D1583460

Probability, Statistics, & Reliability
for Engineers

Probability, Statistics, & Reliability for Engineers

Bilal M. Ayyub
Richard H. McCuen

CRC Press
Boca Raton   New York

11569824

Learning Resources
Centre

Library of Congress Cataloging-in-Publication Data

Ayyub, Bilal M.
    Probability, statistics, and reliability for engineers / Bilal M.
Ayyub, Richard H. McCuen
        p. cm.
    Includes bibliographical references and index.
    ISBN 0-8493-2690-7 (alk. paper)
    1. Engineering mathematics. 2. Reliability (Engineering).
I. McCuen. Richard H., 1941–   . II. Title.
TA330.A99 1997
519.5′02462—dc20
                                    96-41741
                                    CIP

Dedication

To my wife Deena, and our three children Omar, Rami and Samar.

Bilal Ayyub

Preface

In preparing this book, we strove to achieve the following educational objectives: (1) introducing probability, statistics, and reliability methods to engineering students and practicing engineers, (2) emphasizing the practical use of these methods, and (3) establishing the limitations, advantages, and disadvantages of the methods. Although, the book was developed with emphasis on engineering and technological problems, the methods can also be used to solve problems in other fields of sciences.

Problems that are commonly encountered by engineers require decision making under conditions of uncertainty. The uncertainty can be in the definition of a problem, the available information, the alternative solution methodologies and their results, and the random nature of the solution outcomes. Studies show that in the future engineers will need to solve more complex design problems with decisions made under conditions of limited resources, thus necessitating increased reliance on the proper treatment of uncertainty. Therefore, this book is intended to better prepare future engineers, as well as assist practicing engineers, in understanding the fundamentals of probability, statistics, and reliability methods, especially their applications, limitations, and potentials.

STRUCTURE, FORMAT, AND MAIN FEATURES

We have developed this book with a dual use in mind, as both a self-learning guidebook and as a required textbook for a course. In either case, the text has been designed to achieve important educational objectives.

The nine chapters of the book cover of the following subjects: (1) an introduction to the text that covers uncertainty types, decision analysis, and Taylor series expansion; (2) graphical analysis of data, and the computation of important characteristics of sample measurements and basic statistical characteristics; (3) the fundamentals of probability; (4) the joint behavior of random variables and the probabilistic characteristics of functions of random variables; (5) statistical analyses that include parameter estimation, hypothesis testing, confidence-interval estimation, sample-size determination, and probability-model selection; (6) curve fitting or model development based on data using regression analysis; (7) a formal presentation of Monte Carlo simulation; (8) reliability, risk, and decision analysis; and (9) the use of Bayesian methods in engineering. The book was designed for an introductory course in probability, statistics, and reliability with emphasis on applications. In developing the book, a set of educational outcomes as detailed in Chapter 1 motivated the structure and content of this text. Ultimately, serious readers will find the content of the book to be very useful in engineering problem solving and decision making.

One of the most difficult to grasp aspects of probability and statistics is the concept of sampling variation. In engineering practice, an engineer typically has only one sample of data. It is important to recognize that the statistical results would be somewhat different if he or she had collected a different sample, even if that sample were equally likely to have occurred. Simulation is a means of demonstrating the sample-to-sample, or sampling, variation that can be expected. For this reason, we have incorporated a section on simulation at the end of each chapter (Chapters 1 to 6). Performing some simulations is one way of generating a better appreciation for sampling variation that is inherent in statistical problems presented in Chapters 1 to 6. Omitting the sections on simulation does not diminish a reader's understanding of the other sections or chapters.

In each chapter of the book, computational examples are given in the individual sections of the chapter, with more detailed engineering applications given in a concluding section. Also, each chapter includes a set of exercise problems that cover the materials of the chapter. The problems were carefully designed to meet the needs of instructors in assigning homework and the readers in practicing the fundamental concepts.

The book can be covered in one or two semesters depending the level of a course or the time allocated for topics covered in the book. The chapter sequence can be followed as a recommended sequence. However, if needed, instructors can choose a subset of the chapters for courses that do not permit a complete coverage of all chapters or a coverage that cannot follow the presented order. After completing Chapters 1, 2, and 3, the readers will have sufficient background to follow and understand the materials in the following tracks of chapters: Chapter 4; Chapters 5 and 6; Chapters 7 and 8; and Chapter 9 according to the indicated sequence. The accompanying schematic illustrates possible sequences of these chapters in terms of their interdependencies:

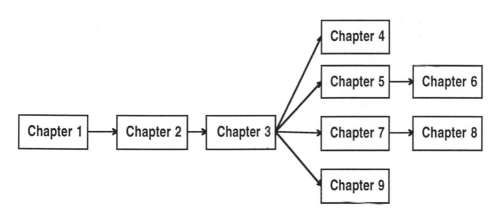

Acknowledgments

This book was developed over several years and draws on our experiences in teaching courses on probability, statistics, and reliability. Drafts of most sections of the book were tested in several courses at the University of Maryland for about four years before its publication. This testing period has proven to be a very valuable tool in establishing its contents and final format and structure.

We would like to acknowledge all the students who participated in the testing of the book for their comments, input, and suggestions. The students who took courses on computational methods in civil engineering in the semesters from 1992 to 1997 contributed to this endeavor. Their feedback was very helpful and greatly contributed to the final product. Also, we would like to acknowledge the instructors who used the book during this testing period, Drs. P. Johnson and R. Muhanna. The assistance of Dr. Ru-Jen Chao, Dr. Maguid Hassan, and Mr. Ibrahim Assakkaf in critically reviewing the problems at the end of the chapters and developing some of the example problems and problem solutions is gratefully acknowledged. The reviewers' comments that were provided to us by publishers were used to improve the book to meet the needs of instructors and enhance the educational process. We acknowledge the publishers and the reviewers for these comments. We also acknowledge the efforts of the editorial staff, artists, and designers at CRC Press LLC for the production of the book.

We would like to invite users of the book, including both readers and instructors, to send us any comments on the book by e-mail at **ayyub@umail.umd.edu**. These comments will be used in developing future editions of the book.

Bilal M. Ayyub
Richard H. McCuen
1997

The Authors

Bilal M. Ayyub, Professor of civil engineering at the University of Maryland at College Park, was born in Palestine and immigrated to the United States in 1980. He completed his B.S. degree in civil engineering in 1980 from the University of Kuwait, and completed both the M.S. (1981) and Ph.D. (1983) in civil engineering at the Georgia Institute of Technology. Dr. Ayyub has an extensive background in risk-based analysis and design, simulation, and marine structural reliability. He is engaged in research work involving structural reliability, marine structures, uncertainty modeling and analysis, and mathematical modeling using the theories of probability, statistics, and fuzzy sets. He has completed various research projects that were funded by the National Science Foundation, the U.S. Coast Guard, the U.S. Army Corps of Engineers, the U.S. Navy, the Department of Defense, the Maryland State Highway Administration, the American Society of Mechanical Engineers, and several engineering companies. He was the general chairman of the first, second and third International Symposia on Uncertainty Modeling and Analysis that were held in 1990, 1993, and 1995. He is the author and editor of eight books and about 250 publications in journals and conference proceedings and reports. Dr. Ayyub is the double recipient of the American Society of Naval Engineers Jimmie Hamilton Award for the best paper in the *Naval Engineers Journal* in 1985 and 1992. Also, he was co-recipient of the American Society of Civil Engineers (ASCE) Outstanding Research Oriented Paper in the *Journal of Water Resources Planning and Management* for 1987, the ASCE Edmund Friedman Award in 1989, and the K.S. Fu Award from the North American Fuzzy Information Processing Society in 1995.

Richard H. McCuen, Professor of civil engineering at the University of Maryland at College Park, received degrees from Carnegie Mellon University and the Georgia Institute of Technology. He is the author of fourteen books and over 175 professional papers. He received the Icko Iben Award from the American Water Resource Association and was co-recipient of the 1988 Outstanding Research Award from the American Society of Civil Engineers Water Resources Planning and Management Division. Topics in statistical hydrology and stormwater management are his primary research interest.

Table of Contents

Chapter 1
Introduction

TABLE OF CONTENTS

1.1. INTRODUCTION

The processes of engineering analysis and design can be systematically performed within a systems framework. Generally, an engineering project can be modeled to include a segment of its environment that interacts significantly with it to define an engineering system. The boundaries of the system are drawn based on the goals and characteristics of the project, the class of performances (including failures) under consideration, and the objectives of the analysis.

The first step in engineering problem solving is to define the architecture of the system. The definition can be based on observations at different system levels that are established based on the goals of the project. The observations can be about the different elements (or components) of the system, interactions among these elements, and the expected behavior of the system. Each level of knowledge that is obtained about an engineering problem defines a system to represent the project. As additional levels of knowledge are added to previous ones, higher epistemological levels of system definition and description are possible which, taken together, form a hierarchy of the system descriptions.

An epistemological hierarchy of systems requires a generalized treatment of uncertainty in both the architecture of the system and the collected information. This treatment can be based, in part, on probability and statistical concepts, as well as other related tools. Therefore, engineering systems should be viewed with an understanding of the knowledge content of collected information including its associated uncertainties. Also, the user should understand the limitations of prediction models that result from inherent insufficiency of models as a result of the assumptions used in developing them. The uncertainty dimension in the analysis of engineering systems can result in valuable insight or information that is necessary to make rational decisions. Such a decision process considers the available information or knowledge, decision choices, alternative decision outcomes, and associated uncertainties. These aspects of decision analysis are discussed in Section 1.1.1.

The need of a proper treatment of uncertainty in engineering systems and the advancement of computer technology and its availability to practicing engineers have impacted engineering education in many aspects. In order to prepare engineering students and practicing engineers for current and future challenges, an introduction to and an emphasis on computer-based methods are needed. Currently, practical engineering problems demand the use of computers to obtain solutions in a timely manner and with an acceptable level of accuracy.

The objective of this book is to introduce the fundamentals of probability, statistics, and reliability to engineering students and practicing engineers in order to deal with uncertainty in engineering systems. The practical aspects of the use of these methods is emphasized throughout the book, with practical applications provided in each chapter. Although the book was developed with emphasis on engineering and technological problems, these methods can be used to solve problems in many other fields. The methods are provided with varying levels of detail and emphasis. A critical presentation of these methods is provided to enhance the readers' understanding.

Probability, statistics, and reliability methods can be used for hand or computer-based computations. The importance of these methods in engineering can be effectively demonstrated in cases dealing with complex problems where analytical solutions cannot be obtained or hand calculations cannot be made. In this book, we use common engineering problems to demonstrate the computational procedures. The examples were intentionally selected with traceable solutions so that readers can reproduce them. It is helpful, but not necessary, for the reader to be familiar with the fundamentals of a computer language.

The use of any computational method without the proper understanding of its limitations and shortcomings can have serious consequences. Before using a method, users should become well versed with these methods in terms of both the computational details and their limitations, shortcomings, and strengths. The methods presented in the book, especially the statistical and reliability methods, should not be used without a complete understanding of the underlying theory and knowledge of computational methods or procedures.

1.1.1. Decision Making in Engineering

Engineering is a profession in which its members are often in decision-making positions. To make the best decisions, engineers must be aware of and fully understand alternative solution procedures. This need has become more important in professional practice as the problems that engineers must address have become more important to society and as the design methods have become more complex. In reference to decisions, the term "best" is multifaceted. Engineers need accurate solutions that are both unbiased and have high precision. The solution must be cost effective and have minimal environmental consequences. The adopted solution should improve the community and be aesthetically appealing. These are just some of the criteria used in making engineering decisions.

When engineers are given a design problem, they typically approach the solution using a systematic procedure. One formulation of the process is as follows: (1) identify the problem, (2) state the objectives, (3) develop alternative solutions, (4) evaluate the alternatives, and (5) implement the best alternative. In this process, engineers should (1) consider the uncertainties associated with the alternative solutions, (2) assess the array of possible outcomes of each alternative with their associated uncertainties, and (3) evaluate all data and prediction models used in the analyses also with their associated uncertainties. Probability, statistics, and reliability methods can be used to help ensure that each of these tasks are properly handled.

Engineering decision problems can be classified into single- and multiple-objective problems. For example, modeling that uses the principle of least squares (as used in Chapter 5) to define a curve fit to data is considered to involve a single objective. A multi-objective problem may seek to minimize the total expected cost, while maximizing safety and minimizing environmental damage. Decision analysis requires all objectives to be clearly stated. Also, the term "expected" in the example objectives means "on the average", which implies the need to properly model the associated uncertainties. For cases of multiple objectives, the objectives need to be stated in the same units. Also, weights that reflect the importance of the objectives and that can be used to combine the objectives need to be assigned. Then, the problem can be formulated in a structure that is suitable for attaining the decision objectives. The development of alternative solutions is the step that is especially critical to success in meeting project objectives. The team of engineers must have a sound technical background and a broad understanding of alternative design methods. To properly evaluate each alternative, a complete understanding of the technical basis of the design method is required. Then, alternatives need to be evaluated and a selection is made.

Computers have increased the number of alternative solution procedures that are available to engineers. Whereas the engineers of past generations were limited to graphical, analytical, and simple empirical methods, engineers can now consider numerical and simulation methods in their design work. The increased complexity of the numerical and simulation methods is believed to improve the accuracy of the solutions, and since these methods are easily implemented with a computer, the increased effort over the less detailed methods is minimal. Thus, the increase in design effort is more than offset by the expected improvement in accuracy. However, to achieve this benefit, the engineer must fully understand the more complex methods so that they can be properly applied.

1.1.2. Expected Educational Outcomes

While a textbook can be the principal medium for someone to individually gain an understanding of a body of knowledge, books such as this one are typically used as part of a course on probability, statistics, and reliability methods. This book has been designed to be used in either capacity, as a self-learning guidebook or as a required textbook for a course. In either case, the text has been designed to produce the following educational outcomes:

1. The reader of the book should be able to (1) collect data on a problem and describe the data using graphical and descriptive measures; (2) develop a probabilistic model for the problem; (3) perform probability operations and evaluations; (4) use discrete and continuous random variables to model some aspects of the problem; (5) assess the joint behavior of random variables with associated correlations; (6) evaluate the probabilistic characteristics of functions of random variables; (7) perform statistical analyses of the data, such as histogram development, univariate analysis, probability model selection, hypothesis testing, parameter estimation, confidence-interval estimation, and the selection of sample sizes; (8) perform correlation and regression analyses for fitting a curve or a model to data; (9) perform random number and random variable generation for the Monte Carlo simulation of selected random variables; (10) evaluate the reliability of a system's component or the reliability of the entire system; (11) perform risk analyses and risk-based decision making; and (12) utilize Bayesian methods for updating probabilities or statistical measures.
2. The reader should be able to select from alternative methods the one method that is most appropriate for a specific problem and state reasons for the selection.
3. The reader should be able to formulate algorithms to solve problems.
4. The reader should understand the limitations of each method, especially the conditions under which they may fail to provide a solution with acceptable accuracy.
5. The reader will become familiar with engineering applications of the various probability, statistical, reliability, and Bayesian methods.

In developing the topics presented in the book, these expected educational outcomes motivated the structure and content of its text. Ultimately, the serious reader will find the material very useful in engineering problem solving and decision making.

1.2. TYPES OF UNCERTAINTY

The analysis of an engineering system often involves the development of a model of the system. The model can be viewed as an abstraction of some aspects of the system. In performing this abstraction, an analyst or engineer must decide which aspects of the system to include and which to leave out. Figure 1-1 shows uncertainties in these aspects that can make model development difficult. Also, depending on the state of knowledge about the system and the background of the analyst or engineer, other aspects of the system might not be known, thus increasing the overall

FIGURE 1-1 Uncertainty types for engineering systems.

uncertainty of the system. In these three categories, i.e., abstracted, nonabstracted, and unknown aspects of the system, several types of uncertainty can be present. Figure 1-1 provides examples of uncertainties within each category. In this book there are some of these uncertainty types that can be dealt with using probability, statistics, reliability, and Bayesian methods (in Chapters 3 to 9). This section provides a brief description of these uncertainty types.

In this description, uncertainties in engineering systems are considered to be mainly attributed to ambiguity and vagueness in defining the variables or parameters of the systems and their relations. The ambiguity component is generally due to noncognitive sources, which include: (1) physical randomness; (2) statistical uncertainty due to the use of limited information to estimate the characteristics of these parameters; and (3) model uncertainties that are due to simplifying assumptions in analytical and prediction models, simplified methods, and idealized representations of real performances. The vagueness-related uncertainty is due to cognitive sources such as: (1) the definition of certain parameters, e.g., structural performance (failure or survival), quality, deterioration, skill and experience of construction workers and engineers, environmental impact of projects, conditions of existing structures; (2) other human factors; and (3) defining the interrelationships among the variables or parameters of the problems, especially for complex systems.

1.2.1. Noncognitive Types

Engineers and researchers currently deal with the ambiguity types of uncertainty in predicting the behavior and designing engineering systems using the theories of probability (Chapters 3 and 4) and statistics (Chapters 5 and 6). Probability distributions are used to model system parameters that are uncertain. Probabilistic methods that include reliability methods, probabilistic engineering mechanics, stochastic finite-element methods, reliability-based design formats, random vibration, and other methods were developed and have been used for this purpose. In this treatment, however, a realization of the presence of a cognitive type of uncertainty was established. Subjective probabilities that are based on mathematics used for the frequency type of probability are used to deal with this uncertainty. Uniform and triangular probability distributions are sometimes used to model this type of uncertainty. The Bayesian techniques have also been used to deal with gaining information about these parameters as described in Chapter 9. The underlying distributions and probabilities were, therefore, updated. Regardless of the nature of the gained information, whether it is cognitive or noncognitive, the same mathematical assumptions and tools were used in these approaches.

1.2.2. Cognitive Types

The cognitive types of uncertainty arise from mind-based abstractions of reality. These abstractions are, therefore, subjective and lack crispness or precision. This vagueness is distinct from ambiguity in source and natural properties. The axioms of probability and statistics are limiting for the proper modeling and analysis of this type, and may not be completely relevant nor completely applicable. The vagueness type of uncertainty can be modeled using the theories of fuzzy sets and possibility. In engineering, the theory has proven to be a useful tool in solving problems that involve the vagueness type of uncertainty. For example, the theory has been successfully used in (1) strength assessment of existing structures and other structural engineering applications; (2) risk analysis and assessment in engineering; (3) analysis of construction failures, scheduling of construction activities, safety assessment of construction activities, decisions during construction, and tender evaluation; (4) the impact assessment of engineering projects on the quality of wildlife habitat; (5) planning of river basins; (6) control of engineering systems; (7) computer vision; and (8) optimization based on *soft* constraints. This type of uncertainty is considered to be beyond the scope of this book.

1.3. TAYLOR SERIES EXPANSION

A Taylor series is commonly used as a basis of approximation in engineering analysis. The objective of this section is to provide a review of the Taylor series.

A Taylor series is the sum of functions based on continually increasing derivatives. For a function $f(x)$ that depends on only one independent variable x, the value of the function at point $x_o + h$ can be approximated by the following Taylor series:

$$f(x_o + h) = f(x_o) + hf^{(1)}(x_o) + \frac{h^2}{2!} f^{(2)}(x_o) + \frac{h^3}{3!} f^{(3)}(x_o)$$

$$+ \cdots + \frac{h^n}{n!} f^{(n)}(x_o) + R_{n+1}$$

(1-1a)

in which x_o is some base value (or starting value) of the independent variable x; h is the distance between x_o and the point x at which the value of the function is needed, that is, $h = x - x_o$; $n! =$ is the factorial of n, i.e., $n! = n(n - 1)(n - 2)\cdots1$; R_{n+1} is the remainder of the Taylor series expansion, and the superscript (n) on the function $f^{(n)}$ indicates the n*th* derivative of the function $f(x)$. The derivatives of $f(x)$ in Equation 1-1a are evaluated at x_o. Equation 1-1a can also be written in the following form:

$$f(x_o + h) = \sum_{k=0}^{\infty} \frac{h^k}{k!} f^{(k)}(x_o)$$

(1-1b)

where $0! = 1$ by convention. Equations 1-1a and 1-1b are based on the assumption that continuous derivatives exist in an interval that includes the points x and x_o.

Equations 1-1a and 1-1b can be used with a finite number of terms to produce approximations for the evaluated function at $f(x_o + h)$. The order of the approximation is defined by the order of the highest derivative that is included in the approximation. For example, the first-order approximation is defined by the terms of the Taylor series expansion up to the first derivative as follows:

$$f(x_o + h) \approx f(x_o) + hf^{(1)}(x_o) \tag{1-2a}$$

Similarly, the second- and third-order approximations are given, respectively, by

$$f(x_o + h) \approx f(x_o) + hf^{(1)}(x_o) + \frac{h^2}{2!}f^{(2)}(x_o) \tag{1-2b}$$

and

$$f(x_o + h) \approx f(x_o) + hf^{(1)}(x_o) + \frac{h^2}{2!}f^{(2)}(x_o) + \frac{h^3}{3!}f^{(3)}(x_o) \tag{1-2c}$$

The accuracy of the approximation improves as the order of the approximation is increased.

The importance of the individual terms of the Taylor series depends on the nature of the function and the distance h. The higher-order terms become more important as the nonlinearity of the function increases and the difference $x - x_o$ increases. If the function is linear, then only the term with the first derivative is necessary. As the separation distance h increases, the importance of the nonlinear terms usually increases. Thus, the error increases as fewer terms of the Taylor series are included.

Example 1-1: Nonlinear Polynomials

Taylor series are used to evaluate a function that is of practical interest, but with an unknown expression. If the expression of the function were known, then the Taylor series approximation would not be necessary. In this example, the true function will be assumed to be known in order to demonstrate the computational aspects of Taylor series and the sizes of the remainder terms in the series. For this purpose, the following fourth-order polynomial is used:

$$f(x) = \frac{1}{8}x^4 - \frac{1}{6}x^3 + \frac{1}{2}x^2 - \frac{1}{2}x + 2 \tag{1-3}$$

The function has the following derivatives:

$$f^{(1)}(x) = 0.5x^3 - 0.5x^2 + x - 0.5 \tag{1-4}$$

$$f^{(2)}(x) = 1.5x^2 - x + 1 \tag{1-5}$$

$$f^{(3)}(x) = 3x - 1 \tag{1-6}$$

$$f^{(4)}(x) = 3 \qquad (1\text{-}7)$$

$$f^{(5)}(x) = 0 \qquad (1\text{-}8)$$

For an assumed base point (x_0) of 1.0, the value of the function and its derivatives are

$$f(x) = 1.95833 \qquad (1\text{-}9a)$$

$$f^{(1)}(x) = 0.5 \qquad (1\text{-}9b)$$

$$f^{(2)}(x) = 1.5 \qquad (1\text{-}9c)$$

$$f^{(3)}(x) = 2 \qquad (1\text{-}9d)$$

$$f^{(4)}(x) = 3 \qquad (1\text{-}9e)$$

$$f^{(5)}(x) = 0 \qquad (1\text{-}9f)$$

Using a separation distance h of 0.5, the values of the individual terms of the Taylor series are shown in Table 1-1(a). For $x = 1.5$, the linear term is dominant. For x from 2 to 3, the second-order term is dominant, with less emphasis given to the linear term and more emphasis given to the cubic term. For $x = 3.5$, the cubic term is the most important, while for $x = 4$, the fourth-order term is most important. These results illustrate the interaction between the importance of the terms and both the nonlinearity of the function and the separation distance. The cumulative values for the terms of the Taylor series are given in Table 1-1(b), and the differences between the cumulative values and the true values are given in Table 1-1(c). The differences show the need to consider the higher-order terms as the separation distance increases. This is clearly evident from Figure 1-2, which shows the cumulative values from Table 1-1(b) as a function of the separation distance h.

1.4. APPLICATIONS

1.4.1. Taylor Series Expansion of the Square Root

The square-root function is used in this section to illustrate the Taylor series expansion. The square-root function can be expressed as

$$f(x) = \sqrt{x} \qquad (1\text{-}10)$$

TABLE 1-1(a) Values of the Individual Terms of the Taylor Series Expansion of Equation 1-1 for the Quadratic Function of Equation 1-3; (b) Cumulative Function of the Taylor Series Expansion; and (c) Difference between the Cumulative Function and the True Value of the Function

Part	x	h	$hf^{(1)}(x_o)$	$\dfrac{h^2}{2!}f^{(2)}(x_o)$	$\dfrac{h^3}{3!}f^{(3)}(x_o)$	$\dfrac{h^4}{4!}f^{(4)}(x_o)$
(a) Individual	1.5	0.5	0.25	0.1875	0.04167	0.0078125
terms	2.0	1.0	0.50	0.7500	0.33333	0.1250000
	2.5	1.5	0.75	1.6875	1.12500	0.6328125
	3.0	2.0	1.00	3.0000	2.66667	2.0000000
	3.5	2.5	1.25	4.6875	5.20833	4.8828125
	4.0	3.0	1.50	6.7500	9.00000	10.1250000
(b) Cumulative	1.5		2.20833	2.39583	2.43750	2.44531
function	2.0		2.45833	3.20833	3.54167	3.66667
	2.5		2.70833	4.39538	5.52083	6.15365
	3.0		2.95833	5.95833	8.62500	10.62500
	3.5		3.20833	7.89583	13.10417	17.98698
	4.0		3.45833	10.20833	19.20833	29.33333
(c) Errors	1.5		−0.23698	−0.04948	−0.00781	0.00000
	2.0		−1.20833	−0.45833	−0.12500	0.00000
	2.5		−3.44531	−1.75781	−0.63281	0.00000
	3.0		−7.66667	−4.66667	−2.00000	0.00000
	3.5		−14.77865	−10.09115	−4.88281	0.00000
	4.0		−25.87500	−19.12500	−10.12500	0.00000

FIGURE 1-2 Taylor series approximation of a fourth-order polynomial.

To evaluate the Taylor series, the derivatives of the function are developed as follows:

$$f^{(1)}(x) = \frac{1}{2}x^{-0.5} \qquad (1\text{-}11a)$$

$$f^{(2)}(x) = -\frac{1}{4}x^{-1.5} \qquad (1\text{-}11b)$$

$$f^{(3)}(x) = \frac{3}{8}x^{-2.5}$$ (1-11c)

Higher-order derivatives can be developed similarly. For a base point $x_o = 1$ and $h = 0.001$, the four terms of the Taylor series produce the following estimate for the square root of 1.001:

$$f(1.001) = \sqrt{1.001} \approx \sqrt{1} + 0.5(0.001)(1)^{-0.5}$$

$$- \frac{1}{4(2!)}(0.001)^2(1)^{-1.5} + \frac{3}{8(3!)}(0.001)^3(1)^{-2.5}$$ (1-12a)

or

$$f(1.001) \approx 1 + 0.5 \times 10^{-3} - 0.125 \times 10^{-6} + 0.625 \times 10^{-10} = 1.0004999$$ (1-12b)

which equals the true value to the number of decimal points shown. Because the interval h is so small, the linear approximation would have been accurate to five decimal places.

1.4.2. Example Taylor Series

In this section, example Taylor series are provided. These series might be familiar to the readers from previous courses in mathematics. For example, the exponential evaluation to the base e of x can be expressed by the following series:

$$e^x = 1 + x + \frac{x^2}{2!} + \frac{x^3}{3!} + \cdots + \frac{x^k}{k!} + \cdots + \sum_{k=0}^{\infty} \frac{x^k}{k!}$$ (1-13)

This equation is valid only for x that is finite. Similarly, the natural logarithm of x can be expressed using a Taylor series as

$$\ln(x) = (x-1) - \frac{(x-1)^2}{2!} + \frac{(x-1)^3}{3!} - \cdots = \sum_{k=0}^{\infty} (-1)^{k-1} \frac{(x-k)^k}{k!} \quad \text{for } 0 < x \leq 2$$ (1-14)

The sine and cosine functions can also be expressed using Taylor series as follows, respectively:

$$\sin(x) = x - \frac{x^3}{3!} + \frac{x^5}{5!} - \cdots = \sum_{k=0}^{\infty} (-1)^k \frac{x^{2k+1}}{(2k+1)!}$$ (1-15)

$$\cos(x) = 1 - \frac{x^2}{2!} + \frac{x^4}{4!} - \cdots = \sum_{k=0}^{\infty} (-1)^k \frac{x^{2k}}{(2k)!} \tag{1-16}$$

Equations 1-15 and 1-16 are valid only for x values that are finite. The reciprocal of $(1 - x)$ where $|x| < 1$ can be expressed by the following Taylor series:

$$\frac{1}{1-x} = 1 + x + x^2 + x^3 + \cdots = \sum_{k=0}^{\infty} x^k \tag{1-17}$$

These Taylor series can be used to evaluate their corresponding functions at any point x using a base value of $x_o = 0$ and increment h.

1.5. INTRODUCTION TO SIMULATION

A central theme of this book is data analysis. Generally, an engineer works with data measured from real engineering systems. For example, transportation engineers frequently use traffic counts at intersections or accident data for various configurations of control signals in designing roadways. Environmental engineers collect water quality data and their analyses of the data are used to decide the type of water quality treatment that is needed. This book emphasizes the analysis of such real-world data. Unfortunately, real-world data may be too limited for reliable decision making. Real-world data may not cover extreme situations that are important in design. For example, stream-flow records often do not include extreme floods that are important in assessing flood risk. For these reasons, engineers often use simulated data to help make decisions.

Simulation is formally covered in depth in Chapter 7. However, it is introduced in all of the earlier chapters to show its usefulness as a supplement to the traditional approaches to data analysis.

1.5.1. Simulation by Coin Flipping

The basics of simulation are easily understood using an example based on the flip of a single coin. Assume that at a particular location on a river, the water quality is considered good about 50% of the time and unacceptable 50% of the time. Also, assume that an environmental engineer tests the quality of the water once a day and the quality on one day is independent of the quality on the previous day. This situation is much like the flip of a coin in that probabilities (i.e., likelihood estimates) are 0.5 and the outcome of the flip of the coin on any one trial is independent of the outcome from the previous flip. Let's also assume that the environmental engineer only has data from the past two weeks. Thus, the sample of 14 values is inadequate to make a decision about the problem. Specifically, the engineer knows that polluted water on two consecutive days is not damaging to the aquatic life, but damage occurs if the water is of unacceptable quality for three or more consecutive days. The actual data for the 14 days were as follows:

AAUAUUAUUAAAUA

in which A indicates acceptable water quality and U unacceptable water quality. Thus, in the engineer's data, the water was of acceptable quality on 57% of the days and there were no instances of aquatic damage. Should the engineer, therefore, believe that aquatic damage will not occur in the future? Of course not! Since the engineer needs to make a decision based on the probability of aquatic damage in the future, the engineer decides to simulate the process and use the simulated probability to make the decision. Flipping a coin 56 times produces the following 8-week sequence:

HHTHTHHHHHTHTHTTHTTTHTHHHTHT

THTTTTHHTHHHTHTHTHTTHHHTTHTH

where H is head and T is tail. If a tail is considered an unacceptable water quality, then two occurrences of unacceptable quality happened in 8 weeks, which represents once every 4 weeks or 13 times a year. Of course, for decision making, the engineer would want to use a longer sequence of simulated data. This example demonstrates the use of simulation in a real-world problem in order to establish a better understanding of the problem, its consequences, and possible solutions.

1.5.2. Generation of Random Numbers

In the example of the previous section, the random occurrence of events was simulated using the flip of a coin. For any one event, only one of the two possible outcomes can occur, H or T, with each having the same likelihood of occurrence. Graphically, this appears as

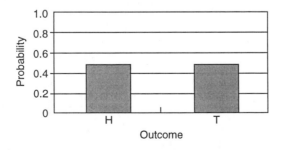

If two coins were flipped and the number of heads counted, one of three possible outcomes could occur: 0, 1, or 2. Letting X be the number of heads, the probability $P(x)$ can be graphed as

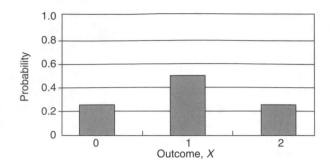

Values of 0, 1, and 2 would result from the following sequences: 0 for the pair (T, T), 1 for the pairs (H, T) or (T, H), and 2 for the pair (H, H).

If random events were generated with the roll of a single die, one of six outcomes is possible. If the die is fair, each outcome y is equally likely and so each would have a probability of 1/6. Graphically, this appears as

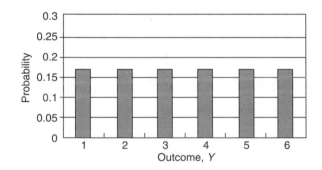

If a pair of dice were rolled simultaneously, the probability of the sum of dots from the two die (Z) would appear graphically as

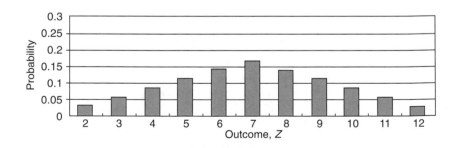

It is left to the reader to show how this would occur.

In each of these cases, only integer values were possible. That is, for example, if a single die is rolled, a value of 3.5 is not possible. Values from the flip of a coin or roll of a die are considered discrete. In many engineering cases, this is desirable, such as if the engineer is interested in the number of traffic fatalities, the number of floods per decade, the number of earthquakes above 6.0 per century, or the number of drums of a hazardous waste.

In many other cases, engineers need values over a continuum, such as the stopping distance of cars, the magnitude of a flood, the compression strength of concrete, or the weight of fertilizer used per acre. Values on a continuum could be generated with a spinner (such as those included in some board games) placed over a 360° protractor. The arrow could be spun and the angle read, with possible values from 0 to 360°. The precision at which the values are recorded would depend on the accuracy of the spinner. Values of the angle (A) to the nearest 5° might be possible for a crude apparatus, while values to eight digits might be possible with an electronic apparatus. Thus, the probability of the outcome of a spin is 1/360 and could be graphed as

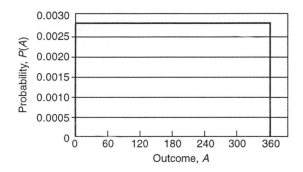

It should be obvious that we could transform the angle A to a new variable B that would take on values from 0 to 1 using the transformation $B = A/360$. This would appear as

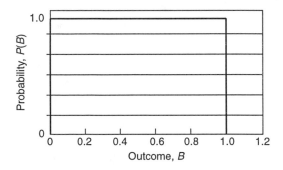

This distribution is very useful and is used frequently.

1.5.3. Computer Generation of Random Numbers

A central element in simulation is a random number generator. In practice, computer packages are commonly used to generate the random numbers used in simulation; however, it is important to understand that these random numbers are generated from a deterministic (well-defined and predictable) process and thus are more correctly called *pseudo random numbers*. Because the random numbers are derived from a deterministic process, it is important to understand the limitations of these generators.

Random-number generators produce numbers that have specific statistical characteristics. Obviously, if the generated numbers are truly random, an underlying population exists, which can be represented by a known probability function using figures as in Section 1.5.2. A single die is the most obvious example of a random-number generator. If we rolled a single die many times, we could tabulate a frequency histogram. If the die were a fair die, we could expect the sample histogram to consist of six bars with almost equal heights. Of course, the histogram for the *population* would consist of six bars of equal height, as shown previously. Other random-number generators would produce random numbers having different distributions, and when a computerized random-number generator is used, it is important to know the underlying population. In this section, the midsquare method for random-number generation is described for illustration purposes. Other methods are described in Chapter 7.

1.5.3.1. Midsquare Method

The midsquare method is one of the simplest but least reliable methods of generating random numbers. However, it is a method that illustrates problems associated with deterministic procedures. The general procedure is as follows:

1. Select at random a four-digit number; this is referred to as the *seed*.
2. Square the number and write the square as an eight-digit number using preceding (lead) zeros if necessary.
3. Use the four digits in the middle as the new random number.
4. Repeat steps 2 and 3 to generate as many numbers as necessary.

As an example, consider the seed number of 2189. This value produces the following sequence of four-digit numbers:

$$
\begin{array}{l}
04\underline{7917}21 \\
62\underline{6788}89 \\
46\underline{0769}44 \\
00\underline{5913}61 \\
34\underline{9635}69 \\
92\underline{8332}25 \\
69\underline{4222}24
\end{array}
$$

At some point one of these numbers must recur, which will begin the same sequence that occurred on the first pass. For example, if the four-digit number of 3500 occurred, the following sequence would result:

12<u>2500</u>00
06<u>2500</u>00
06<u>2500</u>00
06<u>2500</u>00

Such a random-number sequence does not pass statistical tests for randomness. While the procedure could be used for very small samples or five-digit numbers could be used to produce ten-digit squares, the midsquare method has serious flaws that limit its usefulness. However, it is useful for introducing the concept of random-number generation.

The midsquare method cannot be used for actual simulation studies, but numerous methods that are more reliable are available. Other more reliable methods for generating random numbers are discussed in Chapter 7.

1.5.4. Transformation of Random Variables

One example of a transformation was shown in Section 1.5.2. In that case, a value on a continuous scale from 0 to 360 was transformed to a continuous scale from 0 to 1. Transformation is quite frequently necessary and is discussed in detail in Section 7.4. However, the rudiments are discussed herein and used in Chapters 2 to 6.

Assume that a coin was not available, but a die was available. How could we generate flips of a coin with the die? This is easily accomplished by transforming the value of the die to the value of the coin. The die is rolled and an occurrence of a 1, 2, or 3 would indicate a head, while a value of 4, 5, or 6 would represent a tail. Graphically, this transformation appears as

We enter along the ordinate with the outcome of the roll of a die and move horizontally. If the line for the head (H) is intersected, then we assume a head occurred.

If we move horizontally and the line for the tail is first intersected, then we assume a tail occurred. How could we simulate the flip of an unbalanced coin that has a probability of a head of 1/3 rather than 1/2?

For the moment, we will call these graphs, where one value is transformed to another, *transformation graphs*. The transformation graph for the case of the 0 to 360 scale transformed to a 0 to 1 scale is

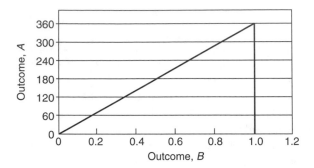

The value of B could be transformed to simulate the outcome (Y) roll of a die using the following transformation graph:

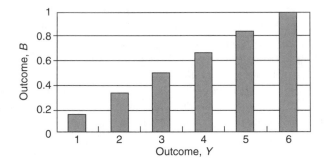

Entering along the ordinate with the value of B, move horizontally until one of the vertical lines is intersected. The transformed variable Y is then read from the abscissa. For example, if B equals 0.42, then Y is 3.

Transformation graphs such as these are widely used in simulation studies. They are central to the discussions of simulation throughout this book.

1.6. PROBLEMS

1-1. Select an engineering system for which you can define different levels of abstractions on the system with different abstraction aspects. Identify the abstracted aspects of the system, the nonabstracted aspects of the system, and the unknown aspects of the system if possible. What are the uncertainty types for the selected system? Describe the uncertainty types using examples.

1-2. Using the Taylor series expansion for cos(x) provided in Section 1.4.2, evaluate the series starting at $x_o = 0$ radian with $h = 0.1$ to $x = 1$ radian. Use one term, two terms, and three terms of the series. Evaluate the error in each value, and discuss the effect of incrementally adding the terms to the series. Compare your results with the true solution.

1-3. Using the Taylor series for sin(x) of Equation 1-15, evaluate the series starting at $x_o = 0$ radian with $h = 0.1$ to $x = 1$ radian. Use one term, two terms, and three terms of the series. Evaluate the error in each value, and discuss the effect of incrementally adding the terms to the series. Compare your results with the true solution.

1-4. Using the Taylor series for e^x provided by Equation 1-13, evaluate the series starting at $x_o = 0$ with $h = 0.1$ to $x = 1$. Use one term, two terms, and three terms of the series. Evaluate the error in each value, and discuss the effect of incrementally adding the terms to the series. Compare your results with the true solution.

1-5. Develop a Taylor series expansion of the following polynomial:

$$f(x) = x^3 - 3x^2 + 5x + 10$$

Use $x = 2$ as the base (or starting) point and h as the increment. Evaluate the series for $h = 0.1, 0.2, 0.3, 0.4, 0.5, 0.6, 0.7, 0.8, 0.9,$ and 1.0. Discuss the accuracy of the method as the terms of the Taylor series are added incrementally.

1-6. Develop a Taylor series expansion of the following polynomial:

$$f(x) = x^5 - 5x^4 + x^2 + 6$$

Use $x = 2$ as the base (or starting) point and h as the increment. Evaluate the series for $h = 0.1, 0.2, 0.3, 0.4, 0.5, 0.6, 0.7, 0.8, 0.9,$ and 1.0. Discuss the accuracy of the method as the terms of the Taylor series are added incrementally.

1-7. Develop a Taylor series expansion of the following function:

$$f(x) = x^2 - 5x^{0.5} + 6$$

Use $x = 2$ as the base (or starting) point and h as the increment. Evaluate the series for $h = 0.1, 0.2, 0.3, 0.4, 0.5, 0.6, 0.7, 0.8, 0.9,$ and 1.0. Discuss the accuracy of the method as the terms of the Taylor series are added incrementally.

1-8. Write a computer program in FORTRAN to evaluate the Taylor series expansion for Problem 1-2.

1-9. Write a computer program in FORTRAN to evaluate the Taylor series expansion for Problem 1-4.

1-10. Write a computer program in FORTRAN to evaluate the Taylor series expansion for Problem 1-6.

1-11. Write a computer program in FORTRAN to evaluate the Taylor series expansion for Problem 1-7.

1-12. Show the development of the probability graph P(Z) vs. Z given in Section 1.5.2.

1-13. Show the transformation graph that transforms the roll of a die (W) to the value of the flip of an unfair coin (C) for which the probability of a tail is 5/6.

1-14. Assume the value of X is random and can take on values from 0 to 1. Assume the following transformation graph relates X to the annual number of fatal accidents (N) at an intersection:

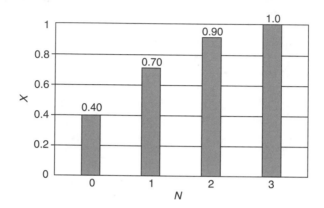

A random number generator yields the following six values of X: 0.37, 0.82, 0.64, 0.25, 0.02, 0.94. How many fatal accidents occurred in each of the 6 years?

1-15. The probabilities of the largest magnitude of an earthquake in any decade are as follows:

Magnitude	3 to 4	4 to 5	5 to 6	6 to 7	7 to 8	8 to 9
Probability	0.78	0.13	0.04	0.03	0.01	0.01

or

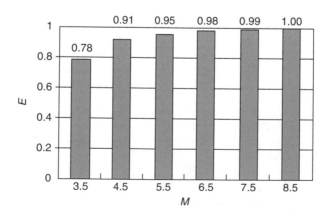

Construct a transformation graph that could transform a random number (E) over the range from 0 to 1 to the magnitude of an earthquake (M) where M takes on the center value for each interval of the magnitude. For each of the following values of E, find the simulated value of M:

$$E = \{0.27, 0.62, 0.13, 0.49, 0.96, 0.06, 0.84\}$$

Chapter 2

Data Description and Treatment

TABLE OF CONTENTS

2.1. INTRODUCTION

It is common in engineering to deal with certain types of dispersion and uncertainty by collecting data and information about some variables that are needed to solve a problem of interest. The data can then be used to establish some understanding about the relations among the different variables. After collecting the data, it is necessary to utilize techniques for describing, treating, and analyzing them. The objective of this chapter is to introduce some of the techniques that are commonly used in engineering to describe and summarize data.

2.2. CLASSIFICATION OF DATA

Graphical analyses are widely used in engineering for both making initial assessments of data characteristics and presenting results to be used by engineers. There are a number of ways of graphically presenting data, each having advantages for specific types of data. Therefore, before discussing the graphical methods, it is necessary to review methods for classifying data.

Data can be measured on one of four scales: nominal, ordinal, interval, and ratio. The four scales are given in order of numerical value. Variables defined on one scale can be reduced to a lower scale of measurement, but cannot be described on a higher scale of measurement. However, when a variable is transformed to a lower scale of measurement, there is an accompanying loss of information.

In addition to the scale of measurement, data can be classified based on their dimensionality. The dimensionality is a measure of the number of axes needed to graphically present the data. Graphical analyses are usually limited to use with one-, two-, and three-dimensional data.

2.2.1. Nominal Scale

The nominal scale of measurement is at the lowest level, because there is no order to the data. Measurements consist of simply identifying the sample as belonging to one of several categories. Nominal measurement scales are both discrete and qualitative. However, numbers may be assigned to the categories for the purpose of coding.

Frequently used examples of variables measured on a nominal scale include (1) gender: female or male; (2) political affiliation: Republican, Democrat, Independent, or other; and (3) college major: engineering, sciences, physical education, or others. Engineering data are sometimes provided using a nominal scale, for example, (1) project failed or did not fail; (2) fatal and nonfatal accidents; and (3) land use: urban, rural, forest, institutional, commercial, or others.

2.2.2. Ordinal Scale

The ordinal scale of measurement is considered to be a higher scale than the nominal scale because it has the added property that there is order among the groups. However, the magnitude of the differences between groups is not meaningful. For example, military ranks are measured on an ordinal scale. The major is above the sergeant and the sergeant is above the private, but we cannot say that a major is two or three times higher than a sergeant.

Variables of interest in engineering that are measured on an ordinal scale include the infiltration potential of soil texture classes and hazard classifications for dam design (high hazard, moderate hazard, low hazard). Soils are classified into one of several categories, such as sand, sandy loam, clay loam, and clay. In this respect, soil texture is measured on a nominal scale. However, if we consider the infiltration potential of the soil, then we can put the soil textures in order according to the infiltration potential, high to low.

2.2.3. Interval Scale

The interval scale of measurement has the characteristics of the ordinal scale, in addition to having a meaningfulness in the separation between any two numbers on the scale. Temperature is defined on the interval scale. We recognize that a difference in temperature of 5°C is less than a difference of 10°C. Values on an interval scale may be treated with arithmetic operators. For example, the mean value of a set of test grades requires addition and division.

Engineering data are frequently recorded on an interval scale. The yield strength of steel, the compression strength of concrete, and the shear stress of soil are variables measured on an interval scale. The annual number of traffic fatalities and the number of lost worker-hours on construction sites due to accidents are also engineering variables recorded on an interval scale.

2.2.4. Ratio Scale

The ratio scale represents the highest level of measurement. In addition to the characteristics of the interval scale, the ratio scale has a true zero point as its origin, not like the interval scale, for which the zero point is set by some standard. For example, in the interval scale the zero point for temperature (°C) is set at the point where water freezes. However, it could have been either set at the point where water boils or based on some other substance.

The standard deviation, which is discussed in Section 2.5.2, is measured on a ratio scale. The zero point is that for which there is no variation. The coefficient of variation and many dimensionless ratios such as the Mach and Reynolds numbers are measured on a ratio scale.

2.2.5. Dimensionality of Data

The dimensionality of data was defined as the number of axes needed to represent the data. Tabular data with one value per classification is an example of one-dimensional data. For example, if a transportation engineer gives the number of fatal traffic accidents for each state in 1991, the variable is described using a one-dimensional, interval scale data. It can be represented in tabular form as a function of the state. It could also be represented pictorially with the number of fatalities as the ordinate (vertical axis) and the state as the abscissa (horizontal axis). In this case, the ordinate is on an interval scale, while the abscissa is on a nominal scale; when presented this way, it appears as a two-dimensional graph.

Two-dimensional plots are very common in engineering. For example, the solution of a dependent variable y of a differential equation could be plotted as a function of the independent variable x. In this case, both variables are expressed on an interval scale. As another example, the corrosion rate of structural steel as a function of the length of time that the steel has been exposed to the corrosive environment is a two-dimensional plot. If we have data for different types of steel (carbon steel, copper steel, and weathered steel), the three relationships can be presented on the same graph, with the steel types identified. In this case, steel type is a nominal variable; so the two-dimensional plot includes two variables on interval scales and one variable on a nominal scale.

2.3. GRAPHICAL DESCRIPTION OF DATA

The first step in data analysis is often a graphical study of the characteristics of the data sample. Depending on the objectives of the analysis and the nature of the problem under consideration, one or more of the following graphical descriptors are commonly used: area charts, pie charts, bar charts, column charts, scatter diagrams, line charts, combinations, and three-dimensional charts. The selection of a graphical descriptor type should be based on (1) its intended readers, (2) the type of data, (3) the dimensionality of the problem, and (4) the ability of the graphical descriptor to emphasize certain characteristics or relations of the parameters of the problem. In this section, examples are used to illustrate the different types.

2.3.1. Area Charts

Area charts are useful for three-dimensional data that include both nominal and interval-independent variables, with the value of the dependent variable measured on an interval scale and cumulated over all values of the nominal variable. The independent variable measured on an interval scale is shown on the abscissa. At any value along the abscissa, the values of the dependent variable are cumulated over the independent variable measured on the nominal scale.

Example 2-1: Area Chart for Traffic Analysis

A traffic engineer is interested in analyzing the traffic at an intersection. An incoming vehicle to the intersection can proceed in one of the following directions: straight, left turn, or right turn (U-turns are assumed to be illegal). The vehicles were counted at the intersection and classified according to direction. The counts were established

FIGURE 2-1 Traffic at an intersection.

for 24 hours in a typical business day. The results are shown in Figure 2-1 in the form of an area chart. The area chart shows that at 0800 approximately 50 vehicles made a right turn, approximately 75 vehicles made either a right or left turn, and a total of approximately 200 vehicles went through the intersection.

2.3.2. Pie Charts

Pie charts are commonly used to graphically present data recorded as fractions, percentages, or proportions. The 100% of the circular pie chart is separated into pie slices based on fractions or percentages. The variable presented in a pie chart is measured on an interval scale.

Example 2-2: Pie Chart for Shipment Breakdown

A shipping company facilitates for customers the transfer of any items from any location to another within the U.S. The company transfers 25, 30, and 45% of the items by air, ground, and sea transportation, respectively. In this case, the variable is the form of transportation, with values recorded as a percentage of all transportation. The breakdown of these items by shipping method is shown in Figure 2-2 in the form of a pie chart.

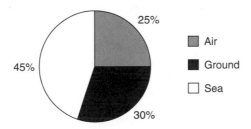

FIGURE 2-2 Shipping methods.

2.3.3. Bar Charts

Bar charts are also useful for data recorded on an interval scale with one or more independent variables recorded on nominal or ordinal scales. The dependent variable can be a magnitude or a fraction. Both one- and two-dimensional bar charts can be used, with a dimension that is based on the number of independent variables.

Example 2-3: Bar Chart for Reinforcing Steel Production

A reinforcing-steel manufacturer provides steel of three different yield strengths: 40, 50, and 60 ksi. The production manager would like to keep track of production for both reinforcing-steel type and the four quarters of a production year. Thus the amount of steel produced is the dependent variable. There are two independent variables, the steel type and the quarters. The dependent variable is on an interval scale, while the independent variables are on an ordinal and nominal scale, respectively. A bar chart, in this case, can meet the requirements of the manager as shown in Figure 2-3. In this case, the bars are shown vertically because the descriptors can be easily included on the abscissa and as a side note.

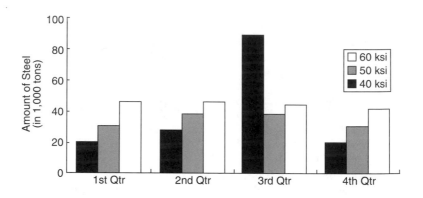

FIGURE 2-3 Reinforcing steel production by yield strength and quarter.

Example 2-4: Bar Chart for Capacity of Desalination Plants

In this case, the dependent variable is the capacity (in million gallons per day, mgd) of water from desalination plants worldwide. The independent variable is the process used in the desalination. Figure 2-4 shows the distribution. The bar chart, in this example, is shown sideways to facilitate presenting the descriptors of the six processes.

FIGURE 2-4 Worldwide desalination capacity (mgd) for available processes.

2.3.4. Column Charts

Column charts are very similar to bar charts, but with the added constraint that the dependent variable is expressed as a percentage (or fraction) of a total. In this case, one of the independent variables is used for the abscissa, while the dependent variable is shown as percentages (or fractions) of the second independent variable. The dependent variable is shown as the ordinate.

Example 2-5: Column Chart for Reinforcing Steel Production

An alternative method for displaying the reinforcing steel production of Example 2-3 is by column charts. Figures 2-5a and 2-5b show example charts. Note that in Figure 2-5a, steel production is displayed as a percentage of the total steel produced for each quarter.

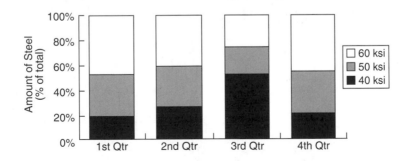

FIGURE 2-5a Reinforcing steel production (as percentage) by yield strength and quarter.

FIGURE 2-5b Reinforcing steel production by yield strength and quarter.

2.3.5. Scatter Diagrams

When both the independent and dependent variables are measured on interval or ratio scales, the data are best presented with scatter plots. Usually, the variable to be predicted (dependent variable) is shown on the ordinate and the independent variable on the abscissa. The ranges for the variables on axes are based on either the minimum and maximum values of the measured data, possible values for the variables, or values that may be expected to occur in the future.

Example 2-6: Scatter Diagram for Yield Strength and Carbon Content

The yield strength of steel depends on several factors, such as the carbon content. An experimental program was designed to understand this relationship. Figure 2-6 shows the measured values. Since we are interested in the yield strength as a function of the carbon content, the yield strength is placed on the ordinate. The limits of the plot were set by the values of the data. An alternative plot with a range of 0 to 200 ksi and 0 to 0.1% carbon content could also be used to display the data. It is evident from this figure that under similar test conditions the yield strength increases as the carbon content increases.

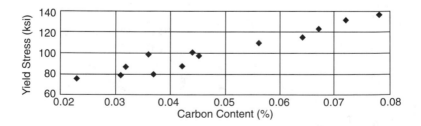

FIGURE 2-6 Yield strength and carbon content.

2.3.6. Line Graphs

Line graphs are used to illustrate mathematical equations. With both variables measured on interval or ratio scales, the variable to be predicted is usually shown as the ordinate. Line graphs are frequently used for design work when the design method is relatively simple.

Example 2-7: Line Chart for Peak Discharge Rates

For one section of the state of Maryland, peak discharge rates (Q in ft³/sec, cfs) can be estimated as a function of drainage area (A in mi²) by the following equations:

$$Q_2 = 55.1A^{0.672} \qquad (2\text{-}1a)$$

$$Q_{10} = 172A^{0.667} \qquad (2\text{-}1b)$$

$$Q_{100} = 548A^{0.662} \qquad (2\text{-}1c)$$

in which the subscripts on Q reflect the return frequency (T in years) of the storm. Thus, Q_{100} is the 100-year peak discharge. Designs with these equations are usually limited to drainage areas of less than 100 mi². Figure 2-7 shows the peak discharge for drainage areas up to 100 mi². The two-dimensional line chart shows the dependent variable on the ordinate, one independent variable as the abscissa, and the second independent variable in the legend taking three values. All three variables are measured on interval scales.

FIGURE 2-7 Peak discharge rate versus drainage area and return period.

Example 2-8: Line Chart for Yield Strength and Carbon Content

To establish the relationship between yield strength and carbon content, two independent laboratories were requested to perform similar tests on similar sets of specimens. The results from the two laboratories are shown in Figure 2-8. The length of the lines connecting the points reflects the magnitude of the variations that are reported based on measurements made at different laboratories.

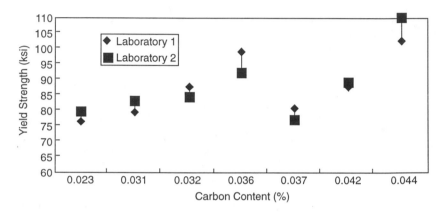

FIGURE 2-8 Line charts for yield strength and carbon content.

2.3.7. Combination Charts

In combination charts, two or more of the previously discussed graphical methods are used to present data. For example, a line graph and bar chart can be combined in the same plot. A combination chart that includes both a scatter plot and a line graph is also commonly used to present experimental data and theoretical (or fitted) prediction equations.

Example 2-9: Operation of a Marine Vessel

The annual number of operation-hours of a high-speed Coast Guard patrol boat is of interest. The annual number of hours was recorded for 10 years. The measured data are shown in Figure 2-9 as vertical bars; they represent sample information. An analytical model is used to fit the data; this is shown in the figure as a solid line representing an assumed frequency model for the population. The combination chart in this case provides an effective presentation tool.

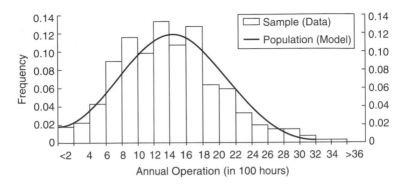

FIGURE 2-9 Combination chart for operational profile.

Example 2-10: Sample Distribution and Probability Function

A combination graph is useful in comparing sample measurements and an assumed probability function. Measurements of daily evaporation (in./day) were collected for each day in a month. A histogram of 30 values is shown in Figure 2-10. An increment of 0.02 in./day is used to form this histogram. The measured evaporation ranges from 0.046 to 0.157 in./day; thus lower and upper bounds of 0.04 and 0.16 in./day are used. Based on the figure, it is reasonable to assume that each increment on the abscissa of the graph is equally likely. Therefore, a uniform frequency of 5, which results from 30 measurements divided by 6 intervals, can be used as the model.

FIGURE 2-10 Combination chart for evaporation data.

Example 2-11: Corrosion Penetration Vs. Exposure Time

Measurements of corrosion were made yearly on a set of 10 steel specimens. The measured data are plotted in the combination graph of Figure 2-11. A power model was fitted to the data (see Chapter 6 for the definition of a power model) and is also shown in Figure 2-11. The fitted power model is

$$\text{Penetration} = 28.075t^{0.39943} \qquad\qquad (2\text{-}2)$$

where t = time exposure. The combination chart of Figure 2-11 is useful because it shows both the measured data and a graph that can be used to make predictions. The scatter of the measured points about the fitted line indicates the accuracy of the model.

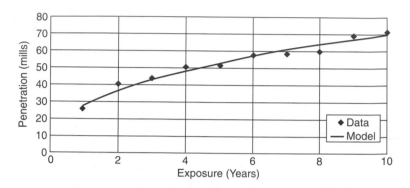

FIGURE 2-11 Combination chart for corrosion prediction.

2.3.8. Three-Dimensional Charts

Any of the charts described in the previous examples can be displayed in three dimensions (three-dimensional). However, three-dimensional charts are commonly used to describe the relationships among three variables. For example, Figure 2-12a shows a three-dimensional pie chart for the data of Example 2-2. Another example is shown in Figure 2-12b, in which the speed (in mph) and number of vehicles per hour passing through an intersection are shown at different times over a 24-h period.

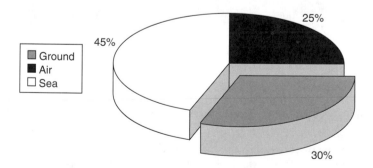

FIGURE 2-12a Three-dimensional pie chart.

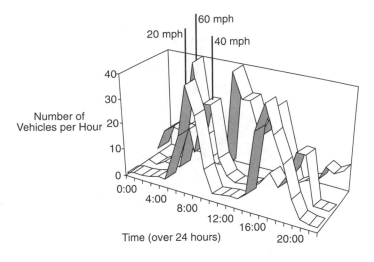

FIGURE 2-12b Three-dimensional surface chart.

2.4. HISTOGRAMS AND FREQUENCY DIAGRAMS

Histograms and frequency diagrams are special types of charts that are commonly used to display and describe data. They are developed from the data for some variables of interest. A histogram is a plot (or a tabulation) of the number of data points vs. selected intervals or values for a parameter. A frequency diagram (or frequency histogram) is a plot (or a tabulation) of the frequency of occurrence vs. selected intervals or values of the parameter. The number of intervals (k) can be subjectively selected depending on the sample size (n). The number of intervals can be approximately determined as

$$k = 1 + 3.3 \log_{10}(n) \qquad (2\text{-}3)$$

Also, the number of intervals can depend on the level of dispersion in the data. The frequency diagrams can be derived from the histogram by dividing the number of data points that correspond to each interval by the sample size. The meaning of these diagrams and their usefulness and development are illustrated in the following example.

Example 2-12: Grades of Students

Students are always interested in the frequency histogram of test scores. This can be developed by determining the number of test scores in various groups, such as the letter grades or intervals of 10 points. If a test is given to 50 students and the numbers of scores are tabulated for each grade level (i.e., A, B, C, D, and F), a

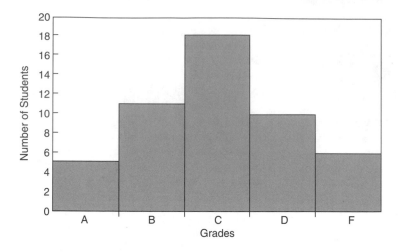

FIGURE 2-13a Histogram of grades.

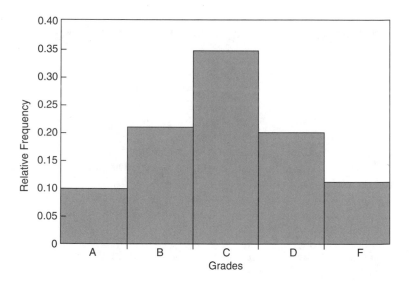

FIGURE 2-13b Frequency histogram of grades.

histogram can be plotted. If the number of students receiving grades of A, B, C, D, and F were 5, 11, 18, 10, and 6, respectively, a graph of the number of students vs. the grade level indicates that the grades have a bell-shaped plot. The resulting histogram and frequency diagrams are shown in Figures 2-13a and 2-13b. Alternatively, the frequency of grades could be computed for intervals, such as 0–10, 11–20, 21–30, ..., 91–100. The frequency could be plotted against these intervals.

The effectiveness of a graphical analysis in identifying the true shape of the frequency diagram depends on the sample size and the intervals selected to plot the abscissa. For small samples, it is difficult to separate the data into groups that provide a reliable indication of the frequency of occurrence. With small samples, the configuration of the histogram may be very different for a small change in the intervals selected for the abscissa; for example, the grade distribution might appear very different if the intervals 0–15, 16–25, 26–35, ..., 76–85, 86–100 are used instead of the 10-point intervals described previously.

2.5. DESCRIPTIVE MEASURES

In engineering it is sometimes desirable to characterize some data by certain descriptive measures. These measures, which take numerical values, can be easily communicated to others and quantify the main characteristics of the data.

Most data analyses include the following three descriptive measures at the fundamental level:

1. Central tendency measures
2. Dispersion measures
3. Percentile measures

In this section an introductory description of these measures is provided. A formal discussion of these measures is provided in probability and statistics textbooks. Additionally, box-and-whisker plots are introduced as a graphical means of presenting these measures.

2.5.1. Central Tendency Measures

A very important descriptor of data is the central tendency measure. The following three types can be used:

1. Average value
2. Median value
3. Mode value

The average value is the most commonly used central tendency descriptor. For n observations, if all observations are given equal weights, the average value is given by

$$\overline{X} = \frac{1}{n} \sum_{i=1}^{n} x_i \tag{2-4}$$

where x_i = a sample point, and $i = 1, 2, ..., n$.

The median value x_m is defined as the point that divides the data into two equal parts, that is, 50% of the data are above x_m and 50% are below x_m. The median value can be determined by ranking the n values in the sample in decreasing order, 1 to n. If n is an odd number, the median is the value with a rank of $(n + 1)/2$. If n is an even number, the median equals the average of the two middle values, that is, those with ranks $n/2$ and $(n/2) + 1$.

The mode value x_d is defined as the point of the highest percent for the frequency of occurrence. This point can be determined with the aid of the frequency histogram.

Although these measures convey certain information about the underlying sample, they do not completely characterize the underlying variables. Two variables can have the same mean, but different histograms. Thus, measures of central tendency cannot fully characterize the data. Other characteristics are also important and necessary.

Example 2-13: Mean Value of Grades

Consider the grades of the students discussed in Example 2-12. Assume the following grade points (levels) that correspond to the letter grades:

A = 4
B = 3
C = 2
D = 1
F = 0

Therefore, the average grade of the class can be computed as follows:

$$\bar{X} = \frac{4+4+4+4+4+3+3+\cdots+3+2+2+\cdots+2+1+1+\cdots+1+0+0+\cdots+0}{50} \tag{2-5}$$

$$= 1.98$$

In this equation, there are 5 of the 4 value, 11 of the 3 value, 18 of the 2 value, 10 of the 1 value, and 6 of the 0 value. Therefore, this equation can be written as

$$\bar{X} = \frac{5 \times 4 + 11 \times 3 + 18 \times 2 + 10 \times 1 + 6 \times 0}{50} = 1.98 \tag{2-6}$$

The average value indicates that on the average the class is at the C level. By inspecting the frequency histogram in Figure 2-13b, the median value is also C, since 32% of the grades are in the A and B levels and 32% of the grades in the D and F levels. Therefore, the C grade divides the grades into two equal percentages. This method of finding the median is proper for values on an ordinal scale, but not on a continuous scale. Again, the C grade is the mode, since it has the highest frequency of 36%. In this example, the average, median, and mode values are the same value. However, in general, these values can be different.

2.5.2. Dispersion Measures

The dispersion measures describe the level of scatter in the data about the central tendency location. The most commonly used measure is the variance and other quantities that are derived from it. For n observations in a sample that are given equal weight, the variance (S^2) is given by

$$S^2 = \frac{1}{n-1} \sum_{i=1}^{n} (x_i - \bar{X})^2 \tag{2-7}$$

The units of the variance are the square of the units of the variable x; for example, if the variable is measured in pounds per square inch (psi), the variance has units of (psi)^2. Computationally, the variance of a sample can be determined using the following alternative equation:

$$S^2 = \frac{1}{n-1} \left[\sum_{i=1}^{n} x_i^2 - \frac{1}{n} \left(\sum_{i=1}^{n} x_i \right)^2 \right] \tag{2-8}$$

Equation 2-8 provides the same answer as Equation 2-7 when computations are made using an appropriate number of significant digits. Equations 2-7 and 2-8 provide an estimate of the variance that is an average of the squared difference between the x values and their average value. In these equations, $(n - 1)$ is used to compute the average, instead of (n), in order to obtain an unbiased estimate of the variance.

There are two commonly used derived measures based on the variance: the standard deviation and the coefficient of variation. By definition the standard deviation (S) is the square root of the variance as follows:

$$S = \sqrt{ \frac{1}{n-1} \left[\sum_{i=1}^{n} x_i^2 - \frac{1}{n} \left(\sum_{i=1}^{n} x_i \right)^2 \right] } \tag{2-9}$$

It has the same units as both the underlying variable and the central tendency measures. Therefore, it is a useful descriptor of the dispersion or spread of a sample of data.

The coefficient of variation (COV or δ) is a normalized quantity based on the standard deviation and the mean. Therefore, it is dimensionless. The COV is defined as

$$COV = \frac{S}{\bar{X}} \tag{2-10}$$

It is also used as an expression of the standard deviation in the form of a percent of the average value. For example, consider $\overline{X}$ and S to be 50 and 20, respectively; therefore, $COV(X) = 0.4$ or 40%. In this case, the standard deviation is 40% of the average value.

Example 2-14: Dispersion Measures of Concrete Strength

A sample of five tests was taken to determine the compression strength (in ksi) of concrete. Test results are 2.5, 3.5, 2.2, 3.2, and 2.9 ksi. Compute the variance, standard deviation, and coefficient of variation of concrete strength. The mean value of concrete strength $\overline{X}$ is given by

$$\overline{X} = \frac{2.5 + 3.5 + 2.2 + 3.2 + 2.9}{5} = 2.86 \text{ ksi} \tag{2-11}$$

The variance of concrete strength is computed using Equation 2-8 as follows:

$$S^2 = \frac{2.5^2 + 3.5^2 + 2.2^2 + 3.2^2 + 2.9^2 - \dfrac{(2.5 + 3.5 + 2.2 + 3.2 + 2.9)^2}{5}}{5 - 1} \tag{2-12}$$

$$= 0.273 \text{ ksi}^2$$

Therefore, the standard deviation is given by

$$S = \sqrt{0.273} = 0.52249 \text{ ksi} \tag{2-13}$$

The coefficient of variation can be computed as follows:

$$\delta \text{ or } COV(X) = \frac{S}{\overline{X}} = \frac{0.52249}{2.86} = 0.183 \tag{2-14}$$

The relatively large coefficient of variation, 18.3%, suggests that the average value is not reliable and that additional measurements might be needed. If, with additional measurements, the coefficient of variation remains large, relatively large factors of safety should be used for projects that use the concrete.

2.5.3. Percentiles

A p percentile value (x_p) for a parameter or variable based on a sample is the value of the parameter such that $p\%$ of the data is less or equal to x_p. On the basis of this definition, the median value is considered to be the 50 percentile value. It is common in engineering to have interest in the 10, 25, 50, 75, and 90 percentile values for a variable.

Example 2-15: Operation of a Marine Vessel

The annual number of operation-hours of a high-speed Coast Guard patrol boat was discussed in Example 2-9. The annual number of hours was recorded for 10 years. An analytical model is used to fit the data; this is shown in Figure 2-9 as a solid line representing an assumed frequency model for the population. The model is also shown in Figure 2-14. Also shown in this figure is the 25 percentile value, which is 702 h. The 75, 90, and 10 percentile values are 1850, 2244, and 578 h, respectively. The calculation of the percentile values, in this example, requires the knowledge of the equation of the model and probabilistic analysis. The model used in Figure 2-14 is called the normal probability density function. The area under this model is a measure of likelihood or probability. The total area under the model is 1. Therefore, the 25 percentile value is the x value such that the area under the model up to this value is 0.25. This value is 702 h, as shown in Figure 2-14.

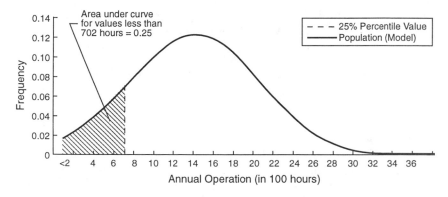

FIGURE 2-14 Percentile values for the operational profile.

2.5.4. Box-and-Whisker Plots

A box-and-whisker plot is a graphical method for showing the distribution of sampled data, including the central tendency (mean and median), dispersion, percentiles (10, 25, 75, and 90 percentiles), and the extremes (minimum and maximum). Additionally, it can be used to show the bias about the standard value and, if the figure includes multiple plots for comparison, the relative sample size.

To construct a box-and-whisker plot, the following characteristics of a data set need to be computed:

1. Mean and median of the sample
2. Minimum and maximum of the sample
3. 90, 75, 25, and 10 percentile values

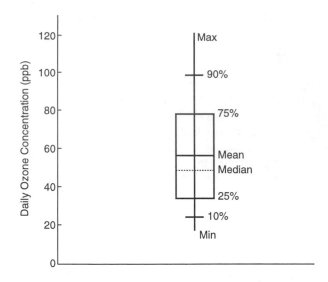

FIGURE 2-15 Box-and-whisker plot for maximum daily ozone concentration (ppb).

The plot consists of a box, the upper and lower boundaries of which define the 75 and 25 percentiles, and upper and lower whiskers, which extend from the ends of the box to the extremes, as shown in Figure 2-15. At the 90 and 10 percentiles, bars, which are one half of the width of the box, are placed perpendicular to the whiskers. The mean and median are indicated by solid and dashed lines that are the full width of the box, respectively.

Figure 2-15 shows the box-and-whisker plot for the maximum daily ozone concentration. The mean and median values are 59 and 52 ppb (parts per billion), respectively. The 10, 25, 75, and 90 percentile points are 24, 36, 79, and 97 ppb, respectively.

If a figure includes more than one box-and-whisker plot and the samples from which each plot is derived are of different sizes, then the width of the box can be used to indicate the sample size, with the width of the box increasing as the sample size increases. Figure 2-16 shows box-and-whisker plots of samples of a toxic chemical analyzed at four different laboratories. Since the samples were of different concentrations, the distributions are presented as the differences between the true concentration and the concentration reported by the laboratory, i.e., error. The difference between the line for zero error and the solid line for the mean in each plot represents the corresponding lab's bias. Lab 1 tends to overpredict by almost 10 ppb. While lab 2 shows a slight positive bias, it shows a skewed distribution of results with many values underestimating the true value. Laboratories 1, 2, 3, and 4 processed the same number of samples each. Therefore, the widths of the boxes are the same, reflecting the equal sample sizes. However, it should be noted that greater accuracy of data descriptors can be expected for the descriptors based on larger sample sizes.

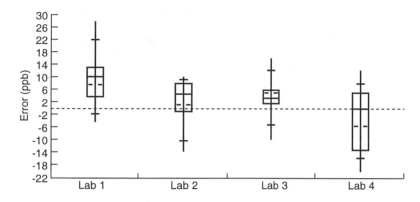

FIGURE 2-16 Display of multiple box-and-whisker plots.

2.6. APPLICATIONS

2.6.1. Two Random Samples

The two random samples shown in Table 2-1 are used to illustrate the meaning of dispersion in data. Both samples have the same mean value of 10. The mean value for sample 1 can be computed as

$$\text{mean of sample } 1 = \frac{15(9) + 15(11)}{30} = 10$$

Similarly, the mean for sample 2 is

$$\text{mean of sample } 2 = \frac{3(5) + 2(6) + 4(7) + \cdots + 2(13) + 2(14) + 2(15)}{30} = 10$$

Although both samples have the same mean and the same sample size, they are different in their levels of scatter or dispersion. Figure 2-17 shows the histograms for the two samples, which show clearly the different levels of scatter. The variances, standard deviations, or coefficients of variation can be computed and used to measure the dispersion. The variance for sample 1 can be computed as

$$\text{variance of sample } 1 = \frac{15(9-10)^2 + 15(11-10)^2}{30-1} = 1.034$$

Similarly, the variance for sample 2 is

$$\frac{3(5-10)^2 + 2(6-10)^2 + 4(7-10)^2 + \cdots + 2(13-10)^2 + 2(14-10)^2 + 2(15-10)^2}{30-1} = 9.379$$

Table 2-1 Two Random Samples

	Number of occurrences	
Sampled value	Sample 1	Sample 2
5	0	3
6	0	2
7	0	4
8	0	0
9	15	3
10	0	4
11	15	2
12	0	6
13	0	2
14	0	2
15	0	2

FIGURE 2-17 Histograms for two samples.

Therefore, the variance of sample 2 is about nine times the variance of sample 1. The standard deviations for samples 1 and 2 are 1.017 and 3.063, respectively, and the coefficient of variations are 0.10 and 0.31, respectively. The larger scatter in sample 2 could reflect natural variation in the value of the variable or a difficulty in making precise measurements. In some cases, variation is not desirable in a sample of data if it reflects a lack of precision. In other cases, sample variation is necessary to ensure that the sample is representative of the full range of the underlying population.

2.6.2. Stage and Discharge of a River

The stage of a river is defined as its flow depth, and the discharge is defined as the volume of flow. For the Little Patuxent River near Guilford, MD, the stage in meters (m) and the discharge in cubic meters per second (cms) were obtained as shown in Table 2-2. The table shows the stage and discharge for the years 1933 to 1989. Descriptive statistics of stage and discharge are shown in Table 2-3. The frequency histogram for the stage is shown in Figure 2-18. The average, which is indicated in Table 2-3, is 2.84 m. The median stage is 2.71 m. The standard deviation and coefficient of variation are 0.71 and 0.25 m, respectively. The maximum and minimum values (extreme values) are 5.60 and 1.69 m, respectively.

TABLE 2-2 Stage and Discharge for the Little Patuxent River, Guilford, MD

Year	Stage (m)	Discharge (cms)	Year	Stage (m)	Discharge (cms)
1933	3.81	119.2	1962	2.78	36.2
1934	2.86	41.9	1963	2.25	23.2
1935	2.35	25.9	1964	2.70	32.8
1936	2.74	37.4	1965	2.35	24.8
1937	3.14	56.6	1966	2.72	34.0
1938	3.08	51.5	1967	2.71	33.4
1939	2.41	27.4	1968	2.86	39.9
1940	3.51	77.6	1969	2.02	19.4
1941	1.87	15.9	1970	2.02	19.4
1942	3.25	61.7	1971	3.49	86.9
1943	2.79	39.6	1972	5.60	351.1
1944	3.19	58.3	1973	3.07	53.2
1945	3.72	107.9	1974	2.65	31.4
1946	2.52	30.6	1975	4.08	152.0
1947	2.44	23.1	1976	3.08	54.1
1948	3.02	43.9	1977	2.14	21.3
1949	2.22	19.9	1978	3.63	103.9
1950	1.95	15.9	1979	3.91	132.5
1951	3.19	55.8	1980	2.65	31.4
1952	4.04	150.1	1981	2.50	28.0
1953	3.11	56.6	1982	2.18	22.0
1954	1.91	18.3	1983	3.33	78.7
1955	3.69	107.3	1984	2.65	42.2
1956	2.55	28.6	1985	2.60	40.5
1957	1.90	17.3	1986	1.69	19.3
1958	2.75	34.8	1987	2.42	34.8
1959	2.14	21.4	1988	2.42	34.8
1960	2.54	28.3	1989	4.01	143.5
1961	2.49	27.4			

TABLE 2-3 Descriptive Statistics of Stage and Discharge for the Little Patuxent River, Guilford, MD

Parameter	Stage (m)	Discharge (cms)
Average	2.84	54.82
Median	2.71	34.8
Mode	2.65	34.8
Standard deviation	0.71	53.78
Sample variance	0.505	2892.5
Coefficient of variation	0.25	0.98
Range	3.91	335.2
Minimum	1.69	15.9
Maximum	5.6	351.1
Count	57	57

FIGURE 2-18 Frequency histogram of the stage of a river.

FIGURE 2-19 Frequency histogram of the discharge of a river.

The frequency histogram for the discharge is shown in Figure 2-19. The average, which is indicated in Table 2-3, is 54.82 cms. The median discharge is 34.8 cms. The standard deviation and coefficient of variation are 53.78 and 0.98 cms, respectively. The maximum and minimum values (extreme values) are 351.10 and 15.90 cms, respectively.

By comparing the two histograms in Figures 2-18 and 2-19, it can be observed that the relative dispersion in discharge is larger than that in stage. The coefficient of variation of discharge is 0.98, which is four times larger than the coefficient of variation of stage (0.25). Also, it can be observed that the frequency histogram for stage is almost bell shaped, whereas the histogram for discharge is highly skewed and appears as an exponential decay function. The closeness of the average stage (2.84 m) to the median stage (2.71 m) is an indication of a symmetric histogram. On the other hand, the average discharge (54.82 cms) is considerably different from the median discharge (34.8 cms), indicating a lack of symmetry in the discharge measurements.

2.7. ANALYSIS OF SIMULATED DATA

During large floods along major rivers, governmental agencies involved in flood preparedness make estimates of maximum flood stages at downstream locations using observed flood stages at upstream locations. For example, when locations on the Upper Mississippi River are at flood stage, people in Memphis, TN, are interested in the maximum flood stage expected in Memphis. A model might use the records of flood stages at Cincinnati on the Ohio River, St. Louis on the Mississippi River, and Kansas City on the Missouri River to simulate the flood profile that might occur in Memphis. The simulated maximum flood stage can allow the local residents to prepare for the flood. After the flood wave has passed through Memphis, those involved in the simulation would compare the simulated and actual flood waves. If the differences are significant, then the information can be used to adjust the simulation model so that the model can produce more accurate predictions of future floods. The point is that it is common practice both to use simulated data in decision making and to compare simulated and actual measured data. As another example, before Apollo astronauts landed on the moon, they went through many simulations in a flight simulator of the landing. These simulations influenced the moon landing program.

Before simulated data are used in decision making, they should be analyzed in much the same way as the measured data, with descriptive measures computed and graphical analyses made. Additionally, the simulated measures should be compared with the descriptive measures and graphs of the actual data. These analyses can be used as an indication of the reasonableness of the simulation.

The histogram of Figure 2-19 suggests that the data follow an exponential decay, which mathematically is

$$f(x) = \frac{1}{b} e^{-x/b} \tag{2-15}$$

where b is a parameter. The exponential distribution is used frequently to represent engineering data. For data that follow this form, the sample mean can be used as an estimate of b. To simulate values, the cumulative function $F_X(x)$ for the density function of Equation 2-15 is as follows:

$$F(x) = 1 - e^{-x/b} \tag{2-16}$$

where values of $F_X(x)$ vary from 0 to 1 and represent the area under $f(x)$ from 0 to x.

Equation 2-16 can be used as a transformation graph to simulate values of the variable x. The cumulative function of the population shown in Figure 2-20 is the graphical representation of the transformation graph. Equation 2-16 can be algebraically rearranged as an expression for x given values of $F_X(x)$ and b as follows:

$$e^{-x/b} = 1 - F_X(x)$$

$$-x/b = \ln\left[1 - F_X(x)\right] \tag{2-17}$$

$$x = -b \ln\left[1 - F_X(x)\right]$$

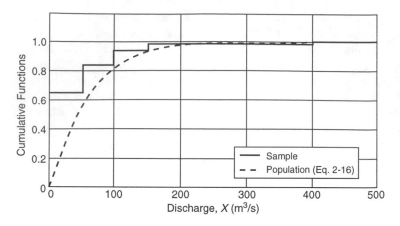

FIGURE 2-20 Cumulative functions for sample and assumed exponential decay model.

Since $F_X(x)$ has a uniform distribution (0 to 1), then $1 - F_X(x)$ also has a uniform distribution (0 to 1). Thus, Equation 2-17 can be rewritten with the definition $F'_X(x) = 1 - F_X(x)$ as follows:

$$x = -b \ln\left[F'_X(x)\right] \qquad (2\text{-}18)$$

Example 2-16: Simulation of River Discharge Rates

The discharge data of Table 2-2, which have the descriptive statistics of Table 2-3, follow an exponential decay (see Figure 2-19). Thus, it would be appropriate to use the exponential function of Equation 2-16 in the simulation of discharge rates. The mean of the sample data can be used as an estimate of the parameter b. Thus, the transformation equation, based on Equation 2-18, is

$$x_i = -54.82 \ln\left(u_i\right) \qquad (2\text{-}19)$$

where u_i is the ith uniform variate (i.e., a random number between 0 and 1) and x_i is the ith simulated discharge.

Values of the discharge x can be simulated using random numbers generated by the midsquare method and the function $F_X(x)$. To illustrate this procedure, a seed value of 8765 for random number generation (r) is used to start the process. A sample of 57 values of x is simulated. The midsquare value is given in column 2 of Table 2-4, with the value transformed to a number between 0 and 1 by dividing it by 10,000 (column 3). The uniform variate is entered into Equation 2-19 to produce the discharge values of column 4 in Table 2-4. Some important descriptive statistics of the simulated discharges are given by

TABLE 2-4 Simulation of Discharges

r^2	r	Random number (u)	Discharge, x (cms)	Rank	r^2	r	Random number (u)	Discharge, x (cms)	Rank
	8,765								
76,825,225	8,252	0.8252	10.5	47	06,594,624	5,946	0.5946	28.5	36
68,095,504	955	0.0955	128.8	9	35,354,916	3,549	0.3549	56.8	25
00,912,025	9,120	0.912	5.0	53	12,595,401	5,954	0.5954	28.4	38
83,174,400	1,744	0.1744	95.7	17	35,450,116	4,501	0.4501	43.8	32
03,041,536	415	0.0415	174.4	5	20,259,001	2,590	0.259	74.1	23
00,172,225	1,722	0.1722	96.4	16	06,708,100	7,081	0.7081	18.9	43
02,965,284	9,652	0.9652	1.9	56	50,140,561	1,405	0.1405	107.6	12
93,161,104	1,611	0.1611	100.1	15	01,974,025	9,740	0.974	1.4	57
02,595,321	5,953	0.5953	28.4	37	94,867,600	8,676	0.8676	7.8	50
35,438,209	4,382	0.4382	45.2	29	75,272,976	2,729	0.2729	71.2	24
19,201,924	2,019	0.2019	87.7	18	07,447,441	4,474	0.4474	44.1	31
04,076,361	763	0.0763	141.1	8	20,016,676	166	0.0166	224.7	1
00,582,169	5,821	0.5821	29.7	35	00,027,556	275	0.0275	197.0	2
33,884,041	8,840	0.884	6.8	51	00,075,625	756	0.0756	141.6	7
78,145,600	1,456	0.1456	105.6	13	00,571,536	5,715	0.5715	30.7	34
02,119,936	1,199	0.1199	116.3	10	32,661,225	6,612	0.6612	22.7	41
01,437,601	4,376	0.4376	45.3	28	43,718,544	7,185	0.7185	18.1	44
19,149,376	1,493	0.1493	104.3	14	51,624,225	6,242	0.6242	25.8	39
02,229,049	2,290	0.229	80.8	19	38,962,564	9,625	0.9625	2.1	55
05,244,100	2,441	0.2441	77.3	20	92,640,625	6,406	0.6406	24.4	40
05,958,481	9,584	0.9584	2.3	54	41,036,836	368	0.0368	181.0	4
91,853,056	8,530	0.853	8.7	49	00,135,424	1,354	0.1354	109.6	11
72,760,900	7,609	0.7609	15.0	46	01,833,316	8,333	0.8333	10.0	48
57,896,881	8,968	0.8968	6.0	52	69,438,889	4,388	0.4388	45.2	29
80,425,024	4,250	0.425	46.9	27	19,254,544	2,545	0.2545	75.0	21
18,062,500	625	0.0625	152.0	6	06,477,025	4,770	0.477	40.6	33
00,390,625	3,906	0.3906	51.5	26	22,752,900	7,529	0.7529	15.6	45
15,256,836	2,568	0.2568	74.5	22	56,685,841	6,858	0.6858	20.7	42
					47,032,164	321	0.0321	188.5	3

Average	64.81 cms
Median	45.25 cms
Standard deviation	57.57 cms
Minimum	1.4 cms
Maximum	224.7 cms
Range	223.3 cms

These values can be compared with the values of the measured data (see Table 2-3). Both the mean and standard deviation of the simulated values are slightly larger than the values of the measured data. Such differences are common and expected, and reflect sampling variation, which are discussed later in the book.

Graphical analyses are also useful for examining simulated data and making comparisons with measured data. Figure 2-21 shows the frequency histograms for the simulated and measured data. The results suggest that the measured data are more skewed than the simulated data, as evident by the slightly higher proportion of small flows and the one event between 350 and 400 cms. However, these differences in the ordinates of the histograms are not large when one considers the relatively small sample size (i.e., $n = 57$). The differences would be less if a larger sample had been used.

FIGURE 2-21 Comparison of frequency histograms for measured and simulated data.

2.8. PROBLEMS

2-1. For each of the measurement scales, identify five variables that are measured with the scale.

2-2. Using age as a variable of interest, identify a function that would be measured on each of the scales: ordinal, interval, and ratio.

2-3. Using the copper content of steel as a variable of interest, identify one function that would be measured for each of the four measurement scales.

2-4. For the eight methods of graphical analysis given in Section 2.3, develop a classification system for distinguishing among them. The classification system should center on basic, yet important, characteristics.

2-5. Using the data from Problem 2-12 and the following population totals, construct an area chart that shows the total and the breakdown of the total population in rural, suburban, and central city.

Year	1900	1910	1920	1930	1940	1950	1960	1970
Total (10^6)	76	92	106	123	132	151	179	203

2-6. The following data are the solid waste (millions of tons) produced annually in the U.S.:

Municipal trash and garbage	150
Industrial	350
Mining	1700
Agriculture	2300

Construct a pie chart to present the data. Discuss the merits of presenting the data in a pie chart vs. the tabular summary given.

2-7. The following data give the age distribution of U.S. citizens as a function of age group. Select a method for graphing the results. Interpret the results. Then combine the values into three meaningful groups and provide a graphical analysis of the distribution. How do the two analyses differ in their information content and emphasis?

Age group	<5	5–9	10–14	15–19	20–24	25–29	30–34
Percentage	7.2	7.4	8.1	9.3	9.4	8.6	7.8

Age group	35–39	40–44	45–49	50–54	55–59	60–64	≥65
Percentage	6.2	5.2	4.9	5.2	5.1	4.5	11.3

2-8. A local highway department compiled the following percentages from accident records according to traffic control method (flashing red light, two-way stop signs, or four-way stop signs) and accident severity (loss of life, major damage, minor damage):

Traffic control method	Loss of life	Major damage	Minor damage	Total
Flashing red light	23	41	36	100
Two-way stop signs	18	39	43	100
Four-way stop signs	12	21	67	100

a. Present the data using three pie charts, one for each of the traffic control methods.
b. Present all the data in a bar chart in a way that emphasizes differences between accident severity.
c. Present all the data in a bar chart in a way that emphasizes differences between the traffic control methods.
d. Present the data as a column chart.
e. Discuss the advantages and disadvantages of each of the above graphical methods with respect to these data.

2-9. Compare the graphical analyses of Figures 2-3, 2-5a, and 2-5b, and identify the circumstances under which each would be the most appropriate for making decisions.

2-10. The following data provide the estimated remaining strippable resources and reserves of bituminous coal in the U.S. (billions of short tons). Create a bar chart to display the data.

Alaska	0.9
Rocky Mountains and North Great Plains	1.1
Interior and Gulf provinces	32.0
Eastern province	27.0

2-11. Create column charts to present the following estimates of U.S. production of bituminous and lignite coal from surface and underground mines:

Year	Surface mines	Underground mines
1940	43	418
1950	123	393
1960	131	285
1970	264	339

Use one column chart to emphasize the temporal variation of total production. Use a second column chart to emphasize the proportion produced from surface mines. Use a third column chart to emphasize the total production from surface vs. underground mines.

2-12. The following percentages indicate the change in rural, suburban, and central city U.S. populations from 1900 to 1970. Present the data graphically to emphasize the decline in the proportion of the population living in rural areas. Present the data graphically to illustrate any association between the change in central city population and the increase in the proportion living in suburban areas.

Year	Rural	Suburban	Central city
1900	58	16	26
1910	55	17	28
1920	51	18	31
1930	46	20	34
1940	46	22	32
1950	45	24	31
1960	40	30	30
1970	34	38	28

2-13. Plot the data shown in the following table using a column chart.

Superstructure type	Number of constructed bridges by year		
	1989	**1990**	**1991**
Steel	5	10	12
Concrete	10	6	7
Prestressed concrete	4	6	5
Total	19	22	24

2-14. Using the bridge data in Problem 2-13, show the number of bridges as a function of year and by bridge type using a line chart, a pie chart, and three-dimensional surface chart.

2-15. The following concrete strength data (in ksi) were collected using an ultrasonic nondestructive testing method at different locations of an existing structure: 3.5, 3.2, 3.1, 3.5, 3.6, 3.2, 3.4, 2.9, 4.1, 2.6, 3.3, 3.5, 3.9, 3.8, 3.7, 3.4, 3.6, 3.5, 3.5, 3.7, 3.6, 3.8, 3.2, 3.4, 4.2, 3.6, 3.1, 2.9, 2.5, 3.5, 3.4, 3.2, 3.7, 3.8, 3.4, 3.6, 3.5, 3.2, 3.6, and 3.8. Plot a histogram and a frequency diagram for concrete strength.

2-16. For the data of Problem 2-15, determine the central tendency measures, that is, the average value, median, and mode.

2-17. For the data of Problem 2-15, determine the dispersion measures, that is, the variance, standard deviation, and coefficient of variation.

2-18. Create a box-and-whisker plot of the data in Problem 2-19. Also create a frequency histogram of the data. Discuss and compare the information content of the two graphical analyses.

2-19. The following data are the maximum daily ozone concentrations for the months indicated. Graph the data in a way that will emphasize the monthly variation in the concentration. Also graph the data in a way that emphasizes the annual variation. Does the effect of monthly variation make it more difficult to assess the importance of the annual variation? Explain.

Year	February	April	June	August	October	December
1980	61	72	77	83	64	55
1981	63	71	78	87	66	58
1982	64	72	78	86	66	59
1983	68	72	80	89	72	63
1984	74	73	85	94	68	66
1985	73	74	86	92	70	64
1986	67	73	85	90	67	64
1987	66	72	83	86	63	59
1988	62	68	81	87	65	57
1989	59	66	82	86	62	51
1990	56	65	80	86	60	50

2-20. Assume a professor gives, on the average, 20% A's, 35% B's, 30% C's, 10% Ds', and 5% F's. Construct a transformation graph that transforms a value from 0 to 1 into a grade. Use the midsquare method (seed = 6348) and the constructed transformation graph to generate grades for a class of 32 students. Construct a pie chart of the simulated grades. Construct a bar chart that shows the proportion of each grade in the simulated class along with the average grades usually given by the professor. Discuss the differences between the simulated and long-term average grades.

2-21. Using a seed of 3597, generate a sample of 25 discharges using the midsquare method and Equation 2-19. Compute the mean and standard deviation and compare them with the values for the measured data (Table 2-3). Graph the frequency histogram of the simulated data and compare it with the histogram for the measured data (Figure 2-19).

Chapter 3
Fundamentals of Probability

TABLE OF CONTENTS

3.1. INTRODUCTION

The analysis of an engineering system should begin with the acquisition of knowledge about the system. This includes its definition, functions, and operational characteristics. The process of knowledge acquisition defines the available information about the system. Equally important is the definition of the lack of information or uncertainty about the system. This process is achieved by characterizing the gained information. These two components are vital for the analysis and decision-making processes.

The treatment of uncertainty requires the understanding of its nature and sources; then appropriate mathematical models can be used to model it. Uncertainties in engineering systems can be attributed mainly to ambiguity and vagueness in defining the variables or parameters of the systems. The ambiguity component is due to (1) physical randomness; (2) statistical uncertainty due to the use of limited information to estimate the characteristics of these parameters; and (3) model uncertainties that are due to simplifying assumptions in analytical and prediction models, simplified methods, and idealized representations of real performances. The vagueness-related uncertainty is common in (1) the definition of certain parameters, e.g., structural performance (failure or survival), quality, deterioration, skill and experience of construction workers and engineers, environmental impact of projects, conditions of existing structures; and (2) defining the interrelationships among the parameters of the problems, especially for complex systems. Probability theory can be used to deal with the ambiguity component of uncertainty.

The objective of this chapter is to introduce the basic concepts of probability theory, relevant computational techniques, and engineering applications.

3.2. SAMPLE SPACES, SETS, AND EVENTS

Sets constitute a fundamental concept in probabilistic analysis of engineering problems. To perform probabilistic analyses of these problems, the definition of the underlying sets is essential for the establishment of a proper model and obtaining realistic results. The goal of this section is to provide the needed set foundation for probabilistic analysis.

3.2.1. Sets

Informally, a set can be defined as a collection of elements or components. Capital letters are usually used to denote sets, e.g., A, B, X, and Y. Small letters are commonly used to denote their elements, e.g., a, b, x, and y.

Example 3-1: Sets

The following are examples of sets:

$$A = \{2, \ 4, \ 6, \ 8, \ 10\} \tag{3-1a}$$

$$B = \{b : b > 0\}; \text{ where ":" means "such that"} \tag{3-1b}$$

$$C = \{\text{Maryland, Virginia, Washington}\} \tag{3-1c}$$

$$D = \{\text{P, M, 2, 7, U, E}\} \tag{3-1d}$$

$$F = \{1, \ 3, \ 5, \ 7, \ 9, \ 11, \ ...\}; \text{ the set of odd numbers} \tag{3-1e}$$

In these example sets, each set consists of a collection of elements. In set A, 2 belongs to A, and 12 does not belong to A. Using mathematical notations, this can be expressed as $2 \in A$ and $12 \notin A$.

Sets can be classified as *finite* and *infinite* sets. For example, sets A, C, and D in Example 3-1 are finite sets; whereas sets B and F are infinite sets. The elements of a set can be either *discrete* or *continuous*. For example, the elements in sets A, C, D, and F are discrete; whereas the elements in set B are continuous. A set without any elements called a null (or empty) set and is denoted as $\emptyset$.

If every element in a set A is also a member of set B, then A is called a subset of B, mathematically expressed as $A \subset B$. Mathematically expressed, A is contained in or equal to B (i.e., $A \subseteq B$), if for every a that belongs to A (i.e., $a \in A$) implies a belongs to B (i.e., $a \in B$). Every set is considered to be a subset of itself. The null set $\emptyset$ is considered to be a subset of every set.

Example 3-2: Subsets

The following are examples of subsets of the sets defined in Example 3-1:

$$A_1 = \{2, \ 4\} \text{ is a subset of } A = \{2, \ 4, \ 6, \ 8, \ 10\} \tag{3-2a}$$

$$B_1 = \{b : 7 < b \le 200\} \text{ is a subset of } B = \{b : b > 0\} \tag{3-2b}$$

$$F = \{1, \ 2, \ 3, \ 4, \ 5, \ 6, \ 7, \ ...\} \text{ is a subset of } F = \{1, \ 2, \ 3, \ 4, \ 5, \ 6, \ 7, \ ...\} \tag{3-2c}$$

Membership (or characteristic) functions are used to describe sets. Let X be a universe, or a set of x values, and let A be a subset of X. Each element, x, is associated with a membership value to the subset A, $\mu_A(x)$. For any set A, the membership function is given by

$$\mu_A(x) = \begin{cases} 1 & \text{if } x \in A \\ 0 & \text{if } x \notin A \end{cases} \tag{3-3}$$

The meaning of this membership function is that there are only two possibilities for an element x, either being a member of A, i.e., $\mu_A(x) = 1$, or not being a member of A, i.e., $\mu_A(x) = 0$. In this case, the set A has sharp (clearly defined) boundaries.

3.2.2. Sample Spaces and Events

In engineering, the set of all possible outcomes of a system (or for an experiment) constitutes the sample space S. A sample space consists of points that correspond to all possible outcomes. Each outcome for the system should constitute a unique element in the sample space. A subset of the sample space is called an event. These definitions are the set-basis of probabilistic analysis. An event without sample points is an empty set and is called the impossible event $\emptyset$. A set that contains all the sample points is called the certain event S. The certain event is equal to the sample space.

Example 3-3: Sample Spaces

The following are examples of sample spaces:

A = {number of cars waiting (queuing) for a left turn at a specified traffic light}
B = {number of units produced by an assembly line}
C = {the strength of concrete delivered at a construction site}
D = {the deformation of a structure under extreme load conditions}

Based on the sample spaces A and D, the following events A_1 and D_1 can be defined:

A_1 = {number of cars waiting (queuing) for a left turn at the specified traffic light between 3:30 p.m. and 6:30 p.m. on a working day}
D_1 = {failure of the structure}

3.2.3. Venn-Euler Diagrams

Events and sets can be represented using spaces that are bounded by closed shapes, such as circles. These shapes are called Venn-Euler (or simply Venn) diagrams. Belonging, nonbelonging, and overlaps between events and sets can be represented by these diagrams.

Example 3-4: Venn Diagrams

In the Venn diagram shown in Figure 3-1, two events (or sets) A and B that belong to a sample space S are represented. The event C is contained in B (i.e., $C \subset B$), and A is not equal to B (i.e., $A \neq B$). Also, the events A and B have an overlap in the sample space S.

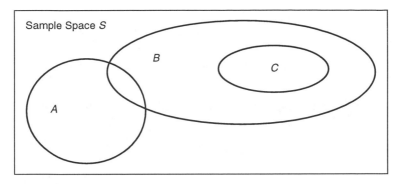

FIGURE 3-1 Events.

3.2.4. Basic Operations

In this section, basic operations that can be used for sets and events are introduced. These operations are analogous to, for example, addition, subtraction, and multiplication in arithmetic calculations.

1. The *union* of events A and B, which is denoted as $A \cup B$, is the set of all elements that belong to A or B or both. Two or more events are called *collectively exhaustive* events if the union of these events results in the sample space.
2. The *intersection* of events A and B, which is denoted as $A \cap B$, is the set of all elements that belong to both A and B. Two events are termed *mutually exclusive* if the occurrence of one event precludes the occurrence of the other event. The term can also be extended to more than two events.
3. The *difference* of events A and B, $A - B$, is the set of all elements that belong to A but not to B.
4. The event that contains all of the elements that do not belong to an event A is called the complement of A, and is denoted $\overline{A}$.

Example 3-5: Operations on Sets and Events

The following are example sets:

$$A = \{2,\ 4,\ 6,\ 8,\ 10\} \tag{3-4a}$$

$$B = \{1,\ 2,\ 3,\ 4,\ 5,\ 6,\ 7,\ 8,\ 9,\ 10\} \tag{3-4b}$$

$$C = \{1,\ 3,\ 5,\ 7,\ 9,\ 11,\ ...\};\ \text{the set of odd numbers} \tag{3-4c}$$

$$F_1 = \{\text{failure of a structure due to an earthquake}\} \tag{3-4d}$$

$$F_2 = \{\text{failure of a structure due to strong winds}\} \tag{3-4e}$$

$$F_3 = \{\text{failure of a structure due to an extreme overload}\} \tag{3-4f}$$

The following operations can be executed for the sets in Equations 3-4a to 3-4f:

$$A \cup B = B = \{1,\ 2,\ 3,\ 4,\ 5,\ 6,\ 7,\ 8,\ 9,\ 10\} \tag{3-5a}$$

$$A \cap B = A = \{2,\ 4,\ 6,\ 8,\ 10\} \tag{3-5b}$$

$$\overline{C} = \{2,\ 4,\ 6,\ 8,\ 10,\ 12,\ ...\};\ \text{the set of even numbers} \tag{3-5c}$$

$$F_1 \cup F_2 = \{\text{failure of the structure due to an earthquake or strong wind}\} \tag{3-5d}$$

$$\overline{F_3} = \{\text{nonfailure of the structure due to an extreme overload}\} \tag{3-5e}$$

The events F_3 and $\overline{F_3}$ can be considered to be mutually exclusive. Table 3-1 shows additional rules based on the above fundamental rules. The validity of these rules can be checked using Venn diagrams.

TABLE 3-1 Additional Operational Rules

Rule Type	Operations
Identity laws	$A \cup \varnothing = A, A \cap \varnothing = \varnothing, A \cup S = S, A \cap S = A$
Idempotent laws	$A \cup A = A, A \cap A = A$
Complement laws	$A \cup \overline{A} = S, A \cap \overline{A} = \varnothing, \overline{\overline{A}} = A, \overline{S} = \varnothing, \overline{\varnothing} = S$
Commutative laws	$A \cup B = B \cup A, A \cap B = B \cap A$
Associative laws	$(A \cup B) \cup C = A \cup (B \cup C), (A \cap B) \cap C = A \cap (B \cap C)$
Distributive laws	$(A \cup B) \cap C = (A \cap C) \cup (B \cap C)$
	$(A \cap B) \cup C = (A \cup C) \cap (B \cup C)$
de Morgan's law	$\overline{(A \cup B)} = \overline{A} \cap \overline{B}, \overline{(E_1 \cup E_2 \cup ... \cup E_n)} = \overline{E}_1 \cap \overline{E}_2 \cap ... \cap \overline{E}_n$
	$\overline{(A \cap B)} = \overline{A} \cup \overline{B}, \overline{(E_1 \cap E_2 \cap ... \cap E_n)} = \overline{E}_1 \cup \overline{E}_2 \cup ... \cup \overline{E}_n$
Combinations of laws	$\overline{(A \cup (B \cap C))} = \overline{A} \cap \overline{(B \cap C)} = (\overline{A} \cap \overline{B}) \cup (\overline{A} \cap \overline{C})$

3.2.5. Cartesian Product

The Cartesian-product set of A and B is the set of all ordered pairs (a,b) such that $a \in A$ and $b \in B$. In mathematical notation, this can be expressed as

$$A \otimes B = \{(a,b): a \in A \text{ and } b \in B\} \tag{3-6}$$

Therefore, the Cartesian product, $\otimes$, of the two sets results in a two-dimensional matrix of all the possible combinations [ordered pairs (a,b)] of the elements of the two sets.

Example 3-6: Production Lines

A precast-concrete factory produces components for floor or roof systems. The manufacturing of these components is performed by three independent production lines. The quality-control department for the factory determines whether the components are defective or nondefective for the lines. Therefore, the following two sets can be defined: (1) the set of production lines $L = \{$Line 1, Line 2, Line 3$\}$, and (2) the set of component quality $Q = \{$defective, nondefective$\}$. Therefore, the Cartesian product of L and Q is given by the following:

	Line 1	Line 2	Line 3
Defective	(Line 1, defective)	(Line 2, defective)	(Line 3, defective)
Nondefective	(Line 1, nondefective)	(Line 2, nondefective)	(Line 3, nondefective)

3.3. MATHEMATICS OF PROBABILITY

The concept of probability has its origin in games of chance. In these games, probabilities are determined based on many repetitions of an experiment and counting the number of outcomes of an event of interest. Then, the probability of the outcome of interest was measured by dividing the number of event occurrences by the total number of repetitions. The objective of this section is to introduce the fundamentals of probability for engineering applications.

3.3.1. Definition of Probability

Upon receiving a graded test from the instructor, a student in a class is interested in three things: (1) the score received, (2) the distribution of the scores in the class or summary characteristics of the distribution, and (3) the grading policy. The scores that the students received are meaningless by themselves, because they fail to tell the students individually both where they stand with respect to others and the opinion of the instructor about how well they understood the material. For example, if the student received a grade of 50 out of possible 100, he or she may or may not consider this to be good. If the class average was 45, the student (with a score of 50) knows that he or she did better than average. However, by itself, the class average does not indicate how well the student performed. The assessment of the performance depends, in part, on the dispersion of the grades. The student should feel better if the grades ranged from 35 to 53 than if the grades ranged from 10 to 90. In the first case, the score of 50 is closer to the highest score (53), and thus the performance is probably better than a student with an average score. Although the closeness of the score with respect to the mean is important, the student is also interested in the grading policy, which indicates the instructor's assessment of how well he or she performed with respect to all the students that the instructor has taught in the past. For example, if the instructor believes that all the test scores were poor in comparison to students of previous years, then the instructor may set the grading policy such that there were no A's or B's; that is, even the student having the highest grade in the class (53) would get a C. Thus, besides the mean and dispersion, the grading policy, which is based on all past students, is important.

Statistics, such as the mean and dispersion, and policy criteria are used extensively in engineering decision making. If the value of 50 were a measurement of a water quality based on a certain pollutant, could we safely conclude that the average level of the pollutant in the river was not different from the level of 45 allowed by the state policy? Similarly, corrosion reduces the thickness of steel beams, but the loss of material due to corrosion is far from uniform over the cross section of a beam. If a 50-mil loss of thickness is measured at one point in a beam and the policy allows for a 45-mil loss, can we conclude that the steel beam is unsafe? Numerous other examples could be presented. The important point made through these examples is that decisions must often be based on sample measurements and data must be analyzed systematically if the decisions are to be universally accepted.

The test score example is useful for introducing some important terminology. In the instructor's viewpoint, the collection of test scores represents *observations* on the *random variable* "test score". The test scores for the class represent a *sample* of observations; the test scores for all the students who have had the test (or similar test) in the past or will have the test in the future represent the *population.* Thus the sample is a random collection or subset of the population. The class average and the range of scores are *sample characteristics*; that is, they are values that provide summary information about the sample. The grading policy is an assessment of the population, and characteristics derived from the population are *population characteristics.*

The student uses the available information, i.e., his or her score, the class distribution of scores, and the grading policy to make inferences about the likelihood of getting a specific final grade at the end of the course. This represents an example of *probabilistic assessment* and *statistical inference.*

In the test score example, a student may be interested in determining the likelihood of his or her final grade. If the deadline for dropping courses is near, the likelihood of receiving a passing grade is of interest. *Probability* is a scale of measurement used to describe the likelihood of an event, where an event is defined as the occurrence of a specified value of the random variable "final grade". It is important to note that there are two random variables: the test score and the final grade. The student must use the value of one random variable to draw inference about the other. The scale on which probability is measured extends from 0 to 1, inclusive, where a value of 1 indicates a certainty of occurrence of the event and a value of 0 indicates certainty of its nonoccurrence.

Quite often, probability is specified as a percentage; for example, when the weather bureau indicates that there is a 30% chance of rain, experience indicates that under similar meteorological conditions it has rained three out of ten times. In this example, the probability was estimated empirically using the concept of relative frequency expressed as

$$P(X = x_o) = \frac{n}{N} \tag{3-7}$$

in which n = number of observations on the random variable X that results in an outcome of interest x_o, and N = total number of observations of X.

In the case of the student who is trying to estimate the probability of getting a passing grade for the course, the student considers the grades that he or she received and how he or she has performed under similar circumstances in the past. If the student has received grades D and F on the two tests taken to date and has failed three out of four courses in which he or she also had a D and F on the first two tests, the likelihood is high (0.75) that the student will also fail this course. Thus, with a high probability of failure, the student would most likely elect to drop the course. This represents decision making under conditions of uncertainty.

Like students, engineers must make decisions under conditions of uncertainty. For example, engineers who have the responsibility of monitoring water quality in our nation's streams and bodies of water estimate pollution levels using samples collected from the water. The samples are then analyzed in a laboratory and the results are used to make a decision. Most sampling programs involve ten or fewer measurements. Uncertainty arises because of the highly variable nature of pollution; that is, the concentration of a pollutant may vary with time, the degree of turbulence in the water, and the frequency with which wastes are discharged into the water. These sources of variation must be accounted for when the engineer makes a decision about water quality.

Traffic engineers must also make decisions under conditions of uncertainty. For example, intersections are frequently the sites of accidents. The traffic engineer knows that accidents can be reduced by installing stop signs or traffic lights. However, there is a cost associated with installing such hardware. Also, traffic controls can cause delay and inconvenience to those that must travel through the intersections. Thus, the traffic engineer must consider numerous factors in making a decision, including the likelihood and severity of accidents at the intersection and the traffic load in each direction. The frequency and severity of accidents can be assessed using data from accidents that have occurred at that intersection in the past. However, these are data of the past, and there is no assurance that they will accurately reflect accident rates in the future. Reduced travel due to an increase in the cost of gasoline may reduce the number of accidents. Data on the traffic volumes originating from each street entering the intersection can be obtained using traffic counters. However, these data may not completely characterize the traffic volumes that will take place in the future. For example, if the traffic volume data are collected during the summer, the opening of schools may alter the relative proportion of traffic volumes on each street entering the intersection. Such sources of variation introduce uncertainty into the decision-making process.

It is clear from these examples that the probability of an event can be defined as the relative frequency of its occurrence or the subjective probability of its occurrence. The type of definition depends on the underlying event. For example, in an experiment that can be repeated N times with n occurrences of the underlying event, the relative frequency of occurrence can be considered as the probability of occurrence. In this case, the probability of occurrence is n/N. However, there are many engineering problems that do not involve large numbers of repetitions, and still we are interested in estimating the probability of occurrence of some event. For example, during the service life of an engineering product, the product either fails or does not fail in performing a set of performance criteria. The events of failure and survival are mutually exclusive and collectively exhaustive of the sample space. The probability of failure (or survival) is considered as a subjective probability. An estimate of this probability can be achieved by modeling the underlying system, its uncertainties, and performances. The resulting subjective probability is expected to reflect the status of our knowledge about the system regarding the true likelihood of occurrence of the events of interest. In this section, the mathematics of probability is applicable to both definitions. However, it is important to keep in mind both definitions, so that results are not interpreted beyond the range of their validity.

3.3.2. Axioms of Probability

In general, an axiomatic approach can be used to define probability as a function from sets to real numbers. The domain is the set of all events within the sample space of the problem, and the range consists of the numbers on the real line. For an event A, the notation $P(A)$ means the probability of occurrence of the event A. The function $P(.)$ should satisfy the following properties:

$$0 \leq P(A) \leq 1 \quad \text{for every event } A \subseteq S \tag{3-8a}$$

$$P(S) = 1 \tag{3-8b}$$

If A_1, A_2, ..., A_n are mutually exclusive events on S, then

$$P(A_1 \cup A_2 \cup ... \cup A_n) = P(A_1) + P(A_2) + ... + P(A_n) \tag{3-8c}$$

The first property states that the probability of any event is inclusively between 0 and 1. Therefore, negative probabilities or probabilities larger than one are not allowed. The second property comes from the definition of the sample space. Since the sample space is the set of *all possible outcomes*, one or more of these outcomes must, therefore, occur resulting in the occurrence of *S*. If the probability of the sample space does not equal 1, this means that the sample space was incorrectly defined. The third property results from the definition of mutually exclusive events. Computational rules can be developed based on these properties. Example rules are given in the following:

$$P(\emptyset) = 0 \tag{3-9}$$

$$P(A \cup B) = P(A) + P(B) - P(A \cap B) \tag{3-10a}$$

$$P(A \cup B \cup C) = P(A) + P(B) + P(C) - P(A \cap B) - P(A \cap C) - P(B \cap C) \tag{3-10b}$$
$$+ P(A \cap B \cap C)$$

$$P(\overline{A}) = 1 - P(A) \tag{3-11}$$

$$\text{If } A \subseteq B, \text{ then } P(A) \le P(B) \tag{3-12}$$

Example 3-7: Probabilities of Events

In the design of structures, the occurrence of damaging winds and earthquakes is of interest. The following events can be defined:

E = {the occurrence of an earthquake in a year}
W = {the occurrence of damaging winds in a year}

Assume that the events E and W have the probabilities $P(E)$ and $P(W)$ of 0.1 and 0.2, respectively. Also, the probability of occurrence of both E and W [i.e., $P(E \cap W)$] is 0.02. Therefore, probability of occurrence of an earthquake or strong winds in a year can be determined as the union of E and W using Equation 3-10a as

$$P(E \cup W) = 0.1 + 0.2 - 0.02 = 0.28 \tag{3-13}$$

The probability of nonoccurrence of an earthquake in a year according to Equation 3-11 is

$$P(\overline{E}) = 1 - 0.1 = 0.9 \tag{3-14}$$

where $\overline{E}$ = the complement of E, i.e., nonoccurrence of an earthquake.

3.3.3. Counting

In experiments that result in finite sample spaces, the processes of identification, enumeration, and counting are essential for the purpose determining the probabilities of some outcomes of interest. The identification process results in defining all possible outcomes and their likelihood of occurrence. The identification of equally likely outcomes is needed to determine any probabilities of interest. The order of occurrence of the outcomes can be important in certain engineering applications, requiring its consideration in the counting process.

The enumeration process can be performed in any systematic form that results in all possible outcomes. The multiplication principle can be used for this purpose. Let events $A_1, A_2, ..., A_n$ have $n_1, n_2, ..., n_n$ elements, respectively. Therefore, the total number of possible outcomes of selecting one element from each of $A_1, A_2, ..., A_n$ is the product $n_1\, n_2\, ... \, n_n$, where the outcomes represent the ways to select the first element from A_1, the second element from A_2, ..., and finally to select the nth element from A_n.

The permutation of r elements from a set of n elements is the number of arrangements that can be made by selecting r elements out of the n elements. The order of selection counts in determining these arrangements. The permutation $P_{r|n}$ of r out of n (where $r \leq n$) is

$$P_{r|n} = \frac{n!}{(n-r)!}$$
(3-15)

where $n!$ = factorial of $n = n(n-1)(n-2)\cdots(2)(1)$. It should be noted that $0! = 1$ by convention. Equation 3-15 results from the fact that there are n ways to select the first element, $(n-1)$ ways to select the second element, $(n-2)$ ways to select the third element, and so on to the last element, i.e., rth element.

The combination of r elements from a set of n elements is the number of arrangements that can be made by selecting r elements out of the n elements. The order of selection in this case does not count in determining these arrangements. One arrangement differs from another, only if the contents of the arrangements are different. The combination $C_{r|n}$ of r out of n (where $r \leq n$) is

$$C_{r|n} = \frac{P_{r|n}}{r!}$$
(3-16)

Therefore, the combination $C_{r|n}$ can be determined as

$$C_{r|n} = \frac{n!}{(r!)(n-r)!}$$
(3-17)

It is very common to use the notation $\binom{n}{r}$ for the combination $C_{r|n}$. It can be shown that the following identity is valid

$$\binom{n}{r} = \binom{n}{n-r}$$

(3-18)

Example 3-8: Counting for Bridge Failure

Consider a simple case of a bridge that is supported by three cables. The failure of interest is the failure of only two cables out of the three cables since it results in the failure of the bridge. What is the number of combinations of $r = 2$ out of $n = 3$ that can result in bridge failure?

$$C_{r|n} = \frac{n!}{r!(n-r)!} = \frac{3!}{2!(3-2)!} = \frac{(3)(2)(1)}{(2)(1)(1)} = 3$$

(3-19a)

This number of combinations can be established by enumeration. The following events can be defined:

$$C_i = \text{failure of cable } i, \text{ where } i = 1, \text{ 2, and 3}$$

(3-19b)

Therefore, the following events result in bridge failure:

$$C_1 \cap C_2$$

(3-20a)

$$C_1 \cap C_3$$

(3-20b)

$$C_2 \cap C_3$$

(3-20c)

The number of events is three, which is the same as the result from Equation 3-19. In Equation 3-21, the order of occurrence of cable failure is not a factor in determining the number of combinations. However, if we assume the order of failure is a factor, then the possible events become

$$C_1 \cap C_2$$

(3-21a)

$$C_1 \cap C_3$$

(3-21b)

$$C_2 \cap C_3$$

(3-21c)

$$C_2 \cap C_1$$

(3-21d)

$$C_3 \cap C_1$$

(3-21e)

$$C_3 \cap C_2 \tag{3-21f}$$

In this case, the number of combinations is six. The same result can be obtained by computing the number of permutations, where order counts, using Equation 3-15 as

$$P_{r|n} = \frac{n!}{(n-r)!} = \frac{3!}{(3-2)!} = 6 \tag{3-22}$$

Now, if we assume that the bridge is supported by $n = 20$ cables, and the failure of $r = 8$ cables results in the failure of the bridge, what is the number of combinations of $r = 8$ out of $n = 20$ that can result in bridge failure? In this case, the number of combinations is

$$C_{r|n} = \frac{n!}{r!(n-r)!} = \frac{20!}{8!12!} = \frac{(20)(19)(18)...(13)}{(8)(7)(6)...(1)} = 125,970 \tag{3-23}$$

For a real bridge, its failure can result from the failure of at least $r = 8$ out of $n = 20$. The number of combinations in this case is

$$\sum_{r=8}^{20} C_{r|n} = \frac{20!}{(8!)(12!)} + \frac{20!}{(9!)(11!)} + \frac{20!}{(10!)(10!)} + \cdots + \frac{20!}{(20!)(0!)} \tag{3-24}$$

3.3.4. Conditional Probability

The probabilities previously discussed are based on and relate to the sample space S. However, it is common in many engineering problems to have interest in the probabilities of occurrence of events that are conditioned on the occurrence of a subset of the sample space. This introduces the concept of conditional probability. For example, the probability of A given that B has occurred, denoted as $P(A|B)$, means the occurrence probability of a sample point that belongs to A given that we know it belongs to B. The conditional probability can be computed as follows:

$$P(A|B) = \frac{P(A \cap B)}{P(B)} \quad \text{if } P(B) \neq 0 \tag{3-25}$$

Clearly, the underlying sample space for the conditional probability is reduced to the conditional event B. The conditional probability satisfies all the properties of probabilities. The following properties can be developed for conditional probabilities:

1. The complement of an event:

$$P(\overline{A}|B) = 1 - P(A|B) \qquad (3\text{-}26)$$

2. The multiplication rule for two events A and B:

$$P(A \cap B) = P(A|B)\,P(B) \quad \text{if } P(B) \neq 0 \qquad (3\text{-}27a)$$

$$P(A \cap B) = P(B|A)\,P(A) \quad \text{if } P(A) \neq 0 \qquad (3\text{-}27b)$$

3. The multiplication rule for three events A, B, and C:

$$P(A \cap B \cap C) = P(A|(B \cap C))\,P(B|C)\,P(C) = P((A \cap B)|C)\,P(C)$$
$$\text{if } P(C) \neq 0 \text{ and } P(B \cap C) \neq 0 \qquad (3\text{-}28)$$

4. For mutually exclusive events A and B:

$$P(A|B) = 0 \qquad (3\text{-}29a)$$

$$P(B|A) = 0 \qquad (3\text{-}29b)$$

5. For statistically independent events A and B:

$$P(A|B) = P(A) \qquad (3\text{-}30a)$$

$$P(B|A) = P(B) \qquad (3\text{-}30b)$$

$$P(A \cap B) = P(A)\,P(B) \qquad (3\text{-}30c)$$

$$A \text{ and } \overline{B} \text{ are independent events} \qquad (3\text{-}30d)$$

$$\overline{A} \text{ and } B \text{ are independent events} \qquad (3\text{-}30e)$$

$$\overline{A} \text{ and } \overline{B} \text{ are independent events} \qquad (3\text{-}30f)$$

Example 3-9: Conditional Probabilities for Production Lines

Consider the three manufacturing lines of Example 3-6. Assume that 20, 30, and 50% of the components are manufactured by lines 1, 2, and 3, respectively. The quality assurance department of the producing factory determined that the probability of having defective products from lines 1, 2, and 3 are 0.1, 0.1, and 0.2, respectively. The following events can be defined in this case:

$$L_1 = \text{Component produced by line 1} \qquad (3\text{-}31a)$$

$$L_2 = \text{Component produced by line 2} \qquad (3\text{-}31b)$$

$$L_3 = \text{Component produced by line 3} \qquad (3\text{-}31c)$$

$$D = \text{Defective component} \qquad (3\text{-}31d)$$

Therefore, the following probabilities are given:

$$P(D|L_1) = 0.1 \qquad (3\text{-}32a)$$

$$P(D|L_2) = 0.1 \qquad (3\text{-}32b)$$

$$P(D|L_3) = 0.2 \qquad (3\text{-}32c)$$

Since these events are not independent, the joint probabilities can be determined as follows:

$$P(D \cap L_1) = P(D|L_1)\,P(L_1) = 0.1(0.2) = 0.02 \qquad (3\text{-}33a)$$

$$P(D \cap L_2) = P(D|L_2)\,P(L_2) = 0.1(0.3) = 0.03 \qquad (3\text{-}33b)$$

$$P(D \cap L_3) = P(D|L_3)\,P(L_3) = 0.2(0.5) = 0.1 \qquad (3\text{-}33c)$$

This example illustrates the importance of using conditional probabilities in determining the probability of the intersection of events.

Sample Space S

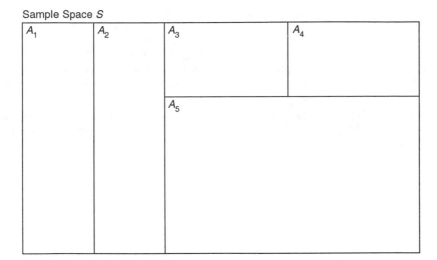

FIGURE 3-2 Partitioned sample space.

Sample Space S

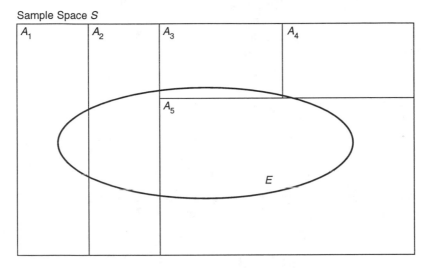

FIGURE 3-3 Theorem of total probability.

3.3.5. Partitions, Total Probability, and Bayes' Theorem

A set of disjoint (i.e., mutually exclusive) events A_1, A_2, ..., A_n form a partition of a sample space if $A_1 \cup A_2 \cup ... \cup A_n = S$. An example partition is shown in Figure 3-2.

If A_1, A_2, ..., A_n represents a partition of a sample space S, and $E \subset S$ represents an arbitrary event as shown in Figure 3-3, the theorem of total probability states that

$$P(E) = P(A_1)\,P(E|A_1) + P(A_2)\,P(E|A_2) + \cdots + P(A_n)\,P(E|A_n) \qquad (3\text{-}34)$$

This theorem is very important in computing the probability of an event E, especially in practical cases where the probability cannot be computed directly, but the probabilities of the partitioning events and the conditional probabilities can be computed.

Bayes' theorem is based on the same conditions of partitioning and events as the theorem of total probability and is very useful in computing the reverse probability of the type $P(A_i|E)$, for $i = 1, 2, ..., n$. The reverse probability can be computed as follows:

$$P(A_i|E) = \frac{P(A_i)\,P(E|A_i)}{P(A_1)\,P(E|A_1) + P(A_2)\,P(E|A_2) + \cdots + P(A_n)\,P(E|A_n)} \qquad (3\text{-}35)$$

The denominator of this equation is $P(E)$, which is based on the theorem of total probability.

Example 3-10: Defective Products in Production Lines

The three manufacturing lines discussed in Example 3-9 are used to illustrate the total probability theorem. The theorem can be used to determine the probability of a defective component as follows:

$$P(D) = P(D|L_1)\,P(L_1) + P(D|L_2)\,P(L_2) + P(D|L_3)\,P(L_3)$$

$$= 0.1(0.2) + 0.1(0.3) + 0.2(0.5) = 0.02 + 0.03 + 0.1 \qquad (3\text{-}36)$$

$$= 0.15$$

This probability includes the three lines. Therefore, on the average, 15% of the components produced by the factory are defective.

3.4. RANDOM VARIABLES AND THEIR PROBABILITY DISTRIBUTIONS

In some engineering applications, the outcomes, say, $E_1, E_2, ..., E_n$, that constitute a sample space S take numerical real values. In other applications, the outcomes can be nonnumeric or combinations of both. It is convenient and sometimes necessary to express all outcomes using numerical values on the real line R. The functions that establish such a transformation to the real line are called random variables. Formally, a *random variable* is defined as a function that assigns a real value to every possible outcome for an engineering system. This mapping can be one-to-one or one-to-many. Based on this definition, the properties of the underlying outcomes, e.g., intersection, union, and complement, are retained in the form of, for example, overlapping ranges of real values, combination of real ranges, and values outside these ranges.

Random variables are commonly classified into two types: *discrete* and *continuous* random variables. A discrete random variable may only take on distinct, usually integer, values; for example, the outcome of a roll of a die may only take on the integer values from 1 to 6 and is, therefore, a discrete random variable. The number of floods per year at a point on a river can only take on integer values, so it is also a discrete random variable. A continuous random variable takes values within a continuum of values. For example, the average of all scores on a test having a maximum possible score of 100 may take on any value including nonintegers, between 0 and 100; thus the class average would be a continuous random variable. A distinction is made between these two types of random variables because the computations of probabilities are different for the two types.

3.4.1. Probability of Discrete Random Variables

The probability of a discrete random variable is given by the *probability mass function*, which specifies the probability that the discrete random variable X equals some value x_i and is denoted by

$$P_X(x_i) = P(X = x_i) \tag{3-37}$$

A capital X is used for the random variable, whereas an x_i is used for the *ith* largest value of the random variable. The probability mass function must satisfy the axioms of probability. Therefore, the probability of an event x_i must be less than or equal to one and greater than or equal to zero, i.e.,

$$0 \le P_X(x_i) \le 1 \tag{3-38}$$

This property is valid for all possible values of the random variable X. Additionally, the sum of all possible probabilities must be equal to one, i.e.,

$$\sum_{i=1}^{N} P_X(x_i) = 1 \tag{3-39}$$

in which N = total number of possible outcomes; for the case of the roll of a die, N equals six.

It is often useful to present the likelihood of an outcome using the *cumulative mass function*, $F_X(x_i)$, which is given by

$$F_X(x_i) = P(X \le x_i) = \sum_{j=1}^{i} P_X(x_j) \tag{3-40}$$

The cumulative mass function is used to indicate the probability that the random variable X is less than or equal to x_i. It is inherent in the definition (Equation 3-40) that the cumulative probability is defined as zero for all the values less than the smallest x_i and one for all values greater than the largest value.

Example 3-11: Roll of a Die

To illustrate the concepts introduced for discrete random variables, assume that the random variable is the outcome of the roll of a die. The *sample space* consists of the collection of all possible values of the random variable, i.e., $x = 1, 2, 3, 4, 5,$ and 6. The probability of each element of the sample space is given by

$$P_X(x) = \frac{1}{6} \quad \text{for } x = 1, 2, 3, 4, 5, \text{ and } 6 \tag{3-41}$$

The probability mass function is shown in Figure 3-4a. The cumulative mass function can be evaluated using Equation 3-40. The results are shown in Figure 3-4b and are given by

$$F_X(x) = P(X \le x) = \frac{x}{6} \quad \text{for } x = 1, 2, 3, 4, 5, \text{ and } 6 \tag{3-42}$$

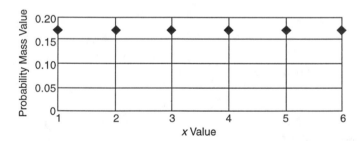

FIGURE 3-4a Probability mass function for the roll of a die.

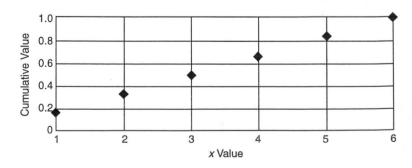

FIGURE 3-4b Cumulative mass function for the roll of a die.

Example 3-12: Pair of Dice

In this example, a random variable is defined as the sum of the dots showing when a pair of dice is rolled. What is the cumulative mass function? What is the probability of getting an even value? What is the probability that for a single roll of the two dice, the sum of the points is equal to 11? What is the probability of obtaining a value of 6 or less?

The sample space consists of the integer values from 2 to 12. If one identifies all the 36 possible outcomes, the relative frequency concept of Equation 3-7 yields the following mass function:

$$P_X(x) = \left\{ \frac{1}{36}, \frac{1}{18}, \frac{1}{12}, \frac{1}{9}, \frac{5}{36}, \frac{1}{6}, \frac{5}{36}, \frac{1}{9}, \frac{1}{12}, \frac{1}{18}, \frac{1}{36} \right\} \tag{3-43}$$

for $x = 2, 3, 4, 5, \ldots, 12$, respectively

It is important to note that the sum of the probabilities equals 1; otherwise, $P_X(x)$ would not be a legitimate mass function. The cumulative mass function is

$$F_X(x) = \left\{ \frac{1}{36}, \frac{1}{12}, \frac{1}{6}, \frac{5}{18}, \frac{15}{36}, \frac{7}{12}, \frac{13}{18}, \frac{5}{6}, \frac{11}{12}, \frac{35}{36}, 1 \right\} \tag{3-44}$$

The probability of the random variable X taking an even value is the sum of the probabilities for the even values of X. Thus

$$P(x \text{ is even}) = \frac{1}{36} + \frac{1}{12} + \frac{5}{36} + \frac{5}{36} + \frac{1}{12} + \frac{1}{36} = \frac{1}{2} \tag{3-45}$$

Also,

$$P(x = 11) = \frac{1}{18} \tag{3-46}$$

and using the probability mass function, we have

$$P(x \leq 6) = \frac{1}{36} + \frac{1}{18} + \frac{1}{12} + \frac{1}{9} + \frac{5}{36} = \frac{15}{36} = 0.4167 \tag{3-47}$$

The same result can be obtained directly from the cumulative mass function.

Equations 3-43 and 3-44 and the values in Example 3-12 apply to the populations that underlie the random variables. They do not give the values obtained from a sample of data. For small samples, probabilities computed using sample results may poorly reflect the true probabilities. For example, if a die is rolled five times with

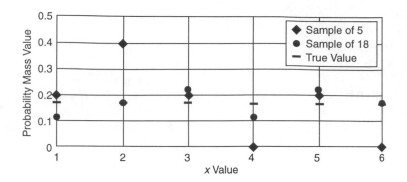

FIGURE 3-5 Probability mass function for two samples.

the following values [3, 2, 5, 1, 2], then the sample mass function would be as shown in Figure 3-5. The sample probability for $x = 2$ is over twice as high as the population probability (1/6). Since $x = 4$ and $x = 6$ did not occur, their sample probabilities of 0 are also poor reflections of the true probabilities. This illustrates the difficulty of using information based on small samples to draw inferences about population probabilities. Larger samples usually provide better estimates of population probabilities than those provided by small samples. Consider the following sample of rolling a single die 18 times: {3, 1, 4, 3, 6, 2, 4, 5, 5, 2, 6, 3, 1, 3, 5, 2, 6, 5}. The sample includes two 1's, three 2's, four 3's, two 4's, four 5's, and three 6's, which gives sample probabilities of 0.111, 0.167, 0.222, 0.111, 0.222, and 0.167. Thus, the larger sample of 18 rolls does not contain the extreme probabilities that were observed in the sample of 5. Note in Figure 3-5 that the probabilities from the sample of 18 are generally closer to the true value of 0.167 than are the sample probabilities based on 5 values.

Example 3-13: Hurricanes

The numbers of hurricanes in the U.S. from 1970/1971 through 1984/1985 are as follows:

1970-71	5	1975-76	3	1980-81	2
1971-72	4	1976-77	1	1981-82	1
1972-73	0	1977-78	3	1982-83	2
1973-74	0	1978-79	1	1983-84	4
1974-75	2	1979-80	6	1984-85	0

The sample probability mass function and the cumulative mass function can be tabulated as shown in Table 3-2. The probability mass values in the table was computed by dividing the number of occurrence by 15, which is the total number of hurricane seasons in the period 1970 to 1985. The cumulative mass value was computed using Equation 3-40. The results of the calculations are also shown in Figures 3-6a and 3-6b.

**TABLE 3-2 Computation of Sample Probability Function
for Hurricanes in Example 3-13**

Number of hurricanes, X	Number of occurrences	Probability mass value	Cumulative mass value
0	3	0.2	0.2
1	3	0.2	0.4
2	3	0.2	0.6
3	2	0.13333	0.73333
4	2	0.13333	0.86667
5	1	0.06667	0.93333
6	1	0.06667	1.
7	0	0.	1.
8	0	0.	1.
.	0	0.	1.
.	0	0.	1.
∞	0	0.	1.

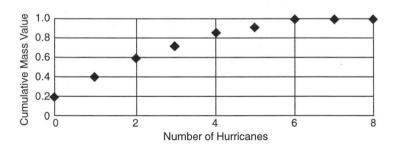

FIGURE 3.6a Sample probability mass function for hurricanes in the U.S.

FIGURE 3-6b Sample cumulative mass function for hurricanes in the U.S.

In spite of the small sample, the mass function shows a consistent trend, with the probability decreasing as the value of the random variable x, i.e., the number of hurricanes per year, increases. It appears that hurricanes do not have the same mass function as the roll of a die. Besides the decreasing trend, the mass function of Table 3-2 does not allow for values greater than six. Even though there was no year in which more than six hurricanes occurred in the 15-year period, the possibility that it could occur exists, so a mass function that allows for more than six occurrences of X should be selected for the population.

3.4.2. Probability for Continuous Random Variables

A *probability density function* (pdf) defines the probability of occurrence for a continuous random variable. Specifically, the probability that the random variable X lies within the interval form x_1 to x_2 is given by

$$P\left(x_1 \le X \le x_2\right) = \int_{x_1}^{x_2} f_X(x)\, dx \qquad (3\text{-}48)$$

in which $f_X(x)$ is the probability density function. If the interval is made infinitesimally small, x_1 approaches x_2 and $P(x_1 \le X \le x_2)$ approaches zero. This illustrates a property that distinguishes discrete random variables from continuous random variables. Specifically, the probability that a continuous random variable takes on a specific value equals zero; that is, probabilities for continuous random variables must be defined over an interval.

It is important to note that the integral of the pdf from $-\infty$ to $+\infty$ equals 1, i.e.,

$$P(-\infty < X < +\infty) = \int_{-\infty}^{+\infty} f_X(x)\, dx = 1 \qquad (3\text{-}49)$$

Also, because of Equation 3-49, the following holds:

$$P\left(X \ge x_o\right) = \int_{x_o}^{+\infty} f_X\left(x_o\right) dx = 1 - P\left(X < x_o\right) \qquad (3\text{-}50)$$

The *cumulative distribution function* (cdf) of a continuous random variable is defined by

$$F_X\left(x_o\right) = P\left(X \le x_o\right) = \int_{-\infty}^{x_o} f_X(x)\, dx \qquad (3\text{-}51a)$$

The cdf is a nondecreasing function in that $P(X \le x_1) \le P(X \le x_2)$ where $x_1 \le x_2$. The cdf equals 0 at $-\infty$ and 1 at $+\infty$. The relationship between $f_X(x)$ and $F_X(x)$ can also be expressed as

$$f_X(x) = \frac{dF_X(x)}{dx} \qquad (3\text{-}51b)$$

Example 3-14: Continuous Random Variables

The continuous random variable X has the following probability density function:

$$f_X(x) = \begin{cases} kx & \text{for } 0 \le x \le 2 \\ 0 & \text{otherwise} \end{cases} \tag{3-52}$$

where k is a constant. Find the value of k that is necessary for $f_X(x)$ to be a legitimate probability density function. Plot both the density and cumulative functions. What is the probability that X equals 1? What is the probability that X takes on a value less than 0.5? What is the probability that X is greater than 1.0 and less than 1.5?

For $f_X(x)$ to be a legitimate pdf, it must satisfy the following constraint:

$$P(-\infty < X < +\infty) = \int_{-\infty}^{+\infty} f_X(x)\, dx = 1 \tag{3-53}$$

Therefore,

$$1 = \int_0^2 kx\, dx = k \int_0^2 x\, dx = \frac{kx^2}{2}\Big|_0^2 \tag{3-54a}$$

and

$$k = 0.5 \tag{3-54b}$$

The cumulative distribution function is given by

$$F_X(x_o) = \int_0^{x_o} 0.5x\, dx = \frac{x^2}{4}\Big|_0^{x_o} = \frac{x_o^2}{4} \tag{3-55}$$

The probability density and cumulative functions appear graphically in Figures 3-7a and 3-7b, respectively. Because probabilities of continuous random variables are defined for regions rather than point values, $P(X = 1) = 0$. The $P(X < 0.5)$ can be determined from the cumulative function; specifically,

$$P(X < 0.5) = \frac{x_o^2}{4}\Big|_{x_o=0}^{x_o=0.5} = \frac{1}{16} \tag{3-56}$$

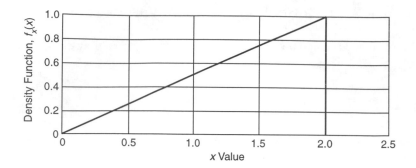

FIGURE 3-7a Probability density function for a continuous random variable.

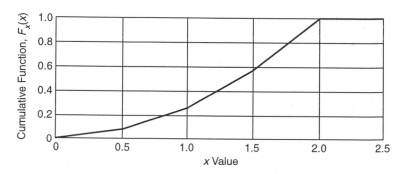

FIGURE 3-7b Cumulative distribution function for a continuous random variable.

Similarly, the cumulative function can be used to find the probability for the following region:

$$P(1.0 < X < 1.5) = P(X < 1.5) - P(X < 1.0)$$

$$= \frac{x_o^2}{4}\bigg|_{x_o=0}^{x_o=1.5} - \frac{x_o^2}{4}\bigg|_{x_o=0}^{x_o=1.0} \tag{3-57}$$

$$= \frac{9}{16} - \frac{1}{4} = \frac{5}{16}$$

Example 3-15: Settlement of Footings

The settlement of a footing (S) depends on the average bearing pressure (B), a dimension of the footing (D), and the soil modulus of compressibility (M). The settlement is assumed to be a random variable that follows the following density function:

$$f_S(s) = \lambda \exp(-\lambda s) \tag{3-58}$$

where λ is a constant depending on the three characteristics: B, D, and M. Integration of the density function yields the following cumulative function:

$$F_S(s) = 1 - \exp(-\lambda s) \qquad (3\text{-}59)$$

If the value of λ is 0.5 for a specified set of conditions (B, D, and M) and the settlement S is in inches, then the probability that the settlement S is less than s_o inches is

$$F_S(s_o) = P(S < s_o) = 1 - \exp(-0.5 s_o) \qquad (3\text{-}60)$$

For example, the probability of 1 in. or less of settlement is

$$F_S(1.0) = P(S < 1.0) = 1 - \exp[-0.5(1.0)] = 0.393 \qquad (3\text{-}61)$$

If damage occurs to the structure where the settlement exceeds 4 in., then the probability of damage is

$$P(S > 4.0) = 1 - F_S(4.0) = 1 - \{1 - \exp[-0.5(4.0)]\} = 0.135 \qquad (3\text{-}62)$$

Therefore for 1000 footings at the specified conditions, damage can be expected with 135 footings.

3.5. MOMENTS

Whether summarizing a data set or attempting to find the population, one must characterize the sample. The moments are useful descriptors of data; for example, the mean, which is a moment, is an important characteristic of a set of test scores. A moment can be referenced to any point on the measurement axis; however, the origin (i.e., zero point) and the mean are the most common reference points.

Although most data analyses use only two moments, it is important for some probabilistic and statistical studies to examine three moments:

1. The *mean*, which is the first moment about the origin
2. The *variance*, which is the second moment about the mean
3. The *skewness*, which is the third moment about the mean

Two characteristics, the mean and variance, were briefly introduced in Chapter 2.

In this section, equations and computational procedures for these moments are introduced. These moments are analogous to the area moments used to compute quantities such as the centroidal distance, the first static moment, and the moment of inertia. For a continuous random variable X, the kth moment about the origin is given by

$$M'_k = \int_{-\infty}^{+\infty} x^k f_X(x)\, dx \tag{3-63}$$

in which X is the random variable, and $f_X(x)$ is its density function. The corresponding equation for a discrete random variable is

$$M'_k = \sum_{i=1}^{n} x_i^k\, P_X(x_i) \tag{3-64}$$

in which n is the number of elements in the underlying sample space of X, and $P_X(x)$ is the probability mass function. The first moment about the origin, i.e., $k = 1$ in Equations 3-63 and 3-64, is called the mean of X, and is denoted μ.

For a continuous random variable, the kth moment about the mean (μ) is

$$M_k = \int_{-\infty}^{+\infty} (x - \mu)^k f_X(x)\, dx \tag{3-65}$$

in which μ is the first moment about the origin, i.e., the mean. The corresponding equation for a discrete random variable is

$$M_k = \sum_{i=1}^{n} (x_i - \mu)^k\, P_X(x_i) \tag{3-66}$$

The above moments are considered as a special case of mathematical expectation. The mathematical expectation of an arbitrary function $g(x)$, which is a function of the random variable X, is defined as

$$E[g(x)] = \int_{-\infty}^{+\infty} g(x) f_X(x)\, dx \tag{3-67}$$

The corresponding equation for a discrete random variable is

$$E[g(x)] = \sum_{i=1}^{n} g(x_i)\, P_X(x_i) \tag{3-68}$$

3.5.1. Mean

The mean value, which was introduced in Chapter 2, can be formally defined as the first moment measured about the origin; it is also the average of all observations on a random variable. It is important to note that the population mean is most often indicated as μ, while the sample mean is denoted by $\overline{X}$. For a continuous random variable, the mean (μ) is computed as

$$\mu = \int_{-\infty}^{+\infty} x\, f_X(x)\, dx \qquad (3\text{-}69)$$

For a discrete random variable, the mean is given by

$$\mu = \sum_{i=1}^{n} x_i\, P_X(x_i) \qquad (3\text{-}70)$$

For n observations, if all observations are given equal weights, i.e., $P_X(x_i) = 1/n$, then the mean for a discrete random variable (Equation 3-70) produces

$$\overline{X} = \frac{1}{n} \sum_{i=1}^{n} x_i \qquad (3\text{-}71)$$

which is the average of the observed values $x_1, x_2, x_3, ..., x_n$.

The mean corresponds to the average value that lies along the measurement axis; this is most easily represented by the density functions shown in Figure 3-8. If the random variable X is the magnitude of a flood, then $f_{X_1}(x_1)$ and $f_{X_2}(x_2)$ could represent the distributions of floods on two rivers. Obviously, the floods of the river characterized by $f_{X_2}(x_2)$ are usually larger than floods on the river characterized by $f_{X_1}(x_1)$. Similarly, if the density functions of Figure 3-8 represent the distribution of grades for two large sections of a class, it is evident that students in section 2 scored higher, on the average, than those in section 1.

Although the mean conveys certain information about a random variable, it does not completely characterize the random variable. The distributions shown in Figure 3-9 are obviously not identical, even though they have the same mean. Thus, the mean, by itself, cannot characterize a distribution. Other characteristics are also important.

Example 3-16: Mean Number of Hurricanes

Consider the data on hurricanes discussed in Example 3-13. The mean annual number of hurricanes in the U.S. can be computed using Equation 3-71 as follows:

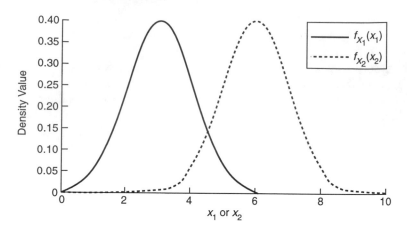

FIGURE 3-8 Mean values for continuous random variables.

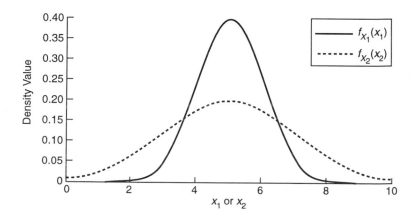

FIGURE 3-9 Random variables with different variances and same mean value.

$$\overline{X} = \frac{5+4+0+0+2+3+1+3+1+6+2+1+2+4+0}{15}$$

$$= 2.27 \text{ hurricanes/year}$$

(3-72)

3.5.2. Variance

The variance is the second moment about the mean. The variance of the population is denoted by σ^2. The variance of the sample is denoted by S^2. The units of the variance are the square of the units of the random variable; for example, if the random variable is measured in pounds per square inch (psi), the variance has units of (psi)^2. For a continuous random variable, the variance is computed as the second moment about the mean as follows:

$$\sigma^2 = \int_{-\infty}^{+\infty} (x - \mu)^2 f_X(x)\, dx \qquad (3\text{-}73)$$

For a discrete variable, the variance is computed as

$$\sigma^2 = \sum_{i=1}^{n} (x_i - \mu)^2 P_X(x_i) \qquad (3\text{-}74)$$

When the n observations in a sample are given equal weight, i.e., $P_X(x_i) = 1/n$, the variance is given by

$$S^2 = \frac{1}{n} \sum_{i=1}^{n} (x_i - \overline{X})^2 \qquad (3\text{-}75)$$

The value of the variance given by Equation 3-75 is biased; an unbiased estimate of the variance is given by

$$S^2 = \frac{1}{n-1} \sum_{i=1}^{n} (x_i - \overline{X})^2 \qquad (3\text{-}76)$$

Equation 3-76, which is the equation used most frequently, provides an unbiased estimate because the average value of many sample estimates of σ^2 would approach the true value of σ^2; whereas the average value of many sample estimates of σ^2 obtained from Equation 3-75 would not approach the true value of σ^2. The concept of unbiased estimators is discussed in Section 5.2.1. Computationally, the variance of a sample can be determined using the following alternative equation which produces the same result as Equation 3-76:

$$S^2 = \frac{1}{n-1} \left[\sum_{i=1}^{n} x_i^2 - n\overline{X}^2 \right] \qquad (3\text{-}77)$$

If $\overline{X}$ needs to be computed, then Equation 3-77 becomes

$$S^2 = \frac{1}{n-1} \left[\sum_{i=1}^{n} x_i^2 - \frac{1}{n} \left(\sum_{i=1}^{n} x_i \right)^2 \right] \qquad (3\text{-}78)$$

Equation 3-77 provides the same answer as Equation 3-78 when computations are made using an appropriate number of significant digits.

TABLE 3-3a Calculation of the Variance for Data Set 1

x	$\overline{X}$	$x - \overline{X}$	$(x - \overline{X})^2$	$\dfrac{1}{n-1}$	$\dfrac{1}{n-1}(x - \overline{X})^2$
9	10	−1	1	0.5	0.5
10	10	0	0	0.5	0
11	10	1	1	0.5	0.5
					Summation = 1

TABLE 3-3b Calculation of the Variance for Data Set 2

x	$\overline{X}$	$x - \overline{X}$	$(x - \overline{X})^2$	$\dfrac{1}{n-1}$	$\dfrac{1}{n-1}(x - \overline{X})^2$
5	10	−5	25	0.5	12.5
10	10	0	0	0.5	0
15	10	5	25	0.5	12.5
					Summation = 25

The variance is an important concept in probabilistic and statistical analyses because many solution methods require some measure of variance. Therefore, it is important to have a conceptual understanding of this moment. In general, it is an indicator of the closeness of the values in sample or a population to the mean. If all values in the sample equal the mean, the sample variance would equal zero. Figure 3-9 illustrates density functions with different variances.

Because of the importance of the concept of variance, it may be worthwhile to show its computation for two samples, each having a sample of size (n) of 3; both sets have a mean of 10, with set 1: {9, 10, 11} and set 2: {5, 10, 15}. Tables 3-3a and 3-3b show the calculations for the two sets. Thus the variance of set 2 is 25 times greater than that of set 1. To illustrate the computational formula (Equation 3-78), the variance of set 1 could be determined as

$$S^2 = \frac{1}{n-1}\left[\sum_{i=1}^{n} x_i^2 - \frac{1}{n}\left(\sum_{i=1}^{n} x_i\right)^2\right]$$

(3-79)

$$= \frac{1}{3-1}\left[(9^2 + 10^2 + 11^2) - \frac{(9+10+11)^2}{3}\right] = 1$$

Although the variance is used in all aspects of probabilistic and statistical analyses, its use as a descriptor is somewhat limited due to its units; specifically, the units of the variance are not the same as those of either the random variable or the mean.

3.5.3. Standard Deviation and Coefficient of Variation

By definition the standard deviation is the square root of the variance. It has the same units as the random variable and the mean; therefore, it is a better descriptor of the dispersion or spread of either a sample of data or a distribution function than the variance. The standard deviation of the population is denoted by σ, while the sample value is denoted by S.

The coefficient of variation (δ or COV) is a dimensionless quantity defined as

$$\delta = \frac{\sigma}{\mu} \tag{3-80}$$

It is also used as an expression of the standard deviation in the form of a proportion of the mean. For example, consider μ and σ to be 100 and 10, respectively; therefore, $\delta = 0.1$ or 10%. In this case the standard deviation is 10% of the mean.

Example 3-17: Variance of Soil Strength

A sample of five tests was taken to determine the unconfined compression strength (in ton/ft²) of soil with the following test results: 1.8, 3.5, 6.1, 3.6, and 4.3 ton/ft². Compute the variance, standard deviation, and coefficient of variation of soil strength.
The mean value of soil strength $\overline{X}$ is given by

$$\overline{X} = \frac{1.8 + 3.5 + 6.1 + 3.6 + 4.3}{5} = 3.86 \text{ ton/ft}^2 \tag{3-81}$$

The variance of soil strength is computed using Equation 3-78 as follows:

$$S^2 = \frac{1.8^2 + 3.5^2 + 6.1^2 + 3.6^2 + 4.3^2 - \dfrac{(1.8 + 3.5 + 6.1 + 3.6 + 4.3)^2}{5}}{5 - 1} \tag{3-82}$$

$$= 2.413 \left(\text{ton/ft}^2\right)^2$$

Therefore, the standard deviation is

$$S = \sqrt{2.413} = 1.553 \text{ ton/ft}^2 \tag{3-83}$$

The coefficient of variation δ is

$$\delta = \frac{S}{\overline{X}} = \frac{1.553}{3.86} = 0.402 \tag{3-84}$$

The large coefficient of variation, i.e., 40.2% of the mean value, suggests that either the mean value is not reliable and that additional measurements might be needed or the underlying random variable has a large dispersion. If with additional measurements the coefficient of variation remains large, relatively large factors of safety should be used for soil strength design.

3.5.4. Skew

The skew is the third moment measured about the mean. Unfortunately, the notation for skew is not uniform from one user to another. The sample skew can be denoted by G while λ can be used to indicate the skew of the population. Mathematically, it is given as follows for a continuous random variable:

$$\lambda = \int_{-\infty}^{+\infty} (x - \mu)^3 \, f_X(x) \, dx \tag{3-85}$$

For a discrete random variable, it can be computed by

$$\lambda = \sum_{i=1}^{n} (x_i - \mu)^3 \, P_X(x_i) \tag{3-86}$$

It has units of the cube of the random variable; thus, if the random variable has units of pounds, the skew has units of (pounds)^3.

The skew is a measure of the lack of symmetry. A symmetric distribution has a skew of zero, while a nonsymmetric distribution has a positive or negative skew depending on the direction of the skewness. If the more extreme tail of the distribution is to the right, the skew is positive; the skew is negative when the more extreme tail is to the left of the mean. Skewed probability distributions are illustrated in Figure 3-10.

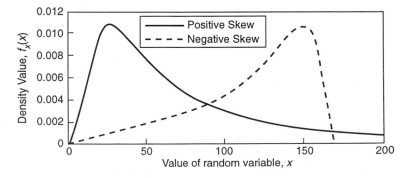

FIGURE 3-10 Skewed distributions.

TABLE 3-4 Computations for the Moments for Example 3-18

x	$P_X(x)$	$x\,P_X(x)$	$(x - \mu)$	$(x - \mu)^2$	$(x - \mu)^2\,P_X(x)$	$(x - \mu)^3\,P_X(x)$
2	1/36	2/36	−5	25	25/36	−125/36
3	1/18	6/36	−4	16	32/36	−64/36
4	1/12	12/36	−3	9	27/36	−27/36
5	1/9	20/36	−2	4	16/36	−8/36
6	5/36	30/36	−1	1	5/36	−1/36
7	1/6	42/36	0	0	0	0
8	5/36	40/36	1	1	5/36	1/36
9	1/9	36/36	2	4	16/36	8/36
10	1/12	30/36	3	9	27/36	27/36
11	1/18	22/36	4	16	32/36	64/36
12	1/36	12/36	5	25	25/36	125/36
Summations						
77	1.00	252/36	0		210/36	0

Example 3-18: Mean and Variance for Discrete Random Variable

The mean, variance, standard deviation, and coefficient of variation for the random variable defined in Example 3-12 are calculated in Table 3-4. Since the probabilities are for the population, the population mean can be determined using Equation 3-70 as

$$\mu = \sum_{i=1}^{n} x_i\,P_X(x_i) = \frac{252}{36} = 7 \tag{3-87}$$

The population variance can be determined using Equation 3-74 as

$$\sigma^2 = \sum_{i=1}^{n} (x_i - \mu)^2\,P_X(x_i) = \frac{210}{36} = 5.833 \tag{3-88}$$

Therefore, the standard deviation is $\sqrt{5.833} = 2.415$, and the coefficient of variation is $\dfrac{2.415}{7} = 0.345$. The skew (λ) is zero indicating a symmetric mass function.

Example 3-19: Mean and Variance for a Continuous Random Variable

The mean, variance, standard deviation, and coefficient of variation for the density function given in Example 3-14 are computed in this example. The mean can be computed using Equation 3-69 as follows:

$$\mu = \int_{-\infty}^{+\infty} x\, f_X(x)\, dx = \int_0^2 x(0.5\,x)\, dx = \frac{4}{3} \qquad (3\text{-}89)$$

The variance can be computed using Equation 3-73,

$$\sigma^2 = \int_{-\infty}^{+\infty} (x-\mu)^2\, f_X(x)\, dx = \int_0^2 \left(x - \frac{4}{3} \right)^2 (0.5\,x)\, dx = \frac{2}{9} \qquad (3\text{-}90)$$

The standard deviation equals $\dfrac{\sqrt{2}}{3}$ = 0.4714, and the coefficient of variation is

$\dfrac{0.4714}{4/3}$ = 0.3771.

3.6. COMMON DISCRETE PROBABILITY DISTRIBUTIONS

Any function that results in probabilities satisfying the axioms of probability is qualified to be a probability distribution. Based on previous experiences and uses of probability distributions, and based on improved understanding of underlying physical processes, several probability distributions have evolved to the common practical use by engineers. In Sections 3.6 and 3.7, these probability distributions are described and discussed. They are classified as continuous and discrete probability distributions. In this section, commonly used discrete probability distributions are described. In Section 3.7, commonly used continuous probability distributions are described.

A probability distribution function is expressed as a real-valued function of the random variable. The location, scale, and shape of the function are determined by its parameters. Distributions commonly have one to three parameters. These parameters take certain values that are specific for the problem and conditions being investigated. The parameters of a distribution can be expressed in terms of its moments, e.g., mean, variance, and skewness, but not necessarily in closed-form expressions.

In this section, the Bernoulli, binomial, geometric, and Poisson distributions are discussed. The first three distributions are based on Bernoulli trials (or sequence), whereas the fourth one is not. An engineering experiment (or system) that consists of N trials is considered to result in a *Bernoulli process* (or sequence) if it satisfies the following conditions: (1) the N trials (or repetitions) are independent; (2) each trial has only two possible outcomes, say, survival (S) or failure (F); and (3) the probabilities of occurrence for the two outcomes remain constant from trial to trial. Also, the negative binomial, Pascal, and hypergeometric distributions are described as other discrete probability distributions in Section 3.6.5.

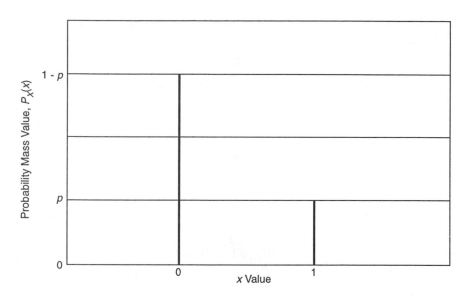

FIGURE 3-11 Probability mass function of the Bernoulli distribution.

3.6.1. Bernoulli Distribution

For convenience, the random variable X is defined as a mapping from the sample space $\{S , F\}$ for each trial of a Bernoulli sequence to the integer values $\{1 , 0\}$, with one-to-one mapping in the respective order, where for example S = success and F = failure. Therefore, the probability mass function is given by

$$P_X(x) = \begin{cases} p & \text{for } x = 1 \\ 1-p & \text{for } x = 0 \\ 0 & \text{otherwise} \end{cases} \tag{3-91}$$

The probability mass function of the Bernoulli distribution is shown in Figure 3-11. The mean and variance for the Bernoulli distribution are, respectively, given by

$$\mu_X = p \tag{3-92}$$

$$\sigma_X^2 = p(1-p) \tag{3-93}$$

Example 3-20: Quality Assurance

The quality assurance department in a structural-steel factory inspects every product coming off its production line. The product either fails or passes the inspection. Past experience indicates that the probability of failure (having a defective product) is 5%. Assuming that failures of the products are independent, determine the average percent of the products that will pass the inspection. What are its variance and coefficient of variation?

The average percent of the products that will pass the inspection is

$$E(X) = p = 1 - 0.05 = 0.95 = 95\% \tag{3-94}$$

Its variance and coefficient of variation (COV) are

$$Var(X) = p(1 - p) = 0.95(0.05) = 0.0475 \tag{3-95}$$

and

$$COV(X) = \frac{\sqrt{Var(X)}}{E(X)} = \frac{\sqrt{0.0475}}{0.95} = 0.229 \tag{3-96}$$

3.6.2. Binomial Distribution

The underlying random variable (X) for this distribution represents the number of successes in N Bernoulli trials. The probability mass function is given by

$$P_X(x) = \begin{cases} \binom{N}{x} p^x (1 - p)^{N-x} & \text{for } x = 0,\ 1,\ 2, ...,\ N \\ 0 & \text{otherwise} \end{cases} \tag{3-97}$$

where $\binom{N}{x}$ can be computed using Equation 3-17. The probability mass and cumulative functions of an example binomial distribution are shown in Figures 3-12a and 3-12b, respectively. The mean and variance for the binomial distribution are, respectively, given by

$$\mu_X = Np \tag{3-98}$$

$$\sigma_X^2 = Np(1 - p) \tag{3-99}$$

A random variable can be represented by the binomial distribution, if the following three assumptions are met:

1. The distribution is based on N Bernoulli trials with only two possible outcomes.
2. The N trials are independent of each other.
3. The probabilities of the outcomes remain constant at p and $(1 - p)$ for each trial.

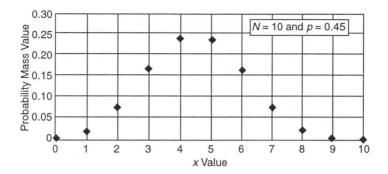

FIGURE 3-12a Probability mass function of the binomial distribution.

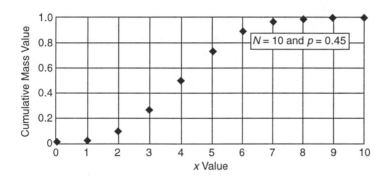

FIGURE 3-12b Cumulative mass function of the binomial distribution.

Therefore, the flip of a coin would meet these assumptions, but the roll of a die would not because there are six possible outcomes.

Example 3-21: Quality Assurance

The production line in Example 3-20 manufactures (on the average) 1000 units of the product daily. What is the expected number of nondefective units? What is the coefficient of variation of the number of nondefective units?

Since there are only two possible outcomes (i.e., defective and nondefective units), a binomial distribution can be used to model this problem, with $N = 1000$ and $p = 0.95$. The characteristics of nondefective units are

$$\text{Expected (mean) value, } E(X) = Np = 1000(0.95) = 950 \text{ units/day} \quad (3\text{-}100a)$$

$$\text{Standard deviation, } \sigma = \sqrt{Np(1-p)} = \sqrt{1000(0.95)(0.05)} = 6.892 \quad (3\text{-}100b)$$

$$\text{Coefficient of variation, } COV(X) = \frac{\sigma}{E(X)} = 0.00725 \quad (3\text{-}100c)$$

The coefficient of variation in this case describes the dispersion of the daily nondefective units produced by a line. The probability of producing 980 nondefective units in a day is

$$P_X(980) = \binom{1000}{980}(0.95)^{980}(1-0.95)^{20} = 4.787 \times 10^{-7}$$

Such a small probability of having 980 nondefective units in a day resulted because of specifying *exactly* 980 units a day in computing the probability.

3.6.3. Geometric Distribution

The underlying random variable for this distribution represents the number of Bernoulli trials that are required to achieve the first success. In this case, the number of trials needed to achieve the first success is neither fixed nor certain. The probability mass function is given by

$$P_X(x) = \begin{cases} p(1-p)^{x-1} & \text{for } x = 1, 2, 3, \dots \\ 0 & \text{otherwise} \end{cases} \tag{3-101}$$

The probability mass function of an example geometric distribution is shown in Figure 3-13. The mean and variance for the geometric distribution are, respectively, given by

$$\mu_X = \frac{1}{p} \tag{3-102}$$

$$\sigma_X^2 = \frac{1-p}{p^2} \tag{3-103}$$

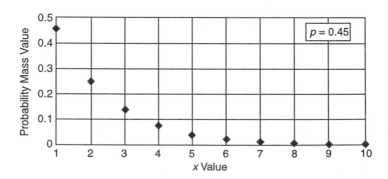

FIGURE 3-13 Probability mass function of the geometric distribution.

Example 3-22: Traffic Accidents

Based on previous accident records, assume that the annual probability of being in a fatal traffic accident is on the average 1.8×10^{-5} per 1000 mi of travel. What is the probability of being in a fatal accident for the first time at 10,000, 100,000, and 500,000 mi of travel? These miles of travel can be considered to correspond to (on the average) 1, 10, and 50 years.

The accident probability at 10,000 mi of travel is $1.8 \times 10^{-5}(1 - 1.8 \times 10^{-5})^{10-1}$ $= 1.8 \times 10^{-5}$. At 100,000 mi, the probability is $1.8 \times 10^{-5}(1 - 1.8 \times 10^{-5})^{100-1} = 1.8 \times 10^{-5}$. At 500,000 mi, the probability is $1.8 \times 10^{-5} (1 - 1.8 \times 10^{-5})^{500-1} = 1.8 \times 10^{-5}$. It is interesting to note that the probability remains the same, since the geometric distribution assumes the accident to be in the last 1000 mi of travel.

3.6.4. Poisson Distribution

The Poisson distribution is commonly used in engineering problems that deal with the occurrence of some random event in the continuous dimension of time or space. For example, the number of occurrences of a natural hazard, such as earthquakes, tornadoes, or hurricanes, in some time interval, such as 1 year, can be considered as a random variable with a Poisson distribution. In these examples, the number of occurrences in the time interval is the random variable. Therefore, the random variable is discrete, whereas its reference space, i.e., the time interval, is continuous. This distribution is considered to be the limiting case of the binomial distribution by dividing the reference space, i.e., time t, into nonoverlapping intervals of size Δt. The occurrence of the event, i.e., a natural hazard, in each interval is considered to constitute a Bernoulli sequence. The number of Bernoulli trials depends on the size of the interval Δt. By considering the limiting case where the size of the interval Δt approaches zero, the binomial distribution becomes the Poisson distribution.

The underlying random variable of this distribution is denoted by X_t, which represents the number of occurrences of an event of interest, and $t = $ the time (or space) interval. The probability mass function for the Poisson distribution is

$$P_{X_t}(x) = \begin{cases} \dfrac{(\lambda t)^x \exp(-\lambda t)}{x!} & \text{for } x = 0, 1, 2, 3, \ldots \\ 0 & \text{otherwise} \end{cases} \qquad (3\text{-}104)$$

The probability mass function of an example Poisson distribution is shown in Figure 3-14. The mean and variance for the Poisson distribution are, respectively, given by

$$\mu_X = \lambda t \qquad (3\text{-}105)$$

$$\sigma_X^2 = \lambda t \qquad (3\text{-}106)$$

The parameter λ of the Poisson distribution represents the average rate of occurrence of the event of interest.

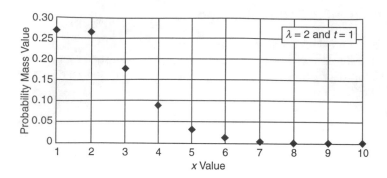

FIGURE 3-14 Probability mass function of the Poisson distribution.

Example 3-23: Traffic Accidents

In Example 3-22, traffic accidents were used to illustrate the use of the geometric distribution. The traffic accident problem can be modeled more adequately using the Poisson distribution. In this case, the annual rate is assumed to equal the annual probability of fatal accidents per 1000 mi, i.e., $\lambda = 1.8 \times 10^{-5}$. The variable t is considered to be the travel distance in thousands of miles. Therefore, the probability of having one fatal accident in 10,000 mi of travel is

$$P_{X_t}(x = 1) = (\lambda t)^x \frac{e^{-\lambda t}}{x!} = \left[1.8 \times 10^{-5} (10)\right]^1 \frac{\exp\left[-1.8 \times 10^{-5} (10)\right]}{1!} \qquad (3\text{-}107)$$

$$= 1.7997 \times 10^{-4}$$

It is important to note that this result is fundamentally different from the corresponding result in Example 3-22. This result gives the probability of one fatal accident in 10,000 mi that can happen at any 1000-mi interval. The result of Example 3-22 is the probability of having a fatal accident in the last 1000-mi interval of 10,000 mi of travel.

3.6.5. Other Discrete Probability Distributions

In this section, a selected group of other discrete probability distributions are described. The group consists of the negative binomial, Pascal distribution, and hypergeometric distribution.

The *negative binomial distribution* is considered a general case of the geometric distribution. Its underlying random variable is defined as the k*th* occurrence of an event of interest on the last trial in a sequence of X Bernoulli trials. The probability of this k*th* occurrence on the last trial is given by the probability mass function of the negative binomial distribution, i.e.,

$$P_X(x) = \begin{cases} \dbinom{x-1}{k-1} p^k(1-p)^{x-k} & \text{for } x = k, k+1, k+2, \ldots \\ 0 & \text{otherwise} \end{cases} \qquad (3\text{-}108)$$

The mean and variance of this distribution are, respectively, given by

$$\mu_X = \frac{k}{p} \qquad (3\text{-}109\text{a})$$

$$\sigma_X^2 = \frac{k(1-p)}{p^2} \qquad (3\text{-}109\text{b})$$

The negative binomial distribution is called the Pascal distribution if k takes on only integer values.

The *hypergeometric distribution* deals with a finite population of size N, with a class of $D \le N$ elements of the population having a property of interest, e.g., defective units or nondefective units. A random sample is selected of size n without replacement, i.e., a sampled element of the population is not replaced before randomly selecting the next element of the sample. The underlying random variable, X, for this distribution is defined as the number of elements in the sample that belong to the class of interest. The probability mass function is given by

$$P_X(x) = \begin{cases} \dfrac{\dbinom{D}{x}\dbinom{N-D}{n-x}}{\dbinom{N}{n}} & \text{for } x = 0, 1, 2, \ldots, \min(n, D) \\ 0 & \text{otherwise} \end{cases} \qquad (3\text{-}110)$$

The mean and variance of this distribution are, respectively, given by

$$\mu_X = n\frac{D}{N} \qquad (3\text{-}111\text{a})$$

$$\sigma_X^2 = n\frac{D}{N}\frac{N-n}{N-1}\left(1 - \frac{D}{N}\right) \qquad (3\text{-}111\text{b})$$

3.7. COMMON CONTINUOUS PROBABILITY DISTRIBUTIONS

In this section, four continuous distributions are discussed: the uniform, normal, lognormal, and exponential distributions. The uniform distribution is very important for performing random number generation in simulation as described in Chapter 7. The normal and lognormal distributions are important due to their common use and applications in engineering. These two distributions also have an important and unique relationship. The importance of the exponential distribution comes from its special relation to the Poisson distribution. Also, the student t-distribution, the chi-squared distribution, and the F-distribution are described as other continuous distributions in Section 3.7.5. They are important because of their use in statistics as discussed in Chapter 5. Also, in this section, extreme value distributions are described.

3.7.1. Uniform Distribution

The density function for the uniform distribution of a random variable X is given by

$$f_X(x) = \begin{cases} \dfrac{1}{b-a} & \text{for } a \le x \le b \\ 0 & \text{otherwise} \end{cases} \qquad (3\text{-}112)$$

where a and b are real values, called parameters, with $a < b$. The density function for the uniform distribution takes a constant value of $\dfrac{1}{b-a}$ in order to satisfy the probability axiom that requires the area under the density function to be one. An example of a uniform density function is shown in Figure 3-15. The mean and variance for the uniform distribution are, respectively, given by

$$\mu_X = \frac{a+b}{2} \qquad (3\text{-}113)$$

$$\sigma_X^2 = \frac{(b-a)^2}{12} \qquad (3\text{-}114)$$

Due to the simple geometry of the density function of the uniform distribution, it can be easily noticed that its mean value and variance, as given by Equations 3-113 and 3-114, respectively, correspond to the centroidal distance and centroidal moment of inertia with respect to a vertical axis of the area under the density function, respectively. This property is valid for other distributions as well. The cumulative function for the uniform distribution is a line with a constant slope and is given by

$$F_X(x) = \begin{cases} 0 & x \le a \\ \dfrac{x-a}{b-a} & a \le x \le b \\ 1 & x \ge b \end{cases} \qquad (3\text{-}115)$$

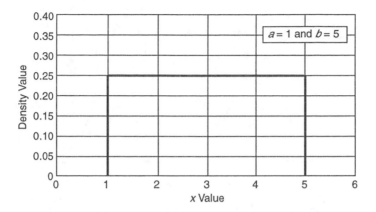

FIGURE 3-15 Probability density function of the uniform distribution.

Example 3-24: Subjective Assessment of Concrete Strength

Subjectively, as well as objectively, determined information is commonly used in engineering for the purpose of decision making. For example, a structural engineer might assess, based on his or her experience, the strength of concrete in an existing bridge to be in the range 3000 to 4000 psi. A uniform probability distribution can be used to estimate probabilities in this case, if the engineer believes that all the values in this range are equally likely. Therefore, the mean value of concrete strength is 3500 psi, and its variance is 83,333 $(psi)^2$, which gives a standard deviation of 288.7 psi. The probability that the strength of concrete X is larger than 3600 psi is

$$P(X > 3600 \text{ psi}) = 1 - F_X(3600)$$

$$= 1 - \frac{3600 - 3000}{4000 - 3000} = 0.4 \tag{3-116}$$

3.7.2. Normal Distribution

The normal distribution (also called the Gaussian distribution) is widely used due to its simplicity and wide applicability. This distribution is the basis for many statistical methods. The normal density function for a random variable X is given by

$$f_X(x) = \frac{1}{\sigma\sqrt{2\pi}} \exp\left(-\frac{1}{2}\left[\frac{x-\mu}{\sigma}\right]^2\right) \qquad -\infty < x < \infty \tag{3-117}$$

It is common to use the notation $X \sim N(\mu, \sigma^2)$ to provide an abbreviated description of a normal distribution. The notation states that X is normally distributed with a

FIGURE 3-16a Probability density function of the normal distribution.

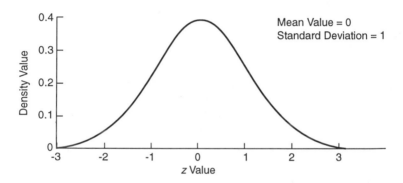

FIGURE 3-16b Probability density function of the standard normal distribution.

mean value μ and variance σ^2. The normal distribution has several important properties that include the following:

1. $f_X(x)$ approaches 0 as x approaches either $-\infty$ or ∞.
2. $f_X(a + \mu) = f_X(-a + \mu)$ for any a, i.e., symmetric density function about the mean μ.
3. The maximum value of $f_X(x)$ (i.e., the mode of the distribution) occurs at $x = \mu$.
4. The inflection points of the density function occurs at $x = \mu \pm \sigma$.
5. The density function has an overall bell shape.
6. The mean value μ and the variance σ^2 are also the parameters of the distribution.

In Figure 3-16a, the normal distribution is used to model the concrete strength of Example 3-24; however, now we are assuming that concrete strength has a normal distribution with a mean = 3.5 ksi and standard deviation = 0.2887 ksi. The density function of another normal distribution is shown in Figure 3-16b. The cumulative distribution function of the normal distribution is given by

$$F_X(x) = \int_{-\infty}^{x} \frac{1}{\sigma\sqrt{2\pi}} \exp\left[-\frac{1}{2}\left(\frac{x-\mu}{\sigma}\right)^2\right] dx \qquad (3\text{-}118)$$

The evaluation of the integral of Equation 3-118 requires numerical methods for each pair (μ, σ^2). This difficulty can be reduced by performing a transformation that results in a standard normal distribution with a mean $\mu = 0$ and variance $\sigma^2 = 1$, denoted as $Z \sim N(0,1)$. Numerical integration can be used to determine the cumulative distribution function of the standard normal distribution. Therefore, by utilizing the transformation between the normal distribution $X \sim N(\mu, \sigma^2)$ and the standard normal distribution $Z \sim N(0,1)$, and the integration results for the standard normal, the cumulative distribution function for the normal distribution can be evaluated. Using the following standard normal transformation:

$$Z = \frac{X - \mu}{\sigma} \qquad (3\text{-}119)$$

The density function, which is shown in Figure 3-16b, and the cumulative distribution function of the standard normal are determined, respectively, as

$$\phi(z) = \frac{1}{\sqrt{2\pi}} \exp\left(-\frac{1}{2} z^2\right) \qquad -\infty < z < \infty \qquad (3\text{-}120)$$

$$\Phi(z) = \int_{-\infty}^{z} \frac{1}{\sqrt{2\pi}} \exp\left(-\frac{1}{2} z^2\right) dz \qquad (3\text{-}121)$$

where $\phi(z)$ = a special notation for the probability density function of the standard normal, and $\Phi(z)$ = a special notation for the cumulative distribution function of the standard normal. The results of the integral $\Phi(z)$ are tabulated in Table A.1 of Appendix A. The tabulated values are for $z \in [0 , 8.2]$. Negative z values can be obtained utilizing the symmetry property of the normal distribution, i.e.,

$$\Phi(-z) = 1 - \Phi(z) \qquad (3\text{-}122)$$

The tabulated Φ values for $z > 0$ are in the range $[0.5, 1.)$. The table can also be used for determining the inverse of the Φ function, i.e., Φ^{-1}. For values of specified $p = \Phi(z)$ that are less than 0.5, Table A-1 can be used with the help of the following relation to obtain the z value:

$$z = \Phi^{-1}(p) = -\Phi^{-1}(1 - |p|) \quad \text{for } p < 0.5 \qquad (3\text{-}123)$$

Equation 3-123 is based on the symmetry property of the normal distribution.

Using the transformation of Equation 3-119, probabilities based on any normal distribution can be easily computed with the aid of the tabulated Φ values. For example, the following cumulative probability for any $X \sim N(\mu, \sigma)$ can be computed

$$P(X \le x) = \int_{-\infty}^{x} \frac{1}{\sigma\sqrt{2\pi}} \exp\left[-\frac{1}{2}\left(\frac{x-\mu}{\sigma}\right)^2\right] dx \tag{3-124}$$

by changing the variable of integration according to Equation 3-119. Therefore, the probability is given by

$$P(X \le x) = \int_{-\infty}^{\frac{x-\mu}{\sigma}} \frac{1}{\sigma\sqrt{2\pi}} \exp\left(-\frac{z^2}{2}\right) \sigma \, dz \tag{3-125}$$

Therefore,

$$P(X \le x) = \int_{-\infty}^{\frac{x-\mu}{\sigma}} \frac{1}{\sqrt{2\pi}} \exp\left(-\frac{z^2}{2}\right) dz = \Phi\left(\frac{x-\mu}{\sigma}\right) \tag{3-126}$$

It can also be shown that

$$P(a < X \le b) = F_X(b) - F_X(a)$$
$$= \Phi\left(\frac{b-\mu}{\sigma}\right) - \Phi\left(\frac{a-\mu}{\sigma}\right) \tag{3-127}$$

The normal distribution has several important and useful properties that include the following:

1. The addition of n normally distributed random variables, $X_1, X_2, ..., X_n$, is a normal distribution as follows:

$$Y = X_1 + X_2 + X_3 + \cdots + X_n \tag{3-128}$$

The mean of Y (i.e., μ_Y) is

$$\mu_Y = \mu_{X_1} + \mu_{X_2} + \mu_{X_3} + \cdots + \mu_{X_n} \tag{3-129}$$

The variance of Y (i.e., σ_Y^2) is

$$\sigma_Y^2 = \sigma_{X_1}^2 + \sigma_{X_2}^2 + \sigma_{X_3}^2 + \cdots + \sigma_{X_n}^2 \tag{3-130}$$

2. *Central limit theorem:* Informally stated, the addition of a number of individual random variables, without a dominating distribution type, approaches a normal distribution as the number of the random variables approaches infinity. The result is valid regardless of the underlying distribution types of the random variables. In practice, the number of the random variables needs to be sufficiently large.

Example 3-25: Subjective Assessment of Concrete Strength

Assume that the structural engineer of Example 3-24 decided to use a normal distribution to model the strength of concrete. The mean and variance are the same as before, i.e., 3500 psi and 83,333 $(psi)^2$, respectively (see Figure 3-16a). Therefore, the probability that the concrete strength is larger than 3600 psi is

$$P(X > 3600 \text{ psi}) = 1 - P(X \le 3600)$$

$$= 1 - \Phi\left[\frac{3600 - 3500}{\sqrt{83,333}}\right] = 1 - \Phi(0.3464) \tag{3-131}$$

Using linear interpolation in Table A.1, $\Phi(0.3464)$ is given by

$$\Phi(0.3464) = 0.633072 + (0.3464 - 0.34)\left(\frac{0.636831 - 0.633072}{0.35 - 0.34}\right) \tag{3-132}$$

$$= 0.635478$$

Therefore,

$$P(X > 3600 \text{ psi}) = 1 - 0.635478 = 0.364522 \tag{3-133}$$

The probability that the concrete strength is larger than 3600 psi was determined in Example 3-24 to be 0.4. The difference between the estimates of 0.4 and 0.36452 reflects the disparity that can result from the selection of the probability distribution. Care should be exercised in selecting a probability distribution since the selection affects the resulting probabilities.

3.7.3. Lognormal Distribution

A random variable X is considered to have a lognormal distribution if $Y = \ln(X)$ has a normal probability distribution, where $\ln(x)$ is the natural logarithm to the base e. The density function of the lognormal distribution is given by

$$f_X(x) = \frac{1}{x\sigma_Y\sqrt{2\pi}}\exp\left[-\frac{1}{2}\left(\frac{\ln x - \mu_Y}{\sigma_Y}\right)^2\right] \quad \text{for } 0 < x < \infty \qquad (3\text{-}134)$$

It is common to use the notation $X \sim LN(\mu_Y, \sigma_Y{}^2)$ to provide an abbreviated description of a lognormal distribution. The notation states that X is lognormally distributed with the parameters μ_Y and $\sigma_Y{}^2$. The lognormal distribution has the following properties:

1. The values of the random variable X are positive, i.e., $x > 0$.
2. $f_X(x)$ is not a symmetric density function about the mean value μ_X.
3. The mean value μ_X and variance $\sigma_X{}^2$ are not equal to the parameters of the distribution, i.e., μ_Y and $\sigma_Y{}^2$. However, they are related to them as follows:

$$\sigma_Y^2 = \ln\left[1 + \left(\frac{\sigma_X}{\mu_X}\right)^2\right] \qquad (3\text{-}135)$$

$$\mu_Y = \ln(\mu_X) - \frac{1}{2}\sigma_Y^2 \qquad (3\text{-}136)$$

These two relations can be inverted as follows:

$$\mu_X = \exp\left(\mu_Y + \frac{1}{2}\sigma_Y^2\right) \qquad (3\text{-}137)$$

$$\sigma_X^2 = \mu_X^2\left[\exp(\sigma_Y^2) - 1\right] \qquad (3\text{-}138)$$

For a relatively small coefficient of variation δ_X, e.g., $\dfrac{\sigma_X}{\mu_X} \le 0.3$, σ_Y is approximately equal to the coefficient of variation δ_X. An example density function of the lognormal distribution is shown in Figure 3-17.

The cumulative distribution function of the lognormal distribution can be determined based on its relationship to the normal distribution. Similar to the normal distribution, the difficulty of performing the integral of the cumulative distribution function can be reduced by performing a transformation that results in a standard normal distribution with a mean $\mu = 0$ and variance $\sigma^2 = 1$, denoted as $Z \sim N(0,1)$. Numerical integration can be used to determine the cumulative distribution function of the standard normal. Therefore, by utilizing the transformation relationship between the lognormal distribution $X \sim LN(\mu_Y, \sigma_Y{}^2)$ and the standard normal distribution $Z \sim N(0,1)$, and the integration results for the standard normal, the cumulative distribution function for the lognormal distribution can be evaluated using the following transformation:

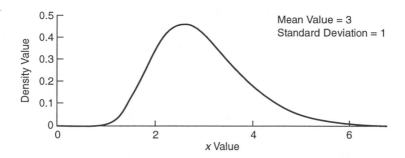

FIGURE 3-17 Probability density function of lognormal distribution.

$$Z = \frac{\ln X - \mu_Y}{\sigma_Y} \tag{3-139}$$

The density function and the cumulative distribution function of the standard normal are given, respectively, by Equations 3-120 and 3-121. They can, therefore, be used to compute probabilities for the lognormal distribution. The results of the integral $\Phi(z)$ are given in Table A.1 of Appendix A. The symmetry property of the standard normal should be used to obtain a complete range of values for the standard variate similar to the normal distribution.

Using the transformation of Equation 3-139, probabilities based on any lognormal distribution can be easily computed with the aid of the tabulated Φ values. For example, the cumulative probability $P(X \leq x)$ for any $X \sim N(\mu_Y, \sigma_Y^2)$ can be computed by changing the variable of integration according to Equation 3-139. Therefore, the cumulative probability is given by

$$P(X \leq x) = \int_{-\infty}^{\frac{\ln x - \mu_Y}{\sigma_Y}} \frac{1}{\sqrt{2\pi}} \exp\left(-\frac{z^2}{2}\right) dz = \Phi\left(\frac{\ln x - \mu_Y}{\sigma_Y}\right) \tag{3-140}$$

It can be also shown that

$$P(a < X \leq b) = F_X(b) - F_X(a)$$

$$= \Phi\left(\frac{\ln b - \mu_Y}{\sigma_Y}\right) - \Phi\left(\frac{\ln a - \mu_Y}{\sigma_Y}\right) \tag{3-141}$$

The lognormal distribution has several important properties that include the following:

1. The multiplication of n lognormally distributed random variables, $X_1, X_2, ..., X_n$, is a lognormal distribution with the following statistical characteristics:

$$W = X_1 X_2 X_3 \cdots X_n \tag{3-142}$$

The first moment of W (i.e., μ_W) is

$$\mu_W = \mu_{Y_1} + \mu_{Y_2} + \mu_{Y_3} + \cdots + \mu_{Y_n} \tag{3-143}$$

The second moment of W (i.e., σ_W^2) is

$$\sigma_W^2 = \sigma_{Y_1}^2 + \sigma_{Y_2}^2 + \sigma_{Y_3}^2 + \cdots + \sigma_{Y_n}^2 \tag{3-144}$$

2. *Central limit theorem:* In this case, the multiplication of a number of individual random variables, without a dominating distribution type, approaches a lognormal distribution as the number of the random variables approaches infinity. The result is valid regardless of the underlying distribution types of the random variables. In practice, the number of the random variables needs to be sufficiently large.

Example 3-26: Subjective Assessment of Concrete Strength

Since the strength of concrete takes on only positive values, someone might argue against using the normal distribution to model concrete strength since it allows for negative values. The lognormal distribution has the property of being defined only for positive values for the underlying random variable. Therefore, let's assume that the structural engineer of Examples 3-24 and 3-25 decides to use a lognormal distribution to model concrete strength. The mean value and variance are the same as before, i.e., 3.5 ksi and 0.083333 (ksi)². Therefore, the parameters of the lognormal distribution can be determined using Equations 3-135 and 3-136 as follows:

$$\sigma_Y^2 = \ln\left(1 + \frac{0.083333}{3.5^2}\right) = 0.00678 \tag{3-145a}$$

$$\mu_Y = \ln(3.5) - \frac{1}{2}(0.00678) = 1.25 \tag{3-145b}$$

The probability that concrete strength is larger than 3.6 ksi can now be determined as

$$P(X > 3.6 \text{ ksi}) = 1 - P(X \le 3.6)$$

$$= 1 - \Phi\left[\frac{\ln(3.6) - 1.25}{\sqrt{0.00678}}\right] \tag{3-146}$$

$$= 1 - \Phi(0.3833) = 0.3507$$

The answer in this case is slightly different from the corresponding value of Example 3-25. It should be noted that this positive property of the random variable of a lognormal distribution should not be used as the only basis for justifying its use. Statistical bases for selecting probability distribution can be used as discussed in Chapter 5.

3.7.4. Exponential Distribution

The importance of this distribution comes from its relationship to the Poisson distribution. For a given Poisson process, the time T between the consecutive occurrence of events has an exponential distribution with the following density function:

$$f_T(t) = \begin{cases} \lambda \exp(-\lambda t) & \text{for } t \geq 0 \\ 0 & \text{otherwise} \end{cases} \tag{3-147}$$

The cumulative distribution function is given by

$$F_T(t) = 1 - \exp(-\lambda t) \tag{3-148}$$

The density and cumulative functions of the exponential distribution with $\lambda = 1$ are shown in Figures 3-18a and 3-18b, respectively. The mean value and the variance are given, respectively, by

$$\mu_T = \frac{1}{\lambda} \tag{3-149}$$

$$\sigma_T^2 = \frac{1}{\lambda^2} \tag{3-150}$$

Based on the means of the exponential and Poisson distributions, the *mean recurrence time* (or *return period*) is defined as $\frac{1}{\lambda}$.

Example 3-27: Earthquake Occurrence

The occurrence of earthquakes is commonly modeled by a Poisson process. Assume that the occurrence of a strong earthquake in a certain region was modeled by a Poisson process with an annual (λ) rate of 0.01. Therefore, the probability of occurrence of exactly one earthquake in 1 year is given by

$$P_{X_t}(x = 1) = (\lambda t)^x \frac{e^{-\lambda t}}{x!} = [(0.01)(1)]^1 \frac{e^{-0.01(1)}}{1!} = 0.00990 \tag{3-151}$$

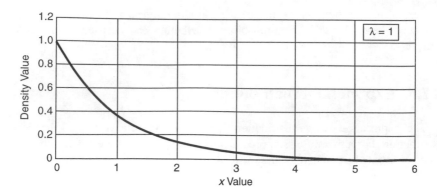

FIGURE 3-18a Probability density function of the exponential distribution.

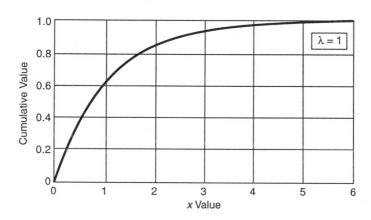

FIGURE 3-18b Cumulative distribution function of the exponential distribution.

The probability of having one or more earthquakes in 1 year is

$$P_{X_t}(x \geq 1) = 1 - P_{X_t}(x = 0)$$

$$= 1 - (\lambda t)^x \frac{e^{-\lambda t}}{x!} \tag{3-152}$$

$$= 1 - [(0.01)(1)]^0 \frac{e^{-0.01(1)}}{0!} = 0.00995$$

The return period for such an earthquake is $\dfrac{1}{\lambda} = 100$ years. A return period of 100 years does not imply that it is only possible to have one earthquake in a 100-year period; in fact, there may be none or any number of earthquakes in any 100-year period. However, on the average, there will be one every 100 years. The probability of having an earthquake for the first time more than 100 years after the last occurrence is computed using the exponential distribution (Equation 3-148) as

$$P(T > 100 \text{ years}) = 1 - P(T \le 100 \text{ years})$$

$$= 1 - \left[1 - e^{-0.01(100)}\right] = 0.3679 \tag{3-153}$$

This result can be generalized to the following probability of an event, that follows a Poisson process, to occur (for the first time) at a time that exceeds *its return period*:

$$P\left(T > \frac{1}{\lambda}\right) = 1 - P\left(T \le \frac{1}{\lambda}\right)$$

$$= 1 - \left[1 - e^{-\lambda(1/\lambda)}\right] = 0.3679 \tag{3-154}$$

3.7.5. Other Continuous Probability Distributions

In statistical analysis, tables of values of the student-*t* distribution, chi-squared distribution, and *F* distribution are commonly used. These distributions and their basic properties are introduced in this section. Also, extreme value distributions are described.

The *student-t (or t) distribution* is a symmetric, bell-shaped distribution with the following density function:

$$f_T(t) = \frac{\Gamma[(k+1)/k]}{(\pi k)^{0.5} \Gamma(k/2)\left[1 + \left(t^2/k\right)\right]^{0.5(k+1)}} \qquad -\infty < t < \infty \tag{3-155}$$

where *k* is a parameter of the distribution and is called the degrees of freedom. The function $\Gamma(.)$ is called the gamma function and is given by

$$\Gamma(n) = \int_0^\infty x^{n-1} \exp(-x) \, dx \qquad \text{for any value } n \tag{3-156}$$

This function is well tabulated in Appendix B. It has the following useful properties:

$$\Gamma(n) = (n-1)\Gamma(n-1) \tag{3-157a}$$

$$\Gamma(n) = (n-1)! \qquad \text{for integer } n \tag{3-157b}$$

For $k > 2$, the mean and variance are, respectively:

$$\mu_T = 0 \qquad (3\text{-}158\text{a})$$

$$\sigma_T^2 = \frac{k}{k-2} \qquad (3\text{-}158\text{b})$$

As k increases toward infinity, the variance of the distribution approaches unity, and therefore the t distribution approaches the standard normal density function. Therefore, the t distribution has heavier tails (with more area) than the standard normal. It is of interest in statistical analysis to determine the percentage points $t_{\alpha,k}$ that correspond to the following probability:

$$\alpha = P\left(T > t_{\alpha,k}\right) \qquad (3\text{-}159\text{a})$$

or

$$\alpha = \int_{t_{\alpha,k}}^{\infty} f_T(t)\, dt \qquad (3\text{-}159\text{b})$$

where α is called the level of significance. These percentage points are tabulated; e.g., see Table A-2. Due to the symmetry of the t distribution, it is common to tabulate values for only the upper tail, i.e., nonnegative t values. For the lower tail, the following relationship can be used to determine the percentage points:

$$t_{1-\alpha,k} = -t_{\alpha,k} \qquad (3\text{-}160)$$

The *chi-squared* (χ^2) *distribution* is encountered frequently in statistical analysis, where we deal with the sum of squares of k random variables with standard normal distributions, i.e.,

$$\chi^2 = C = Z_1^2 + Z_2^2 + \cdots + Z_k^2 \qquad (3\text{-}161)$$

where $C =$ random variable with chi-square distribution, and Z_1 to Z_k are normally (standard normal) and independently distributed random variables. The probability density function of the chi-square distribution is

$$f_C(c) = \frac{1}{2^{0.5k}\, \Gamma\!\left(\dfrac{k}{2}\right)} c^{0.5k-1} \exp\!\left(\frac{-c}{2}\right) \quad \text{for } c > 0 \qquad (3\text{-}162)$$

where $\Gamma(.)$ is the gamma function as defined in Equation 3-157. The distribution is defined only for positive values and has the following mean and variance, respectively:

$$\mu_C = k \qquad\qquad (3\text{-}163a)$$

$$\sigma_C^2 = 2k \qquad\qquad (3\text{-}163b)$$

The parameter of the distribution, k, represents the degrees of freedom. This distribution is positively skewed with a shape that depends on the parameter k. It is of interest in statistical analysis to determine the percentage points $c_{\alpha,k}$ that correspond to the following probability:

$$\alpha = P\left(C > c_{\alpha,k}\right) \qquad\qquad (3\text{-}164a)$$

$$\alpha = \int_{c_{\alpha,k}}^{\infty} f_C(c)\,dc \qquad\qquad (3\text{-}164b)$$

where α is called the level of significance. These percentage points are tabulated, e.g., see Table A-3.

The *F distribution* is used quite frequently in statistical analysis. It is a function of two shape parameters, $v_1 = k$ and $v_2 = u$, and has the following density function:

$$f_F(f) = \frac{\Gamma\left(\dfrac{u+k}{2}\right)\left(\dfrac{k}{u}\right)^{\frac{k}{2}}(f)^{\frac{k}{2}-1}}{\Gamma\left(\dfrac{k}{2}\right)\Gamma\left(\dfrac{u}{2}\right)\left[\dfrac{fk}{u}+1\right]^{\frac{u+k}{2}}} \quad \text{for } f > 0 \qquad\qquad (3\text{-}165)$$

where $\Gamma(.)$ is the gamma function as given by Equation 3-157. For $u > 2$, the mean and variance of this distribution are

$$\mu_F = \frac{u}{u-2} \qquad\qquad (3\text{-}166a)$$

$$\sigma_F^2 = \frac{2u^2(u+k-2)}{k(u-2)^2(u-4)} \quad \text{for } u > 4 \qquad\qquad (3\text{-}166b)$$

This distribution is positively skewed with a shape that depends on the parameters k and u. It is of interest in statistical analysis to determine the percentage points $f_{\alpha,k,u}$ that correspond to the following probability:

$$\alpha = P\left(F > f_{\alpha,k,u}\right) \tag{3-167a}$$

$$= \int_{f_{\alpha,k,u}}^{\infty} f_F(x)dx = \alpha \tag{3-167b}$$

where α is called the level of significance. These percentage points are tabulated, e.g., see Table A-4. The F-distribution has a unique property that allows tabulating values for the upper tail only. For the lower tail, the following relationship can be used to determine the percentage points:

$$f_{1-\alpha,k,u} = \frac{1}{f_{\alpha,u,k}} \tag{3-167c}$$

It should be noted that the parameters of this distribution cannot be transposed, i.e.,

$$f_{\alpha,u,k} \neq f_{\alpha,k,u} \tag{3-167d}$$

Extreme value distributions are a class of commonly used distributions in engineering and sciences. These distributions are described in the remaining part of this section. The extreme value distributions are of three types.

Two forms of the type I extreme value distribution can be used, the largest and smallest extreme value. These two types are described in this section. The density function for the largest distribution of a random variable X_n is given by

$$f_{X_n}(x) = \alpha_n e^{-\alpha_n(x-u_n)} \exp\left[-e^{-\alpha_n(x-u_n)}\right] \tag{3-168a}$$

The density function for the smallest distribution of a random variable X_1 is given by

$$f_{X_1}(x) = \alpha_1 e^{\alpha_1(x-u_1)} \exp\left[-e^{\alpha_1(x-u_1)}\right] \tag{3-168b}$$

where u_n = the location parameter for X_n, α_n = the shape parameter of X_n, u_1 = the location parameter for X_1, and α_1 = the shape parameter of X_1. The cumulative function for the largest distribution is given by

$$F_{X_n}(x) = \exp\left[-e^{-\alpha_n(x-u_n)}\right] \tag{3-168c}$$

The cumulative function for the smallest extreme is given by

$$F_{X_1}(x) = 1 - \exp\left[-e^{\alpha_1(x-u_1)}\right] \tag{3-168d}$$

For the largest extreme, the mean (μ) and variance (σ^2) for the distribution are, respectively, given by

$$\mu_{X_n} = u_n + \frac{\gamma}{\alpha_n} \qquad (3\text{-}168e)$$

$$\sigma^2_{X_n} = \frac{\pi^2}{6\alpha_n^2} \qquad (3\text{-}168f)$$

where $\pi = 3.14159$, and $\gamma = 0.577216$. For the smallest extreme, the mean (μ) and variance (σ^2) for the distribution are, respectively, given by

$$\mu_{X_1} = \mu_1 - \frac{\gamma}{\alpha_1} \qquad (3\text{-}168g)$$

$$\sigma^2_{X_1} = \frac{\pi^2}{6\alpha_1^2} \qquad (3\text{-}168h)$$

Two forms of the type II extreme value distribution can be used, the largest and smallest extreme value. The two types are described in this section, although only the largest distribution has practical value. The density function for the largest extreme of a random variable X_n is given by

$$f_{X_n}(x) = \frac{k}{v_n} \left(\frac{v_n}{x} \right)^{k+1} \exp\left[-\left(\frac{v_n}{x} \right)^k \right] \qquad (3\text{-}168i)$$

The density function for the smallest extreme of a random variable X_1 is given by

$$f_{X_1}(x) = -\frac{k}{v_1} \left(\frac{v_1}{x} \right)^{k+1} \exp\left[-\left(\frac{v_1}{x} \right)^k \right] \qquad (3\text{-}168j)$$

where v_n = the location parameter for X_n, v_1 = the location parameter for X_1, and k = the shape parameter of X_1 and X_n. The cumulative function for the largest distribution is given by

$$F_{X_n}(x) = \exp\left[-\left(\frac{v_n}{x} \right)^k \right] \qquad (3\text{-}168k)$$

The cumulative function for the smallest extreme is given by

$$F_{X_1}(x) = 1 - \exp\left[-\left(\frac{v_n}{x}\right)^k\right]$$

(3-168l)

where $x \le 0$, and $v_1 > 0$. For the largest extreme, the mean (μ) and variance (σ^2) for the distribution are, respectively, given by

$$\mu_{X_n} = v_n \Gamma\left(1 - \frac{1}{k}\right)$$

(3-168m)

$$\sigma^2_{X_n} = v_n^2\left[\Gamma\left(1 - \frac{2}{k}\right) - \Gamma^2\left(1 - \frac{1}{k}\right)\right] \quad \text{for } k \ge 2$$

(3-168n)

where the $\Gamma(.)$ = the gamma function that is given by

$$\Gamma(t) = \int_0^\infty r^{t-1} \exp(-r) \, dr$$

It should be noted that

$$\Gamma(t) = (t-1)\Gamma(t-1)$$

For an integer n, the gamma function becomes the factorial as follows:

$$\Gamma(n) = (n-1)!$$

Appendix B contains tabulated values of the gamma function. The coefficient of variation (δ) based on Equations 3-168m and 3-168n is

$$\delta^2_{X_n} = \left[\frac{\Gamma\left(1 - \frac{2}{k}\right)}{\Gamma^2\left(1 - \frac{1}{k}\right)}\right] - 1$$

(3-168o)

For the smallest extreme, the mean (μ) and variance (σ^2) for the distribution are, respectively, given by

$$\mu_{X_1} = v_1 \Gamma\left(1 - \frac{1}{k}\right)$$

(3-168p)

$$\sigma_{X_1}^2 = v_1^2\left[\Gamma\left(1-\frac{2}{k}\right)-\Gamma^2\left(1-\frac{1}{k}\right)\right] \quad \text{for } k \geq 2 \qquad (3\text{-}168\text{q})$$

The coefficient of variation (δ) is

$$\delta_{X_1}^2 = \left[\frac{\Gamma\left(1-\dfrac{2}{k}\right)}{\Gamma^2\left(1-\dfrac{1}{k}\right)}\right]-1 \qquad (3\text{-}168\text{r})$$

Two forms of the type III (Weibull) extreme value distribution can be used, the largest and smallest extreme value. These two types are described in this section. The density function for the largest extreme random variable X_n is given by

$$f_{X_n}(x) = \frac{k}{\omega-n}\left(\frac{\omega-x}{\omega-u}\right)^{k-1}\exp\left[-\left(\frac{\omega-x}{\omega-u}\right)^k\right] \quad \text{for } x \leq \omega \qquad (3\text{-}168\text{s})$$

The density function for the smallest extreme random variable X_1 is given by

$$f_{X_1}(x) = \frac{k}{u-\omega}\left(\frac{x-\omega}{u-\omega}\right)^{k-1}\exp\left[-\left(\frac{x-\omega}{u-\omega}\right)^k\right] \quad \text{for } x \geq \omega \qquad (3\text{-}168\text{t})$$

where $u > 0$ and $k > 0$, u = scale parameter, k = shape parameter, and ω = upper or lower limit on x for the largest and smallest extreme, respectively. The cumulative distribution function for the largest extreme random variable X_n is given by

$$F_{X_n}(x) = \exp\left[-\left(\frac{\omega-x}{\omega-u}\right)^k\right] \quad \text{for } x \leq \omega \text{ and } k > 0 \qquad (3\text{-}168\text{u})$$

The cumulative distribution function for the smallest extreme random variable X_1 is given by

$$F_{X_1}(x) = 1-\exp\left[-\left(\frac{x-\omega}{u-\omega}\right)^k\right] \quad \text{for } x \geq \omega \qquad (3\text{-}168\text{v})$$

For the largest extreme, the mean (μ) and variance (σ^2) for the distribution are, respectively, given by

$$\mu_{X_n} = \omega-(\omega-u)\Gamma\left(1+\frac{1}{k}\right) \qquad (3\text{-}168\text{w})$$

$$\sigma^2_{X_n} = (\omega - u)^2 \left[\Gamma\left(1+\frac{2}{k}\right) - \Gamma^2\left(1+\frac{1}{k}\right) \right] \qquad (3\text{-}168\text{x})$$

For the smallest extreme, the mean (μ) and variance (σ^2) for the distribution are, respectively, given by

$$\mu_{X_1}(x) = \omega + (u - \omega)\left[\Gamma\left(1+\frac{1}{k}\right) \right] \qquad (3\text{-}168\text{y})$$

$$\sigma^2_{X_1} = (u - \omega)^2 \left[\Gamma\left(1+\frac{2}{k}\right) - \Gamma^2\left(1+\frac{1}{k}\right) \right] \qquad (3\text{-}168\text{z})$$

3.8. APPLICATIONS

3.8.1. Water Supply and Water Quality

The water for a residential unit is pumped from a well, then passed through a neutralization unit (neutralizer), and then through a water filter. The annual probability of failure of the pump (which is electrically operated) is 0.1. The annual failure probabilities of the neutralizer and filter are 0.2 and 0.1, respectively. The probability of failure of both the neutralizer and filter at the same time is 0.02. The failure of the pump results in the loss of the water supply, whereas the failure of the neutralizer or filter results in reducing the water quality below acceptable limits without the loss of water supply. The failure events of the pump, neutralizer, and filter can be assumed to be independent. What is the probability of having water of acceptable quality at the residential unit?

The following events are defined:

$$P = \text{pump failure} \qquad (3\text{-}169\text{a})$$

$$N = \text{neutralizer failure} \qquad (3\text{-}169\text{b})$$

$$F = \text{filter failure} \qquad (3\text{-}169\text{c})$$

The probability of receiving water of acceptable quality is

$$P(\bar{P} \cap \bar{N} \cap \bar{F}) = P[\bar{P} \cap \overline{(N \cup F)}] \qquad (3\text{-}170\text{a})$$

Probability of acceptable water $= (1-0.1)[1-(0.2+0.1-0.02)] = 0.648 \quad (3\text{-}170\text{b})$

If low-quality water is received at the residential unit, the probability that the low-level quality is due to the failure of the neutralizer is determined using Equation 3-25 as

$$P[N|(N \cup F)] = \frac{P(N \cap (N \cup F))}{P(N \cup F)} = \frac{P(N)}{P(N \cup F)} \qquad (3\text{-}171)$$

Therefore, the probability is

$$P[N|(N \cup F)] = \frac{0.2}{0.2 + 0.1 - 0.02} = 0.714 \qquad (3\text{-}172)$$

This result indicates that the neutralizer is responsible for 71.4% of the failures in water quality.

If the neutralizer is replaced with a more reliable unit with $P(N) = 0.1$ and $P(N \cap F) = 0.01$, then the conditional probability of Equation 3-172 becomes

$$P[N|(N \cup F)] = \frac{0.1}{0.1 + 0.1 - 0.01} = 0.526 \qquad (3\text{-}173)$$

3.8.2. Earthquakes and Structures

A structure is located in an earthquake zone, where ten damaging earthquakes of intensity (VI) have occurred in the last 100 years. For an earthquake of this intensity, the probability of failure of the structure is 0.01. The occurrence of earthquakes can be assumed to follow a Poisson distribution. Therefore, the time between consecutive earthquakes follows an exponential distribution. The rate of earthquake occurrence is

$$\lambda = \frac{10}{100} = 0.1 \text{ earthquakes/year} \qquad (3\text{-}174)$$

The probability of an earthquake occurring in the next year is computed using Equation 3-148 as

$$P(T \le 1) = 1 - \exp[-0.1(1)] = 0.09516 \qquad (3\text{-}175)$$

where T = the time between consecutive earthquakes. The probability that an earthquake will not occur in the next 5 years is

$$P(T < 5) = 1 - \{1 - \exp[-0.1(5)]\} = 0.6065 \qquad (3\text{-}176)$$

The probability of occurrence of X_t = two earthquakes in the next year is computed using the probability mass function of the Poisson distribution (Equation 3-104) as

$$P_{X_t}(2) = \frac{[(0.1)(1)]^2}{2!} \exp[-(0.1)(1)] = 0.004524 \tag{3-177}$$

The probability of failure, P_f, of the structure during its design life of 50 years, due to earthquakes, can be computed using the theorem of total probability. Partitioning the sample space using the number of earthquakes (x) in 50 years results in the following failure probability for a given number of earthquakes (x):

$$P(\text{structural failure}|x \text{ earthquakes}) = [P(\text{structural failure}|\text{an earthquake})]^x \tag{3-178a}$$

Equation 3-178a is based on the assumption of independence between structural-failure events due to the x earthquakes. Using the theorem of total probability (Equation 3-34), the failure probability of the structure can be computed as

$$P_f = \sum_{x=1}^{\infty} P(\text{earthquakes})[P(\text{structural failure}|\text{an earthquakes})]^x \tag{3-178b}$$

In Equations 3-178a and 3-178b, a surviving structure after an earthquake is assumed to sustain no damage. Therefore, Equation 3-178b can be rewritten as

$$P_f = \sum_{x=1}^{\infty} \frac{[0.1(50)]^x}{x!} \exp[-0.1(50)](0.01)^x \tag{3-179}$$

Utilizing the following relationship for infinite series:

$$\exp(x) = 1 + x + \frac{x^2}{2!} + \frac{x^3}{3!} + \cdots \tag{3-180}$$

P_f can be expressed as

$$P_f = \exp[-0.1(50)]\{\exp[0.1(50)(0.01)] - 1\} = 0.0003455 \tag{3-181}$$

3.8.3. Foundation of a Structure

The foundation of a column in a structure supports a concentrated axial force with a normally distributed magnitude, N(100,000 lb, $10,000^2$ lb^2). The supporting foundation is 5×5 ft^2; therefore, its area is 25 ft^2. Since the foundation is concentrically loaded, the soil bearing pressure, P_b, can be assumed to a have a uniform distribution over the foundation area with a mean and variance computed, respectively, as

$$\overline{P}_b = \frac{100,000}{25} = 4000 \text{ psf} \tag{3-182}$$

$$\text{Var}(P_b) = \frac{(10,000)^2}{(25)^2} = 160,000 \text{ (psf)}^2 \tag{3-183}$$

The standard deviation of the bearing pressure is $\sqrt{160,000} = 400$ psf, and its coefficient of variation is $400/4000 = 0.1$. The bearing strength of soil, P_s, was determined from test borings to be normally distributed N(10,000 psf; 2500^2 psf^2). The probability of bearing failure, P_f, of the soil is

$$P_f = P(P_b > P_s) = P(P_s - P_b < 0) \tag{3-184}$$

This probability can be computed by defining the following random variable:

$$M = P_s - P_b \tag{3-185}$$

where M is called the safety margin. Therefore, the mean and variance of M are, respectively:

$$\overline{M} = \overline{P}_s - \overline{P}_b = 10,000 - 4000 = 6000 \text{ psf} \tag{3-186}$$

$$\text{Var}(M) = \text{Var}(P_s) + \text{Var}(P_b) = (2500)^2 + 160,000 = 6,410,000 \text{ (psf)}^2 \tag{3-187}$$

According to Equations 3-125 and 3-126, M is normally distributed, i.e., N(6000 psf, 6,410,000 psf^2). Therefore, the probability of failure, P_f, is

$$P_f = P(M < 0) \tag{3-188a}$$

$$= \Phi\left[\frac{0 - \overline{M}}{\sigma_M}\right] \tag{3-188b}$$

$$= \Phi\left[\frac{0 - 6000}{2,532}\right] \tag{3-188c}$$

$$= 1 - \Phi(2.37) = 0.0089 \qquad\qquad \text{(3-188d)}$$

Table A.1 for the standard normal variate was used to determine the cumulative probability $\Phi(2.37)$.

3.8.4. Failure of Highway Drainage

A ditch alongside a roadway has a capacity of 4 m³/min, and based on its cross-sectional geometry and grade, the average water-flow velocity is 0.5 m/sec with a standard deviation of 0.25 m/sec. The maximum permissible velocity, so erosion is not a problem, is 1 m/sec. Historical rainfall records indicate that the ditch is expected to have a flow with a mean and standard deviation of 2 and 1 m³/min, respectively. The water-flow volume rate and velocity can be assumed to be normally distributed. Therefore, the probability, P_e, that the water-flow velocity exceeds the erosion limit is

$$P_e = P(V > 1) \qquad\qquad \text{(3-189a)}$$

$$= 1 - \Phi\left[\frac{1 - 0.5}{0.25}\right] = 1 - \Phi(2) = 0.0228 \qquad\qquad \text{(3-189b)}$$

where V = velocity of water flow. Table A.1 for the standard normal variate was used to determine the cumulative probability $\Phi(2)$. The probability, P_Q, of exceeding the volume-rate capacity of the ditch is

$$P_Q = P(Q > 4) \qquad\qquad \text{(3-190a)}$$

$$= 1 - \Phi\left[\frac{4 - 2}{1}\right] = 1 - \Phi(2) = 0.0228 \qquad\qquad \text{(3-190b)}$$

where Q = volume-rate of the water flow.

3.8.5. Floods and Coffer Dams

A specific coffer dam has a probability of 0.05 of being overtopped by a flood in any 1 week. We are interested in the probability that the first overflow will not occur until the xth week. Since there are only two possible outcomes, overtopped or not overtopped, and the probability of overtopping remains constant from week to week, the geometric probability distribution of Equation 3-101 can be used to approximate the probability that the first overtopping occurs in week x = 1, 2, 3, ..., 7 as

$$P(X = 1) = (0.05)(0.95)^0 = 0.0500 \qquad\qquad (3\text{-}191\text{a})$$

$$P(X = 2) = (0.05)(0.95)^1 = 0.0475 \qquad\qquad (3\text{-}191\text{b})$$

$$P(X = 3) = (0.05)(0.95)^2 = 0.0451 \qquad\qquad (3\text{-}191\text{c})$$

$$P(X = 4) = (0.05)(0.95)^3 = 0.0429 \qquad\qquad (3\text{-}191\text{d})$$

$$P(X = 5) = (0.05)(0.95)^4 = 0.0407 \qquad\qquad (3\text{-}191\text{e})$$

$$P(X = 6) = (0.05)(0.95)^5 = 0.0387 \qquad\qquad (3\text{-}191\text{f})$$

$$P(X = 7) = (0.05)(0.95)^6 = 0.0368 \qquad\qquad (3\text{-}191\text{g})$$

In this case, the probability of the first occurrence remains fairly constant from week to week, for relatively a small number of weeks. On the average, the first overtopping occurs on the $\dfrac{1}{0.05}$ week = 20th week (see Equation 3-102) with a standard deviation of first occurrence, S_o, given by Equation 3-103 as

$$S_o = \sqrt{\frac{1 - 0.05}{0.05^2}} = 19.5 \text{ weeks} \qquad\qquad (3\text{-}192)$$

The probability of at least one overtopping in the next 10 weeks is computed using the binomial distribution of Equation 3-97 as

$$P_X(\text{at least } 1) = 1 - P_X(0) \qquad\qquad (3\text{-}193\text{a})$$

$$= 1 - \frac{10!}{10!(0!)}(0.05)^0 (0.95)^{10-0} = 0.40126 \qquad (3\text{-}193\text{b})$$

The probability that the dam will not be overtopped in a 2-year period (104 weeks) is

$$P_X(0) = \binom{104}{0}(0.05)^0 (0.95)^{104-0} = 0.0048 \qquad\qquad (3\text{-}194)$$

The probability of one overtopping is

$$P_X(1) = \binom{104}{1}(0.05)^1 (0.95)^{104-1} = 0.0264 \qquad\qquad (3\text{-}195)$$

The probability of more than one overtopping is

$$P_X(x \text{ more than } 1) = 1 - 0.0048 - 0.0264 = 0.9688 \qquad (3\text{-}196)$$

The probability of the dam not being overtopped can be increased by increasing the height of the dam. If the height of the dam is increased so that the probability of being overtopped in a 1-week period is decreased to 0.02, the probability of the dam not being overtopped in a 104-week period increases to

$$P_X(0) = \binom{104}{0}(0.02)^0 (0.98)^{104-0} = 0.1223 \qquad (3\text{-}197)$$

Thus, the probability of the dam not being overtopped during the 104-week period increased 25 times when the probability of overtopping during 1 week was decreased from 0.05 to 0.02. Increasing the height of the dam requires additional resources but has the return that the probability of overtopping, which can result in damages, is reduced. Therefore, these probabilities can be used in a trade-off analysis, such as a benefit-cost study of alternative dam heights.

The binomial and geometric distributions were used in this example for the purpose of illustration. Since we are dealing with the random occurrence of an event in time, the Poisson distribution can be used to compute the above probabilities.

3.9. SIMULATION AND PROBABILITY DISTRIBUTIONS

Simulation offers a unique opportunity to understand several important elements related to probability distributions and probabilistic analysis. First, it clearly shows the concept of sampling variation, which is random variation in statistics that characterize the difference between a sample value of a statistic and the corresponding population value. This is the underlying basis for probability distributions. Sampling variation was evident in the simulated discharge data of Example 2-16.

Second, simulation of random samples provides the opportunity to examine the effect of sample size on the accuracy of sample statistics. Small samples yield statistics that generally show greater deviation from the population value than provided by large samples. Thus, the spread of a probability distribution of a random variable, on the average, decreases as the sample size increases.

Third, simulation can be used to show why it is important to assess the probability function of the random variable prior to the simulation. Different probability functions can lead to different decisions. Furthermore, random variables can be discrete or continuous, and the type of problem dictates the selection.

Fourth, measured data are often very limited, and making decisions with small sample sizes increases the risk of an incorrect decision. Using measured data for decision making is subject to the problems of (1) not having extreme values in the sample that enable the response of the system to be studied under extreme conditions and (2) outlying measurements that distort sample statistics. Simulation enables the effect of either of these factors on the probability distribution to be studied.

The uniform distribution (see Section 3.7.1) is of special importance in simulation because many generators produce random numbers that are uniformly distributed. The uniform distribution was used as an input to the transformation equation of Equation 2-18. Random numbers from other distributions are often transformed from numbers generated with a uniform distribution. The midsquare method introduced in Chapter 1 produces uniformly distributed random numbers from 0 to 9999. Multiplying such numbers by 0.0001 will essentially produce numbers over the range from 0 to 1.

Since the midsquare method is not an especially reliable generator, generated numbers should be tested prior to their use. Consider the 28 random numbers given in column 4 of Table 3-5. Figure 3-19 shows histograms developed from these random numbers using different cell widths: 0.1, 0.2, and 0.25. The histograms suggest that the 28 generated numbers are approximately uniformly distributed. It should be noted that for cases with cell widths that have small numbers of events per cell, such as Figure 3-19a, the graph shows greater deviation from the population. For large cell widths, the frequencies are greater and the sample provides a better representation of the population. Considerable sampling variation can occur for small sample sizes.

3.9.1. Simulation of Discrete Random Variables

In Chapter 1, values of the flip of a coin and a roll of a die were simulated; both of these are discrete random variables. In Chapter 2, values for an exponential distribution were simulated. The discharge values of Example 2-16 are continuous variates. Both discrete and continuous variates can be simulated.

When using uniform variates to simulate discrete variates, the transformation graph uses a continuous scale for the uniform variate and a discrete scale on the axis for the discrete random variable. In such cases, the transformation graph can be portrayed as a series of spikes as shown in Chapter 1 or as a cumulative histogram as shown in Figure 2-20. They are used in the same way.

Example 3-28: Simulation of Pump Reliability

A mechanical engineering company produces hydraulic pumps. Over a long period of time, they have found that one in eight of the pumps is returned because of defects. If the company has orders to deliver six pumps per week for a period of 28 weeks, they would expect 1/8 of the 168 pumps, or 21, to be returned over the 28-week period. To assess the likelihood that multiple pumps will be returned in any week, the company decides to simulate a 28-week sequence.

Since a pump can only be defective or not defective and since defects in pumps are independent events, then the sequence can be represented as a binomial process, with the probability of a defective pump being 1/8. Therefore, the probabilities for pumps being defective in a group of six are as follows:

Number of defective pumps	Binomial probability	Cumulative probability	
0	$\binom{6}{0}\left(\frac{1}{8}\right)^0\left(\frac{7}{8}\right)^6 = 0.448795$	0.448795	(3-198a)
1	$\binom{6}{1}\left(\frac{1}{8}\right)^1\left(\frac{7}{8}\right)^5 = 0.384682$	0.833477	(3-198b)
2	$\binom{6}{2}\left(\frac{1}{8}\right)^2\left(\frac{7}{8}\right)^4 = 0.137386$	0.970863	(3-198c)
3	$\binom{6}{3}\left(\frac{1}{8}\right)^3\left(\frac{7}{8}\right)^3 = 0.026169$	0.997032	(3-198d)
4	$\binom{6}{4}\left(\frac{1}{8}\right)^4\left(\frac{7}{8}\right)^2 = 0.002804$	0.999836	(3-198e)
5	$\binom{6}{5}\left(\frac{1}{8}\right)^5\left(\frac{7}{8}\right)^1 = 0.000160$	0.999996	(3-198f)
6	$\binom{6}{6}\left(\frac{1}{8}\right)^6\left(\frac{7}{8}\right)^0 = 0.000004$	1.000000	(3-198g)

The cumulative probability function can be used as the transformation function. Any random number less than 0.448795 would indicate zero defects. A value between 0.448795 and 0.833477 would indicate one defect.

To simulate a 28-week period, random numbers are generated using the midsquare method. The results are given in Table 3-5. A seed of 1941 was used. The sequence (column 5 of Table 3-5) indicates that only four pumps were returned in the first 10 weeks. However, in the last 10 weeks, a total of nine pumps were returned because of defects. In the 28-week period, a total of 18 pumps were returned. Thus, the sample probability of a pump being returned was 18/[28(6)] = 0.1071, which is slightly less than the long-term average of 0.125. In 3 of the 28 weeks, more than one pump was returned. The simulation indicates that the probability of multiple pumps being returned in 1 week is 0.1071 (i.e., 3/28). Based on the long-term probability of 1/8, the expected rate can be computed from the cumulative function of Equations 3-198a to 3-198g. The probability of more than one pump being returned equals $1-F_X(1)$, which would be $1-0.833477$ or 0.166523. If a simulation period longer than 28 weeks was used, the sample probability would be closer to the true probability. Again, the difference reflects sampling variation.

Example 3-29: Simulation of Traffic Flow I

To illustrate the generation of values that have a discrete distribution, uniform variates (continuous) are used to generate Poisson variates (discrete). The Poisson distribution was introduced in Section 3.6.4. Assume that the number of cars N arriving at a left-turn-only signal is Poisson distributed with a rate $\lambda = 4$. A traffic

TABLE 3-5 Simulation of Normal, Binomial, and Poisson Variates

(1)	(2)	(3)	(4)	(5)	(6)	(7)
				Number of pumps returned	Normal random	
			Random	returned	random	Poisson
Week	X^2	X	number	(binomial)	number	variate
		1,941	Seed			
1	03,767,481	7,674	0.7674	1	0.7300	5
2	58,890,276	8,902	0.8902	2	1.2274	7
3	79,245,604	2,456	0.2456	0	−0.6884	3
4	06,031,936	0,319	0.0319	0	−1.8538	1
5	00,101,761	1,017	0.1017	0	−1.2718	2
6	01,034,289	0,342	0.0342	0	−1.8225	1
7	00,116,964	1,169	0.1169	0	−1.1905	2
8	01,366,561	3,665	0.3665	0	−0.3411	3
9	13,432,225	4,322	0.4322	0	−0.1708	3
10	18,679,684	6,796	0.6796	1	0.4667	5
11	46,185,616	1,856	0.1856	0	−0.8946	2
12	03,444,736	4,447	0.4447	0	−0.1390	4
13	19,775,809	7,758	0.7758	1	0.7580	5
14	60,186,564	1,865	0.1865	0	−0.8908	2
15	03,478,225	4,782	0.4782	1	−0.0548	4
16	22,867,524	8,675	0.8675	2	1.1148	6
17	75,255,625	2,556	0.2556	0	−0.6569	3
18	06,533,136	5,331	0.5331	1	0.0830	4
19	28,419,561	4,195	0.4195	0	−0.2031	3
20	17,598,025	5,980	0.5980	1	0.2482	4
21	35,760,400	7,604	0.7604	1	0.7077	5
22	57,820,816	8,208	0.8208	1	0.9185	6
23	67,371,264	3,712	0.3712	0	−0.3287	3
24	01,778,944	7,789	0.7789	1	0.6683	5
25	60,668,521	6,685	0.6685	1	0.4358	5
26	44,689,225	6,892	0.6892	1	0.4936	5
27	47,499,664	4,996	0.4996	1	−0.0010	4
28	24,960,016	9,600	0.9600	2	1.7511	8

engineer is trying to design the length of the left-turn-only lane; cars wishing to turn left in excess of the design capacity would have to wait in the left lane, which represents a traffic hazard. For this simulation, a simplifying assumption, that no cars remain from the previous cycle of the traffic signal, is made. That is, all cars are able to turn left prior to the light turning red. The proportion of time that the left lane is occupied by cars wishing to use the left-turn-only lane is of interest to the traffic engineer.

The uniformly distributed random numbers for 28 cycles of the traffic signal were generated with the midsquare method, with the resulting values given in column 4 of Table 3-5. The cumulative Poisson distribution for $\lambda = 4$ is given in Table 3-6. The uniform random numbers were entered into the cumulative function and the value of N found (see column 7 of Table 3-5). For example, for the first cycle of the traffic light the uniform variate is 0.7674. Entering Table 3-6, this corresponds

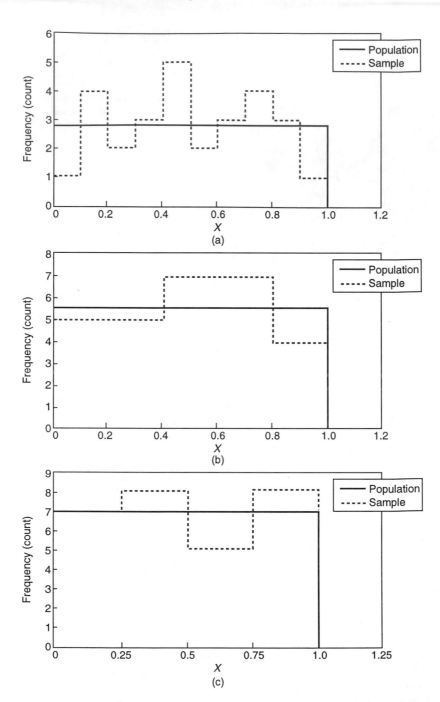

FIGURE 3-19 Histograms of uniformly generated random numbers for cell widths. (a) Cell width = 0.10; (b) cell width = 0.20; and (c) cell width = 0.25.

TABLE 3-6 Cumulative Values of the Poisson
Distribution with $\lambda = 4$ and $\lambda = 4.5$

(1)	(2)	(3)
	$\sum_{x=0}^{N} P_{X_t}(x)$	$\sum_{x=0}^{N} P_{X_t}(x)$
N	with $\lambda = 4$	with $\lambda = 4.5$
0	0.0183	0.0111
1	0.0916	0.0611
2	0.2381	0.1736
3	0.4335	0.3423
4	0.6288	0.5321
5	0.7851	0.7029
6	0.8893	0.8311
7	0.9489	0.9134
8	0.9786	0.9597
9	0.9919	0.9829
10	0.9972	0.9933
11	0.9991	0.9976
12	0.9997	0.9992
13	0.9999	0.9997
14	1.0000	0.9999
15	1.0000	1.0000

to $N = 5$ since it is greater than 0.6288 and less than 0.7851. For the second cycle, a uniform variate of 0.8902 yields a Poisson variate of $N = 7$. The 28 values of N are given in column 7 of Table 3-5. A sample frequency histogram is tabulated in column 2 of Table 3-7. The sample cumulative distribution is tabulated in column 5 of Table 3-7. If the engineer set the length of the left-turn-only lane at five cars, then 14.3% of the time the lane capacity would be exceeded. For $\lambda = 4$, the true proportion is 0.2149 (1–0.7851 from Table 3-6). The difference between the sample and the population is the result of the small sample size (i.e., sampling variation).

Example 3-30: Simulation of Traffic Flow II

The above example does not illustrate the true value of simulation since the population was assumed to be Poisson with $\lambda = 4$. Thus, it was not really necessary to perform the simulation because the probability of N could be determined from the population. The example was also not overly realistic because an assumption was made that all cars in the queue were able to turn left before the light turned red. Anyone who has used a left-turn-only lane knows that this is not a realistic assumption. The value of simulation becomes more evident if random numbers are used for both the number of cars entering the left-turn-only lane and the number of cars that turn left during one cycle of the traffic light. Both of these are discrete random variables and two random variates are needed to simulate one cycle of the traffic light.

TABLE 3-7 Sample Frequency Histogram and Simulated
Frequency Distribution for Example 3-29

(1)	(2) Sample frequency or count (f)	(3) Nf	(4) Simulated proportion	(5) Cumulative proportion
N				
0	0	0	0	0
1	2	2	0.071	0.071
2	4	8	0.143	0.214
3	6	18	0.214	0.428
4	5	20	0.179	0.607
5	7	35	0.250	0.857
6	2	12	0.071	0.928
7	1	7	0.036	0.964
8	1	8	0.036	1.000
Total	28	110	1.000	

To illustrate the process, the problem of Example 3-29 can be modified by allowing a queue to form in the left-turn-only lane. The number of cars that turn during one cycle of the light is a discrete random variable, which is assumed to be of a Poisson distribution with $\lambda = 4.5$, with the cumulative distribution as given in column 3 of Table 3-6. A simulation run of 28 traffic-light cycles is given in Table 3-8. A separate midsquare transformation is used for the two random variables. A seed of 1941 is used for cars entering the queue, and a seed of 5287 is used for simulating the number of cars that are able to make the left turn before the light turns red. A random variate is used to decide on the initial number of cars in the left-turn-only lane. The seed of 5287 leads to the first generated variate of 0.9523. This is used to compute the initial number of cars waiting to make a left turn. Entering Table 3-6 (column 3) yields an initial queue length of eight cars. For the first traffic-light cycle, the random variate of 0.7674 indicates five additional cars enter the queue (column 2 of Table 3-6). The random variate of 0.6875 indicates that time and traffic from the opposing direction would permit five cars to make a left turn. Thus, during the first cycle, five vehicles entered the queue and the five cars at the front of the queue made the left turn, which leaves eight cars in the queue. Column 9 of Table 3-8 gives the number in the left-turn-only lane at the end of the green-light part of the cycle. If we assume that those cars that make the left turn do so at the end of the cycle, then the number of cars in both the left-turn-only and left lanes would be the number at the start of the cycle plus the number of arrivals. This queue length is given in column 4 of Table 3-8. These values show if the design engineer only allowed for five cars in the left-turn-only lane, cars would always be on hold in the left lane. Thus, a longer left-turn-only lane is warranted.

3.9.2. Simulation of Continuous Random Variables

In Section 1.5.4, a transformation graph was shown for transforming uniform variates on the scale of 0 to 360 to a uniform variate on the scale from 0 to 1. The

TABLE 3-8 Left-Turn-Only Simulation with Random Variates for Both Car Arrivals and Left Turns

(1)	(2)	(3)	(4)	(5)	(6)	(7)	(8)	(9)
						Number		
4-digit value (X)	Uniform variate	Car arrivals	Number in lane	4-digit value (X)	Uniform variate	eligible for turn	Number turning	Number in lane
1941				5287				
				9523	0.9523			8
7674	0.7674	5	13	6875	0.6875	5	5	8
8902	0.8902	7	15	2656	0.2656	3	3	12
2456	0.2456	3	15	543	0.0543	1	1	14
319	0.0319	1	15	2948	0.2948	3	3	12
1017	0.1017	2	14	6907	0.6907	5	5	9
342	0.0342	1	10	7066	0.7066	6	6	4
1169	0.1169	2	6	9283	0.9283	8	6	0
3665	0.3665	3	3	1740	0.1740	3	3	0
4322	0.4322	3	3	276	0.0276	1	1	2
6796	0.6796	5	7	761	0.0761	2	2	5
1856	0.1856	2	7	5791	0.5791	5	5	2
4447	0.4447	4	6	5356	0.5356	5	5	1
7758	0.7758	5	6	6867	0.6867	5	5	1
1865	0.1865	2	3	1556	0.1556	2	2	1
4782	0.4782	4	5	4211	0.4211	4	4	1
8675	0.8675	6	7	7325	0.7325	6	6	1
2556	0.2556	3	4	6556	0.6556	5	4	0
5331	0.5331	4	4	9811	0.9811	9	4	0
4195	0.4195	3	3	2557	0.2557	3	3	0
5980	0.5980	4	4	5382	0.5382	5	4	0
7604	0.7604	5	5	9659	0.9659	9	5	0
8208	0.8208	6	6	2962	0.2962	3	3	3
3712	0.3712	3	6	7734	0.7734	6	6	0
7789	0.7789	5	5	8147	0.8147	6	5	0
6685	0.6685	5	5	3736	0.3736	4	4	1
6892	0.6892	5	6	9576	0.9576	8	6	0
4996	0.4996	4	4	6997	0.6997	5	4	0
9600	0.9600	8	8	9580	0.958	8	8	0

graph itself was a continuous line, with both axes being continuous scales. This is the character of a transformation graph for transformation of one continuous variate to a second continuous variate.

Normally distributed random numbers are frequently used in simulation studies. Many statistical methods assume an underlying normal population. Therefore, normally distributed random numbers are widely used in statistically based simulations. Normally distributed random numbers can be generated by transforming uniformly distributed random numbers. In Chapter 7, a mathematical approach to making the transformation is provided. To provide a conceptual understanding of the transforming of one continuous variable to another, it will be presented using the tabular representation of the normal distribution (Table A-1 in Appendix A).

Assuming that a random number generator that provides uniformly distributed numbers from 0 to 1 is available, values of the standard normal deviate Z can be obtained by entering Table A-1 using the uniform variate U as the probability and reading the value of Z that corresponds to the cumulative probability of U. Values of U less than 0.5 produce negative values of Z. Consider a U value of 0.4814. Entering Table A-1 with a probability of 0.4814 gives a Z value of –0.0467, assuming that linear interpolation is acceptable.

Example 3-31: Generation of Standard Normal Variates

The uniformly distributed numbers in column 4 of Table 3-5 were transformed to a set of 28 numbers that have a standard normal distribution (see column 6 of Table 3-5). A histogram of simulated values is shown in Figure 3-20. The values appear to closely match the standard normal distribution.

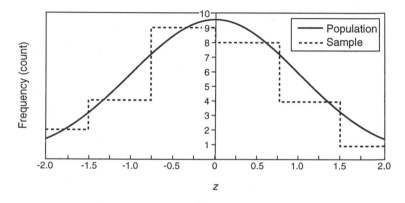

FIGURE 3-20 Comparison of random sample of standard normal variate and N(0,1).

3.10. PROBLEMS

3-1. A construction manager needs to procure building materials for the construction of a bridge. The following sources are identified:

Material type	Sources
Concrete	Sources A and B
Reinforcing steel	Sources C and D
Timber	Sources D and E
Structural steel	Sources C and F
Hardware	Sources F, G, and H

Define the sample space of all possible combinations of sources supplying the construction project, assuming that each material type can be procured from one source only, but a source may supply more than one material type at the same time.

3-2. A construction tower crane can operate up to a height H of 300 ft, a range (radius) R of 50 ft, and an angle ϕ of $\pm 90°$ in a horizontal plane. Sketch the sample space of operation of the crane. Sketch the following events:

Event	Definition
A	$30 < H < 80$ and $R < 30$
B	$H > 50$ and $0° < \phi < 50°$
C	$H < 40$ and $R > 60$
D	$H > 80$ and $-30° < \phi < 50°$

3-3. Construct Venn diagrams for each of the following:
   a. A deck of playing cards
   b. The roll of a die
   c. Letter grades on a test assuming equal probabilities for each grade
   d. Letter grades on a test assuming the following probabilities:
      (A) 15%; (B) 25%; (C) 30%; (D) 20%; and (F) 10%
   e. Options at an intersection with the following probabilities: (left turn) 20%; (straight ahead) 40%; (right turn) 25%; (U-turn) 10%; and (remain stopped) 5%.

3-4. For the data and events of Problem 3-2, sketch the following events: $A \cup B$, $A \cap B$, $C \cup D$, $C \cap D$, $A \cup C$, $A \cup (B \cap C)$, $\overline{A}$, and $\overline{A} \cap B$.

3-5. The traffic that makes a left turn at an intersection consists of two types of vehicles, types A and B. A type A vehicle is twice the length of type B. The left-turn lane can accommodate eight vehicles of type B, four of type A, or combinations of A and B. Define the sample space of all possible combinations of vehicles waiting for a left turn at the intersection. Also define the following events: (1) at least one vehicle of type A waiting for left turn, (2) two vehicles of type B waiting for left turn, and (3) exactly one of type A and one of type B waiting for left turn.

3-6. Construct a Venn diagram for a deck of playing cards (4 suits, 13 cards per suit). Show the following events:
   a. A = all diamonds and all aces.
   b. B = all face cards.
   c. C = the intersection of red cards and face cards.
   d. D = the union of black cards and cards with values of 4 or smaller.

3-7. What is the difference between the population and a sample? A construction engineer obtains five specimens of steel reinforcing bars from a shipment to a construction site. What is the corresponding population? How could the engineer use the sample to characterize the population?

3-8. The scores on a test were distributed as shown below for a class of 60 students. Convert the frequency histogram to a probability distribution. What is the probability that a student had a score of at least 90? What is the probability that a student failed the test (F grade), assuming that a score less than 70 was considered an F?

Range	50–59	60–69	70–74	75–79	80–84	85–89	90–94	95–100
Number	4	9	7	11	13	7	6	3

3-9. Two loads A and B are applied on a structural column and are mutually exclusive. The probability that the beam is safe against load A is 0.7 and against load B is 0.9. What is the probability that the structure is safe assuming that the "safe" events for the two loads are independent?

3-10. A concrete beam may fail by shear or flexure. The failure probability in shear is equal to failure probability in flexure, and the probability of failure in shear when the beam is loaded beyond its flexural capacity (i.e., the beam already failed in flexure) is 80%. Find the failure probability of the beam in flexure given that the probability of failure of the beam is 0.2.

3-11. Materials for a construction site are supplied by two sources A and B. Upon arrival of all materials from both sources, the construction is started after the arrival of workers. If the probability that materials would be supplied by A on time is 0.8 and by B is 0.75, and if the probability that the workers arrive on time for work is 0.95, find the probability that the job is started on time.

3-12. A shipping company facilitates for customers the transfer of any items from any location to another within the U.S. The company transfers 20, 50, and 30% of the items by air, ground, and sea transportation, respectively. The breakdown of these items by shipping method is shown in the accompanying figure. Assume that 5% of the items shipped by air get damaged during shipping, 10% of the items shipped by ground get damaged during shipping, and 15% of the items shipped by sea get damaged during shipping. Determine the following probabilities:

  a. The probability of receiving a damaged item sent by air, ground, or sea.

  b. If an item was received damaged, what is the probability that it was sent by (a) air; (b) by ground; and (c) by sea?

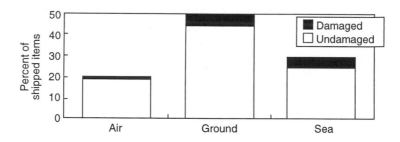

PROBLEM 3-12 Shipping methods.

3-13. Find the following conditional probabilities:

  a. On a single roll of a die, the probability of a 3 given that the value is an odd integer (draw the sample space for the conditional event)

  b. The probability that the outcome of a single draw from a card deck is a red ace given that it is not a face card

  c. The probability that the outcome of a single draw from a card deck is an ace given that five cards, including one ace, have already been drawn and discarded from the deck

3-14. A construction site receives fill material from three sources, with sources A, B, and C providing 15, 25, and 60% of the total, respectively. On the average, fill material from sources A, B, and C do not have the proper moisture characteristics 2, 6, and 3% of the time, respectively. If one dump truck is sampled at random and the fill material does not have the proper moisture, what is the probability that it came (a) from source A? (b) from source B? (c) from source C?

3-15. Given the following cumulative mass function, derive the mass function

$$F_X(x) = \begin{cases} 0.2 & x = 2 \\ 0.2 & x = 3 \\ 0.2 & x = 4 \\ 0.5 & x = 5 \\ 0.7 & x = 6 \\ 0.8 & x = 7 \\ 0.9 & x = 8 \\ 1.0 & x = 9 \end{cases} \text{ for }$$

Graph both the probability and cumulative mass functions.

3-16. Given the following probability mass function:

$$P_X(x) = \begin{cases} i & x = 1 \\ 2i & x = 2 \\ 3i & x = 3 \\ 0 & \text{otherwise} \end{cases} \text{ for }$$

determine the value of i that results in a legitimate mass function. Graph both the probability and cumulative mass functions.

3-17. For the probability mass function of Problem 3-15, determine the following probabilities: (a) $P(X = 4)$; (b) $P(X = 5)$; (c) $P(X \le 5)$; (d) $P(X < 5)$; (e) $P(4 < X \le 7)$; and (f) $P(X \ge 7)$.

3-18. The number of years of drought in the last nine decades are:

1900 to 1909	2
1910 to 1919	0
1920 to 1929	3
1930 to 1939	5
1940 to 1949	4
1950 to 1959	1
1960 to 1969	1
1970 to 1979	2
1980 to 1989	2

Show the sample mass function and cumulative function for the number of draught years in a decade. What is the sample space?

3-19. Find the value k that is necessary to make the following a legitimate density function:

$$f_X(x) = \begin{cases} kx & \text{for } 0 \le x \le 1 \\ k & \text{for } 1 < x \le 2 \end{cases}$$

Graph both the density and the cumulative functions.

3-20. The duration of a construction activity was estimated to be in the range [2,6] days with a most likely duration of 4 days. A construction engineer used the following symmetric density function to model the duration:

$$f_X(x) = \begin{cases} bx & \text{for } 2 \le x \le 4 \\ a - bx & \text{for } 4 < x \le 6 \end{cases}$$

Determine the necessary constants a and b to have a legitimate density function. Graph both the density and cumulative functions.

3-21. For the probability mass function of Problem 3-15, determine the mean, variance, and standard deviation.

3-22. For the probability mass function of Problem 3-16, determine the mean, variance, standard deviation, coefficient of variation, and skewness.

3-23. For the probability density function of Problem 3-19, determine the mean, variance, and standard deviation.

3-24. For the probability density function of Problem 3-20, determine the mean, variance, standard deviation, coefficient of variation, and skewness.

3-25. Calculate the mean, variance, standard deviation, coefficient of variation, and skewness for the following data sample:

$$X = \{3, 5, 8, 6, 8\}$$

3-26. Ten steel specimens were tested for their yield strength. Determine the mean, variance, standard deviation, coefficient of variation, and skewness for the tested steel based on the following test results:

$$\{38, 36, 34, 37, 38, 39, 35, 38, 40, 36\} \text{ ksi}$$

3-27. The probability of a flood in any 1 year is 0.1. In a 10-year period, what is the probability of (a) no floods; (b) two or fewer floods; (c) from one to three floods?

3-28. Piles are used to support the foundation of a structure. The failure probability of a pile during proof testing is 0.1. Determine the following:
   a. The probability of three failed piles out of ten tested piles
   b. The probability of no failures in 20 tested piles
   c. The probability of ten tested piles to obtain the first failure
   d. The probability of three consecutive pile failures

3-29. A fair coin is tossed ten times. What is the probability of getting (a) exactly five heads; (b) exactly two tails; (c) no heads; (d) two or fewer heads; (e) five or more tails; and (f) at least two but not more than six heads?

3-30. The probability that a flood of a specified magnitude occurs in any 1 year is 0.05. What is the probability that in the next 10 years (a) exactly two such floods will occur; (b) no more than one such flood will occur; (c) no such floods will occur; (d) at least four floods will occur; and (e) at least three but no more than six such floods will occur?

3-31. Over the past 15 years, the numbers of hurricanes per year to cause damage in a certain city were 4, 2, 1, 3, 0, 2, 1, 3, 5, 2, 3, 1, 1, 2, and 0. What is the probability of not having hurricane damage in any 1 year? What are the average and variance of the number of hurricanes in 1 year?

3-32. The overload occurrence of an offshore facility due to extreme waves can be modeled by a Poisson process with a rate $\lambda = 0.5$. What is the probability of occurrence of two overloads in a year? What is the probability of no overloads in a year? What is the probability of one or two overloads in the design life of the structure of 20 years?

3-33. Derive Equations 3-113 and 3-114 using Equations 3-69 and 3-73.

3-34. Assume that daily evaporation rates (E) have a uniform distribution with $a = 0$ and $b = 0.35$ in./day. Determine the following probabilities: (a) $P(E > 0.1)$; (b) $P(E < 0.22)$; (c) $P(E = 0.2)$; and (d) $P(0.05 < E < 0.15)$.

3-35. Using the test grades of Problem 3-8, fit a uniform distribution to the data.

3-36. The compressive strength of concrete specimen follows normal distribution with a mean value (μ) of 2.8 ksi and coefficient of variation (COV) of 0.1. If the applied stress is 2.5 ksi, find the probability of failure.

3-37. The average annual precipitation for Washington, D.C. is 43 in., with a standard deviation of 6.5 in. If the amounts can be assumed to have a normal distribution, find the probability that the precipitation in any 1 year is (a) greater than 55 in.; (b) less than 35 in.; (c) either less than 32 in. or greater than 51 in.; and (d) between 38 and 48 in. Find the precipitations that have (e) 5% chance of being exceeded and (f) 1% chance of not being exceeded.

3-38. A concrete delivery truck shuttles between a concrete mixing plant and a construction site for the entire 10-h working day. The one-way trip takes on the average 1 h with a standard deviation of 0.2 h. Assuming that the travel time can be modeled by a normal distribution, what is the probability of completing a two-way (complete) trip within 1.5 hr? within 2.5 hr? Assuming that the time needed to unload the truck is also a normal variate with a mean of 0.5 h and standard deviation of 0.2 h, what are the mean and standard deviation of the needed time for a complete two-way trip with unloading?

3-39. The Reynolds number for a pipe flow was found to follow lognormal distribution with a mean value and standard deviation of $\mu_{R_e} = 2650$ and $\sigma_{R_e} = 700$, respectively. The flow can be classified as laminar flow if $R_e < 2700$ and as turbulent flow if $R_e > 4000$. Find the probability that the flow is in transitional state.

3-40. The random variable $Y = \ln(X)$ has a normal distribution with a mean of 5 and a standard deviation of 1. Determine the mean, variance, standard deviation, mode, median, and skewness coefficient of X.

3-41. Over the past 15 years, the numbers of hurricanes per year to cause damage in a certain city were given in Problem 3-31 as 4, 2, 1, 3, 0, 2, 1, 3, 5, 2, 3, 1, 1, 2, and 0. What is the probability that the time between consecutive hurricanes exceeds 1 year? What is the return period of hurricanes for this city?

3-42. A random variable X with a mean of 2.5 follows an exponential distribution. Find the following probabilities: (a) $P(X > 2.5)$; (b) $P(X < 1)$; and (c) $P(0.5 < X < 1.5)$. Find the values of x_0, if (d) $P(X < x_0) = 0.1$; and (e) $P(X > x_0) = 0.05$.

3-43. Find the following probabilities for the t statistic: (a) $P(t > 2.45 \mid k = 6)$; (b) $P(t < 2.72 \mid k = 11)$; and (c) $P(t < -1.75 \mid k = 16)$.

3-44. Find the values of t_α for a random variable that has the t distribution such that: (a) $P(t < t_\alpha \mid k = 15) = 0.05$; (b) $P(t > t_\alpha \mid k = 8) = 0.025$; and (c) $P(t > t_\alpha \mid k = 24) = 0.90$.

3-45. Find the following probabilities for the chi-square (C) statistic: (a) $P(C > 5.024 | k = 1)$; (b) $P(C < 0.831 | k = 5)$; (c) $P(C > 2.555 | k = 10)$; and (d) $P[(C < 6.251 \text{ or } C > 10.864) | k = 18]$.

3-46. Find the values of c_α for each of the following: (a) $P(C > c_\alpha | k = 4) = 0.020$; and (b) $P(C < c_\alpha | k = 7) = 0.01$.

3-47. Find the values of the F statistic (f_α) for each of the following: (a) $P[F > f_\alpha | (k = 8, u = 5)] = 0.05$; (b) $P[F > f_\alpha | (k = 15, u = 10)] = 0.01$.

3-48. The extreme wave height that is used in the design of an offshore facility for a design life of 30 years is a random variable with a mean of 25 ft and standard deviation of 5 ft. Determine the probability that the extreme wave height will exceed 30 ft in 30 years. Assume that the extreme wave height follows the largest extreme value distribution of type I.

3-49. Redo Problem 3-49 assuming that the extreme wave height follows the largest extreme value distribution of type II.

3-50. Redo Problem 3-49 assuming that the extreme wave height follows the smallest extreme value distribution of type III.

3-51. Use the midsquare method with a seed of 1937 to generate 30 uniform random numbers (0 to 1). Plot frequency histograms using cell widths of 0.1, 0.2, and 0.25. Comment on: (a) the degree to which the numbers follow a uniform distribution; (b) the effect of cell size on the ability of the sample to appear to have a uniform distribution.

3-52. Construct a transformation curve that transforms uniform variates on a scale from 0 to 1 to uniform variates on a scale from 10 to 20. Use the 30 variates of Problem 3-51 and the transformation curve to obtain 30 uniform variates on a scale from 10 to 20. Discuss the results of the transformation.

3-53. Use the 30 uniform variates from Problem 3-51 to generate standard normal variates (z). Compute the mean and standard deviation of the z values. Compare them with the population values, and discuss the differences.

Chapter 4
Multiple Random Variables

TABLE OF CONTENTS

4.1. INTRODUCTION

The discussion in Chapter 3 is limited to single random variables. However, in engineering, it is common to deal with two or more random variables simultaneously in solving problems. In general, multiple random variables are encountered in these problems in the following two forms:

1. Joint occurrences of multiple random variables that can be correlated or uncorrelated
2. Random variables that are known in terms of their functional relationship with other basic random variables

The objective of this chapter is to provide analytical tools to deal with these two forms of multiple random variables.

4.2. JOINT RANDOM VARIABLES AND THEIR PROBABILITY DISTRIBUTIONS

In some engineering applications, the outcomes, say, $E_1, E_2, ..., E_n$, that constitute a sample space S are mapped to an n-dimensional (n-D) space of real numbers. The functions that establish such a transformation to the n-D space are called multiple random variables (or random vectors). This mapping can be one-to-one or one-to-many.

Multiple random variables are commonly classified into two types: discrete and continuous random vectors. A discrete random vector may only take on distinct, usually integer, values, whereas a continuous random vector takes on values within a continuum of values. A distinction is made between these two types of random vectors because the computations of probabilities depend on their type.

4.2.1. Probability for Discrete Random Vectors

The probability of a discrete multiple random variable or random vector $X = (X_1, X_2, ..., X_n)$ is given by a *joint probability mass function*. A joint mass function specifies the probability that the discrete random variable X_1 is equal to some value x_1, X_2 is equal to some value x_2, ..., X_n is equal to some value x_n and is denoted by

$$P_X(x) = P(X_1 = x_1, X_2 = x_2, ..., X_n = x_n) \tag{4-1}$$

where X = random vector that includes the random variables $(X_1, X_2, ..., X_n)$, and x = some specified values for the random vectors, $(x_1, x_2, ..., x_n)$. The probability mass function must satisfy the axioms of probability. Therefore, the probability of an event $(X_1 = x_1, X_2 = x_2, ..., X_n = x_n)$ must be less than or equal to one, and it must be greater than or equal to zero, i.e.,

$$0 \leq P(X_1 = x_1, X_2 = x_2, ..., X_n = x_n) \leq 1 \tag{4-2}$$

This property is valid for all possible values of all of the random variables. Additionally, the sum of all possible probabilities must equal one.

It is often useful to present the likelihood of an outcome using the *cumulative mass function*, which is given by

$$F_X(x) = P(X_1 \leq x_1, X_2 \leq x_2, ..., X_n \leq x_n) = \sum_{\text{all } (X_1 \leq x_1, X_2 \leq x_2, ..., X_n \leq x_n)} P_X(x_1, x_2, x_3, ..., x_n) \tag{4-3}$$

The cumulative mass function is used to indicate the probability that the random variable X_1 is less than or equal x_1, X_2 is less than or equal x_2, ..., and X_n is less than or equal x_n. The cumulative distribution function must satisfy the following properties:

1. $F_X(all\ x \rightarrow -\infty) = 0$
2. $F_X(x_1, x_2, ..., x_i \rightarrow -\infty, ..., x_n) = 0$, for any $i = 1, 2, ..., n$
3. $F_X(x_1, x_2, ..., x_i \rightarrow -\infty, ..., x_k \rightarrow -\infty, ..., x_n) = 0$, for any values of $x_i, ..., x_k$
4. $F_X(x_1, x_2, ..., x_i \rightarrow +\infty, ..., x_n) = F_{Xj}(x_j: j = 1, 2, ..., n$ and $j \neq i)$, called the *marginal distribution* of all the random variables except X_i
5. $F_X(x_1, x_2, ..., x_i \rightarrow +\infty, ..., x_k \rightarrow +\infty, ..., x_n) = F_{Xj}(x_j: j = 1, 2, ..., n$ and $j \neq i$ to $k)$, called the *marginal distribution* of all the random variables except X_i to X_k
6. $F_X(all\ x \rightarrow +\infty) = 1$
7. $F_X(x)$ is a nonnegative and nondecreasing function of x

The first, second, and third properties define the limiting behavior of $F_X(x)$; as one or more of the random variables approach $-\infty$, $F_X(x)$ approaches zero. The fourth and fifth properties define the possible marginal distributions as one or more of the random variable approaches $+\infty$. The sixth property is based on the probability axiom that states the probability of the sample space is one. The seventh property is based on the cumulative nature of $F_X(x)$.

The presentation of the materials in the remaining part of this section is limited to two random variables. The presented concepts can be generalized to n random variables. Based on the definition of conditional probabilities, the conditional probability mass function $P_{X_1|X_2}(x_1|x_2)$, for two random variables X_1 and X_2, is given by

$$P_{X_1|X_2}(x_1|x_2) = \frac{P_{X_1X_2}(x_1,x_2)}{P_{X_2}(x_2)} \tag{4-4}$$

where $P_{X_1|X_2}(x_1|x_2)$ results in the probability of $X_1 = x_1$ given that $X_2 = x_2$; $P_{X_1X_2}(x_1,x_2)$ = joint mass function of X_1 and X_2, and $P_{X_2}(x_2)$ = marginal mass function for X_2 that is not equal to zero. In this case, the marginal distribution is given by

$$P_{X_2}(x_2) = \sum_{all\ x_1} P_{X_1X_2}(x_1,x_2) \tag{4-5}$$

Similarly, the conditional probability mass function $P_{X_2|X_1}(x_2|x_1)$, for two random variables X_1 and X_2, is given by

$$P_{X_2|X_1}(x_2|x_1) = \frac{P_{X_1X_2}(x_1,x_2)}{P_{X_1}(x_1)} \tag{4-6}$$

where the marginal mass function $P_{X_1}(x_1)$ is

$$P_{X_1}(x_1) = \sum_{\text{all } x_2} P_{X_1 X_2}(x_1, x_2) \tag{4-7}$$

The definitions provided by Equations 4-4 to 4-7 can be generalized for the n-D case. Based on the definition of conditional probabilities, it can be stated that, if X_1 and X_2 are statistically uncorrelated random variables, then

$$P_{X_1 | X_2}(x_1 | x_2) = P_{X_1}(x_1) \tag{4-8a}$$

and

$$P_{X_2 | X_1}(x_2 | x_1) = P_{X_2}(x_2) \tag{4-8b}$$

Therefore, using Equations 4-4 or 4-6, the following important relationship can be obtained:

$$P_{X_1 | X_2}(x_1, x_2) = P_{X_1}(x_1) P_{X_2}(x_2) \tag{4-9}$$

Example 4-1: Reliability Engineering

A reliability engineer evaluates each unit produced at the end of a manufacturing line in a factory for mechanical and electrical defects. The evaluation is based on an accelerated test for a simulated life of the manufactured product. Based on a sample of 100 units, the following numbers of defects were determined:

X = number of mechanical defects

		0	1	2	3	4
Y =	0	70	4	2	1	0
number of	1	4	2	3	1	1
electrical	2	2	1	1	1	2
defects	3	2	1	0	0	0
	4	1	1	0	0	0

$$\tag{4-10a}$$

In this table, two random variables are defined: X = number of mechanical defects and Y = number of electrical defects. The table entries can be used to obtain likelihood measures of combination occurrences of defects by dividing the entries by the total number of units, i.e., 100 units. The result can be considered as the joint mass function of X and Y, i.e., $P_{XY}(x,y)$, as

X = number of mechanical defects

		0	1	2	3	4	
Y =	0	0.70	0.04	0.02	0.01	0	
number of	1	0.04	0.02	0.03	0.01	0.01	(4-10b)
electrical	2	0.02	0.01	0.01	0.01	0.02	
defects	3	0.02	0.01	0	0	0	
	4	0.01	0.01	0	0	0	

The marginal mass distributions for X and Y, i.e., $P_X(x)$ and $P_Y(y)$, can be determined by performing summations over all Y and X values, respectively, of $P_{XY}(x,y)$. The resulting marginal distributions are

x	0	1	2	3	4	
$P_X(x)$	0.79	0.09	0.06	0.03	0.03	(4-10c)

and

y	0	1	2	3	4	
$P_Y(y)$	0.77	0.11	0.07	0.03	0.02	(4-10d)

The marginal distribution of a random variable provides a description of its probabilistic characteristics while the other random variable remains random. For example, $P_X(x)$ provides the probability mass function for mechanical defects without regard to the occurrences of the electrical defects, i.e., with random occurrence of electrical welds. The probability distribution of x mechanical defects given that there are y electrical defects, i.e., $P_{X|Y}(x|y)$, is given by

For $y = 0$

$x|y$	0|0	1|0	2|0	3|0	4|0	
$P_{X|Y}(x|y)$	70/77	4/77	2/77	1/77	0/77	(4-10e)
or	0.909	0.052	0.026	0.013	0.000	

For $y = 1$

$x|y$	0|1	1|1	2|1	3|1	4|1	
$P_{X|Y}(x|y)$	4/11	2/11	3/11	1/11	1/11	(4-10f)
or	0.364	0.182	0.273	0.091	0.091	

For $y = 2$

$x\vert y$	0|2	1|2	2|2	3|2	4|2	
$P_{X\vert Y}(x\vert y)$	2/7	1/7	1/7	1/7	2/7	(4-10g)
or	0.286	0.143	0.143	0.143	0.286	

For $y = 3$

$x\vert y$	0|3	1|3	2|3	3|3	4|3	
$P_{X\vert Y}(x\vert y)$	2/3	1/3	0/3	0/3	0/3	(4-10h)
or	0.667	0.333	0.000	0.000	0.000	

For $y = 4$

$x\vert y$	0|4	1|4	2|4	3|4	4|4	
$P_{X\vert Y}(x\vert y)$	1/2	1/2	0/2	0/2	0/2	(4-10i)
or	0.5	0.5	0.0	0.0	0.0	

The conditional distributions enable the reliability engineer to investigate the probabilistic characteristics of a random variable as the other random variable is maintained constant at some specified value. For example, $P_{X\vert Y=4}(x\vert y = 4)$ provides the probability mass function for mechanical defects for the case that we have four electrical defects. Since $P_{XY}(x,y)$ is not equal to the product $P_X(x) P_Y(y)$, the random variables X and Y are statistically correlated.

4.2.2. Probability for Continuous Random Vectors

A *joint probability density function* (pdf) is used to define the likelihood of occurrence for a continuous random vector. Specifically, the probability that the random vector $X = (X_1, X_2, ..., X_n)$ is within the interval from $x^l = \left(x_1^l, x_2^l, x_3^l, ..., x_n^l\right)$ to $x^u = \left(x_1^u, x_2^u, x_3^u, ..., x_n^u\right)$ is

$$P\left(x^l \leq X \leq x^u\right) = \int_{x_1^l}^{x_1^u} \int_{x_2^l}^{x_2^u} ... \int_{x_n^l}^{x_n^u} f_X(x) \, dx_1 dx_2 ... dx_n \qquad (4-11)$$

in which $f_X(x)$ is the joint density function. It is important to note that the multiple integral of the joint pdf from $-\infty$ to $+\infty$ equals 1, i.e.,

$$P(-\infty < X < +\infty) = \int\limits_{-\infty}^{+\infty}\int\limits_{-\infty}^{+\infty}\cdots\int\limits_{-\infty}^{+\infty} f_X(x)\, dx_1 dx_2 \ldots dx_n = 1 \qquad (4\text{-}12)$$

The *cumulative distribution function* (cdf) of a continuous random variable is defined by

$$F_X(x) = P(X \le x) = \int\limits_{-\infty}^{x_1}\int\limits_{-\infty}^{x_2}\cdots\int\limits_{-\infty}^{x_n} f_X(x)\, dx_1 dx_2 \ldots dx_n \qquad (4\text{-}13)$$

The cumulative distribution function must satisfy the following properties:

1. $F_X(all\ x = -\infty) = 0$
2. $F_X(x_1, x_2, \ldots, x_i \to -\infty, \ldots, x_n) = 0$, for any $i = 1, 2, \ldots, n$
3. $F_X(x_1, x_2, \ldots, x_i \to -\infty, \ldots, x_k \to -\infty, \ldots, x_n) = 0$, for any values of $x_i, \ldots, x_k$
4. $F_X(x_1, x_2, \ldots, x_i \to +\infty, \ldots, x_n) = F_{X_j}(x_j, j = 1, 2, \ldots, n$ and $j \ne i)$, called the *marginal distribution* of all the random variables except X_i
5. $F_X(x_1, x_2, \ldots, x_i \to +\infty, \ldots, x_k \to +\infty, \ldots, x_n) = F_{X_j}(x_j, j = 1, 2, \ldots, n$ and $j \ne i$ to $k)$, called the *marginal distribution* of all the random variables except X_i to X_k
6. $F_X(all\ x \to +\infty) = 1$
7. $F_X(x)$ is a nonnegative and nondecreasing function of x

These properties are similar to the properties of $F_X(x)$ for the discrete random variables as was discussed in Section 4.2.1. The joint density function can be obtained from a given joint cumulative distribution function by evaluating the partial derivative as follows:

$$f_X(x) = \frac{\partial^n F_X(x)}{\partial X} \qquad (4\text{-}14a)$$

That is,

$$f_{X_1 X_2 \ldots X_n}(x_1, x_2, \ldots, x_n) = \frac{\partial^n F_{X_1 X_2 \ldots X_n}(x_1, x_2, \ldots, x_n)}{\partial X_1 \partial X_2 \ldots \partial X_n} \qquad (4\text{-}14b)$$

The presentation of the materials in the remaining part of this section is limited to two random variables. The presented concepts can be generalized to n random variables. Based on the definition of conditional probabilities, the conditional probability density function $f_{X_1|X_2}(x_1|x_2)$ for two random variables X_1 and X_2 is given by

$$f_{X_1|X_2}(x_1|x_2) = \frac{f_{X_1 X_2}(x_1, x_2)}{f_{X_2}(x_2)} \qquad (4\text{-}15)$$

where $f_{X_1X_2}(x_1,x_2)$ = joint density function of X_1 and X_2, and $f_{X_2}(x_2)$ = marginal density function for X_2 that is not equal to zero. In this case, the marginal distribution is given by

$$f_{X_2}(x_2) = \int_{-\infty}^{+\infty} f_{X_1X_2}(x_1, x_2)\, dx_1 \tag{4-16}$$

Similarly, the conditional probability density function $f_{X_2|X_1}(x_2|x_1)$ for two random variables X_1 and X_2 is given by

$$f_{X_2|X_1}(x_2|x_1) = \frac{f_{X_1X_2}(x_1, x_2)}{f_{X_1}(x_1)} \tag{4-17}$$

where the marginal density function $f_{X_1}(x_1)$ is

$$f_{X_1}(x_1) = \int_{-\infty}^{+\infty} f_{X_1X_2}(x_1, x_2)\, dx_2 \tag{4-18}$$

The definitions as given by Equations 4-15 to 4-18 can be generalized for the n-D case, but they are not discussed herein. Based on the definition of conditional probabilities, it can be stated that if X_1 and X_2 are statistically uncorrelated random variables, then

$$f_{X_1|X_2}(x_1|x_2) = f_{X_1}(x_1) \tag{4-19}$$

and

$$f_{X_2|X_1}(x_2|x_1) = f_{X_2}(x_2) \tag{4-20}$$

Therefore, using Equations 4-15 or 4-17, the following important relationship can be obtained:

$$f_{X_1X_2}(x_1, x_2) = f_{X_1}(x_1) f_{X_2}(x_2) \tag{4-21}$$

Example 4-2: Times to Failure

An automated control system consists of two components with random times to failure X and Y. Previous experience with these components indicates that their times to failure are exponentially distributed and that the x time to failure is three times larger than the y time to failure. The following joint density function was suggested in this case:

$$f_{XY}(x,y) = \begin{cases} c\exp[-(x+3y)] & x\geq 0 \text{ and } y\geq 0 \\ 0 & \text{otherwise} \end{cases} \tag{4-22}$$

where c = constant. The magnitude of c can be determined based on the following condition:

$$\text{The volume under } f_{XY}(x,y) = \int_0^\infty \int_0^\infty f_{XY}(x,y)\,dx\,dy = 1 \tag{4-23a}$$

$$= \int_0^\infty \int_0^\infty c\exp[-(x+3y)]\,dx\,dy \tag{4-23b}$$

$$= \int_0^\infty \int_0^\infty c\exp(-x)\exp(-3y)\,dx\,dy \tag{4-23c}$$

$$= \int_0^\infty \left([-c\exp(-3y)\exp(-x)]\Big|_{x=0}^{x=\infty} \right) dy \tag{4-23d}$$

$$= \frac{-c\exp(-3y)}{3}\Bigg|_{y=0}^{y=\infty} = \frac{c}{3} = 1 \tag{4-23e}$$

Therefore, $c = 3$.

The marginal distributions of X and Y are

$$f_X(x) = \int_0^\infty c\exp[-(x+3y)]\,dy = \exp(-x) \tag{4-24}$$

$$f_Y(y) = \int_0^\infty c\exp[-(x+3y)]\,dx = 3\exp(-3y) \tag{4-25}$$

The marginal distribution of a random variable provides a description of its probabilistic characteristics while the other random variable remains random. For example, $f_X(x)$ provides the probability density function for X without regard to the occurrence of Y, i.e., random occurrence of Y. Since $f_{XY}(x,y) = f_X(x) f_Y(y)$, the random variables X and Y are statistically uncorrelated.

The probability that X as a random variable is larger than Y as a random variable is

$$P(X > Y) = \int_0^\infty \int_0^x f_{XY}(x, y) \, dy \, dx \qquad (4\text{-}26a)$$

$$= \int_0^\infty \exp(-x)[-\exp(-3x) + 1] \, dx \qquad (4\text{-}26b)$$

$$= \left(\frac{\exp(-4x)}{4} - \exp(-x) \right)\Bigg|_0^\infty = \frac{3}{4} \qquad (4\text{-}26c)$$

This result is expected since the problem states that the X time to failure is three times larger than the Y time to failure.

4.2.3. Conditional Moments, Covariance, and Correlation Coefficient

In general, moments can be computed using the concept of mathematical expectation. For a continuous random vector X, the kth moment about the origin is given by

$$M'_k = \int_{-\infty}^{+\infty} \int_{-\infty}^{+\infty} \cdots \int_{-\infty}^{+\infty} x_1^k \, x_2^k \cdots x_n^k \, f_{X_1 X_2 \ldots X_n}(x_1, x_2, \ldots, x_n) \, dx_1 dx_2 \cdots dx_n \qquad (4\text{-}27)$$

in which $\{X_1, X_2, \ldots, X_n\}$ is the random vector and $f_{X_1 X_2 \ldots X_n}(x_1, x_2, \ldots, x_n)$ is its joint density function. The corresponding equation for a discrete random vector X is

$$M'_k = \sum_{\text{all } x} x_1^k \, x_2^k \cdots x_n^k \, P_{X_1 X_2 \ldots X_n}(x_1, x_2, \ldots, x_n) \qquad (4\text{-}28)$$

in which $P_{X_1 X_2 \ldots X_n}(x_1, x_2, \ldots, x_n)$ is the joint probability mass function.

The above moments are commonly considered special cases of mathematical expectation. The mathematical expectation of an arbitrary function $g(X)$, a function of the random vector X, is given by

$$E[g(X)] = \int_{-\infty}^{+\infty} \int_{-\infty}^{+\infty} \cdots \int_{-\infty}^{+\infty} g(x) f_{X_1 X_2 \ldots X_n}(x_1, x_2, \ldots, x_n) \, dx_1 dx_2 \cdots dx_n \qquad (4\text{-}29)$$

The corresponding equation for a discrete random vector X is

$$E[g(X)] = \sum_{\text{all } x} g(x) P_{X_1 X_2 \dots X_n}(x_1, x_2, \dots, x_n) \tag{4-30}$$

For the two-dimensional case, X_1 and X_2, the conditional mean value for X_1 given that X_2 takes a value x_2, denoted $\mu_{X_1|x_2}$, is defined in terms of the conditional mass and density functions for the discrete and continuous random variables, respectively. The conditional mean for the continuous case is

$$\mu_{X_1|x_2} = E(X_1|x_2) = \int_{-\infty}^{+\infty} x_1 \, f_{X_1|X_2}(x_1|x_2) \, dx_1 \tag{4-31}$$

where $f_{X_1|X_2}(x_1|x_2)$ = the conditional density function of X_1 at a given (or specified) value of X_2 (see Equation 4-15). In this case, the conditional mean is the average value of the random variable X_1 given that the random variable X_2 takes the value x_2. For a discrete random variable, the conditional mean is given by

$$\mu_{X_1|x_2} = E(X_1|x_2) = \sum_{\text{all } x_1} x_1 \, P_{X_1|X_2}(x_1|x_2) \tag{4-32}$$

where $P_{X_1|X_2}(x_1|x_2)$ = the conditional mass function of X_1 at a given (or specified) value of X_2 (see Equation 4-4).

For statistically uncorrelated random variables X_1 and X_2, the conditional mean of a random variable is the same as its mean, i.e.,

$$\mu_{X_1|x_2} = E(X_1|x_2) = E(X_1) \tag{4-33}$$

and

$$\mu_{X_2|x_1} = E(X_2|x_1) = E(X_2) \tag{4-34}$$

Also, it can be shown that the expected value with respect to X_2 of the conditional mean $\mu_{X_1|X_2}$ is the mean of X_1, i.e.,

$$E_{X_2}(\mu_{X_1|X_2}) = E(X_1) \tag{4-35}$$

where E_{X_2} = the expected value with respect to X_2, i.e., the variable of integration (or summation) for computing the expected value is x_2. In Equation 4-35, the quantity $\mu_{X_1|X_2}$ is treated as a random variable, since conditioning is performed on the random variable X_2 (not a specified value x_2).

As previously discussed, the variance is the second moment about the mean. For two random variables X_1 and X_2, the conditional variance $\sigma^2_{X_1|x_2}$ [or Var($X_1|x_2$)] is computed as follows:

$$\text{Var}(X_1|x_2) = \int_{-\infty}^{+\infty} \left(x_1 - \mu_{X_1|x_2}\right)^2 f_{X_1|X_2}(x_1|x_2)\, dx_1 \tag{4-36}$$

For a discrete variable, the conditional variance is computed by

$$\text{Var}(X_1|x_2) = \sum_{\text{all } x_1} \left(x_1 - \mu_{X_1|x_2}\right)^2 P_{X_1|X_2}(x_1|x_2) \tag{4-37}$$

The variance of the random variable X_1 can also be computed using the conditional variance as follows:

$$\text{Var}(X_1) = E_{X_2}\left[\text{Var}(X_1|X_2)\right] + \text{Var}_{X_2}\left[E(X_1|X_2)\right] \tag{4-38}$$

where E_{X_2} = the expected value with respect to X_2, and Var_{X_2} = variance with respect to X_2, i.e., the variable of integration (or summation) for computing the variance is x_2. In Equation 4-38, the quantity Var($X_1|X_2$) is treated as a random variable, since the conditioning is performed on the random variable X_2 (not a value x_2).

The covariance (Cov) of two random variables X_1 and X_2 is defined in terms of mathematical expectation as

$$\text{Cov}(X_1, X_2) = E\left[\left(X_1 - \mu_{X_1}\right)\left(X_2 - \mu_{X_2}\right)\right] \tag{4-39}$$

It is common to use the notation $\sigma_{X_1 X_2}$, σ_{12}, or Cov(X_1,X_2) for the covariance of X_1 and X_2. The covariance for two random variables can also be determined using the following equation that results from Equation 4-39:

$$\text{Cov}(X_1, X_2) = E(X_1 X_2) - \mu_{X_1}\mu_{X_2} \tag{4-40}$$

where the expected value of the product ($X_1 X_2$) is given by

$$E(X_1 X_2) = \int_{-\infty}^{+\infty}\int_{-\infty}^{+\infty} x_1 x_2\, f_{X_1 X_2}(x_1, x_2)\, dx_1 dx_2 \tag{4-41}$$

Equation 4-40 can be derived from Equation 4-39 based on the definition of mathematical expectation, and by separating terms of integration. If X_1 and X_2 are statistically uncorrelated, then

$$\mathrm{Cov}(X_1, X_2) = 0 \qquad\qquad (4\text{-}42a)$$

and

$$\mathrm{E}(X_1 X_2) = \mu_{X_1}\mu_{X_2} \qquad\qquad (4\text{-}42b)$$

The correlation coefficient is defined as a normalized covariance with respect to the standard deviations of X_1 and X_2 and is given by

$$\rho_{X_1 X_2} = \frac{\mathrm{Cov}(X_1, X_2)}{\sigma_{X_1}\sigma_{X_2}} \qquad\qquad (4\text{-}43)$$

The correlation coefficient ranges inclusively between -1 and $+1$, i.e.,

$$-1 \le \rho_{X_1 X_2} \le +1 \qquad\qquad (4\text{-}44)$$

If the correlation coefficient is zero, then the two random variables are described to be uncorrelated. From the definition of correlation, in order for $\rho_{X_1 X_2}$ to be zero, the $\mathrm{Cov}(X_1, X_2)$ must be zero. Therefore, X_1 and X_2 are statistically uncorrelated. However, the converse of this finding does not hold. The correlation coefficient can also be viewed as a measure of the degree of linear association between X_1 and X_2. The sign ($-$ or $+$) indicates the slope for the linear association. It is important to note that the correlation coefficient does not give any indications about the presence of nonlinear relationship between X_1 and X_2 (or the lack of it).

Example 4-3: Covariance and Correlation

The sample space for two random variables, X and Y, is defined by the region between the two curves shown in Figure 4-1. Assume that any pair (x,y) in this region is equally likely to occur, i.e., a uniform distribution is assumed over the region between the two curves. The two curves are given by the following equations:

$$y = x^{\frac{1}{n}} \quad \text{for } 0 \le x \le 1 \qquad\qquad (4\text{-}45)$$

and

$$y = x^n \quad \text{for } 0 \le x \le 1 \qquad\qquad (4\text{-}46)$$

Therefore, the joint density value for these random variables can be viewed on a third axis that is perpendicular to the plane of the curves in Figure 4-1. The joint

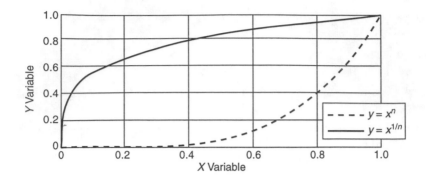

FIGURE 4-1 The two curves in Example 4-3.

density function in this case takes on a constant value over this region. The range of y for both curves is $[0,1]$. Therefore, the value of the density function can be determined based on the following condition:

$$\int_0^\infty \int_0^\infty f_{XY}(x,y) \, dx \, dy = 1 \qquad (4\text{-}47a)$$

where $f_{XY}(x,y)$ = constant (c). Therefore,

$$1 = \left[\int_0^1 x^{\frac{1}{n}} \, dx - \int_0^1 x^n \, dx \right] c \qquad (4\text{-}47b)$$

Solving for c, the following expression can be obtained:

$$c = \frac{n+1}{n-1} \qquad (4\text{-}48)$$

Therefore, the density function is given by

$$f_{XY}(x,y) = \begin{cases} \dfrac{n+1}{n-1} & \text{for } 0 \le x \le 1 \text{ and } x^n \le y \le x^{1/n} \\ 0 & \text{otherwise} \end{cases} \qquad (4\text{-}49)$$

The marginal density functions of X is given by

$$f_X(x) = c \int_{x^n}^{x^{\frac{1}{n}}} dy = \frac{n+1}{n-1}\left(x^{\frac{1}{n}} - x^n \right) \qquad \text{for } 0 \le x \le 1 \qquad (4\text{-}50)$$

Similarly, the marginal density functions of Y is given by

$$f_Y(y) = \frac{n+1}{n-1}\left(y^{\frac{1}{n}} - y^n\right) \quad \text{for } 0 \le y \le 1 \tag{4-51}$$

Therefore, $f_Y(y)$ is similar to $f_X(x)$. Thus, the expected value of X is equal to the expected value of Y and is given by

$$E(X) = E(Y) = \int_0^1 x\, f_X(x)\, dx \tag{4-52a}$$

$$= \frac{n+1}{n-1}\left[\frac{n}{1+2n} - \frac{1}{n+2}\right] \tag{4-52b}$$

Also, the second moments of X and Y are equal and given by

$$E(X^2) = E(Y^2) = \int_0^1 x^2\, f_X(x)\, dx \tag{4-53a}$$

$$= \frac{n+1}{n-1}\left[\frac{n}{1+3n} - \frac{1}{n+3}\right] \tag{4-53b}$$

Therefore, the variances of X and Y are

$$\text{Var}(X) = E(X^2) - [E(X)]^2 \tag{4-54a}$$

$$\text{Var}(Y) = E(Y^2) - [E(Y)]^2 \tag{4-54b}$$

The expected value of the product XY is

$$E(XY) = c\int_0^1 \int_{x^n}^{x^{\frac{1}{n}}} xy\, dy\, dx \tag{4-55a}$$

$$= \frac{1}{2}\frac{n+1}{n-1}\left[\frac{n}{2+2n} - \frac{1}{2n+2}\right] = \frac{1}{4} \tag{4-55b}$$

Therefore, the covariance of X and Y is

$$\text{Cov}(X,Y) = E(XY) - E(X)E(Y) \tag{4-56}$$

For $n = 2$, these moments take the following values:

$$E(X) = E(Y) = \frac{9}{20} \tag{4-57}$$

$$E(X^2) = E(Y^2) = \frac{9}{35} \tag{4-58}$$

$$\text{Var}(X) = \text{Var}(Y) = 0.0546 \tag{4-59}$$

$$E(XY) = \frac{1}{4} \tag{4-60}$$

$$\text{Cov}(X,Y) = 0.0475 \tag{4-61}$$

Therefore, the correlation coefficient is

$$\rho_{XY} = \frac{\text{Cov}(X,Y)}{\sqrt{\text{Var}(X)\text{Var}(Y)}} = \frac{0.0475}{0.0546} = 0.87 \tag{4-62}$$

For $n = 3$, the correlation coefficient ρ_{XY} is 0.71. Figure 4-2 shows selected probability descriptors, including ρ_{XY}, as functions of n. It is interesting to note that as the power order n approaches 1, the area between the two curves diminishes and the correlation coefficient approaches 1. Also, by increasing n, the area between the two curves increases, approaching a limiting case where it covers the entire area of Figure 4-2 and the correlation coefficient approaches zero.

4.2.4. Example Joint Probability Distributions

There are a few joint probability distributions that are used in engineering. They include the multinomial distribution, a discrete distribution, and the joint bivariate normal, a continuous distribution. In this section, these two distributions are described.

The *multinomial distribution* is a generalization of the binomial distribution. In this case, the distribution is not based on a Bernoulli sequence (i.e., trials with two possible outcomes); it is based on trials with more than two possible outcomes, say

FIGURE 4-2 Probability descriptors as functions of n.

n outcomes. The probabilities of occurrence for these outcomes are assumed to be $p_1, p_2, ..., p_n$. The following condition is imposed on these probabilities:

$$p_1 + p_2 + \cdots + p_n = 1 \tag{4-63}$$

The trials are assumed to be independent. The random variables, in this case, $X_1, X_2, ..., X_n$, are the numbers of occurrences of the respective outcomes in N identical and independent trials. The joint probability mass function is

$$P_{X_1 X_2 ... X_n}(x_1, x_2, ..., x_n) = \frac{N!}{x_1! x_2! \cdots x_n!} (p_1)^{x_1} (p_2)^{x_2} \cdots (p_n)^{x_n} \tag{4-64}$$

The following condition should be satisfied:

$$N = x_1 + x_2 + \cdots + x_n \tag{4-65}$$

The mean and variance of the respective random variables, $X_1, X_2, ..., X_n$, are

$$\mu_i = NP_i \quad \text{for } i = 1, 2, ..., n \tag{4-66a}$$

$$\sigma_i^2 = NP_i(1 - p_i) \quad \text{for } i = 1, 2, ..., n \tag{4-66b}$$

The *joint bivariate normal distribution* is a generalization of the normal probability distribution. This distribution describes the joint probabilistic characteristics of random variables. For the case of two random variables, X_1 and X_2, the joint density function is given by

$$f_{X_1X_2}(x_1, x_2) = \frac{1}{2\rho\sigma_1\sigma_2\sqrt{1-\rho^2}}$$

$$\exp\left\{-\frac{1}{2(1-\rho^2)}\left[\left(\frac{x_1-\mu_1}{\sigma_1}\right)^2 - 2\rho\left(\frac{x_1-\mu_1}{\sigma_1}\right)\left(\frac{x_2-\mu_2}{\sigma_2}\right) + \left(\frac{x_2-\mu_2}{\sigma_2}\right)^2\right]\right\}$$

(4-67)

where $-\infty < x_1 < +\infty$, $-\infty < x_2 < +\infty$, ρ is the correlation coefficient for the two random variables, μ is the mean, and σ is the standard deviation. The marginal distributions for X_1 and X_2 are the same as the probability density function of the normal distribution (i.e., Equation 10-117) and are given, respectively, by

$$f_{X_1}(x_1) = \frac{1}{\sigma_1\sqrt{2\pi}}\exp\left[-\frac{1}{2}\left(\frac{x_1-\mu_1}{\sigma_1}\right)^2\right] \quad \text{for } -\infty < x_1 < \infty \quad (4\text{-}68)$$

$$f_{X_2}(x_2) = \frac{1}{\sigma_2\sqrt{2\pi}}\exp\left[-\frac{1}{2}\left(\frac{x_2-\mu_2}{\sigma_2}\right)^2\right] \quad \text{for } -\infty < x_2 < \infty \quad (4\text{-}69)$$

The following moments can be determined for the two random variables:

$$E(X_1) = \mu_1 \quad (4\text{-}70a)$$

$$E(X_2) = \mu_2 \quad (4\text{-}70b)$$

$$\text{Var}(X_1) = \sigma_1^2 \quad (4\text{-}70c)$$

$$\text{Var}(X_2) = \sigma_2^2 \quad (4\text{-}70d)$$

It can be shown that the conditional random variables $X_1|x_2$ and $X_2|x_1$ have normal probability distributions with the following means, respectively:

$$E(X_1|x_2) = \mu_1 + \frac{\sigma_1}{\sigma_2}(x_2-\mu_2)\rho \quad (4\text{-}71a)$$

$$E(X_2 | x_1) = \mu_2 + \frac{\sigma_2}{\sigma_1}(x_1 - \mu_1)\rho \qquad (4\text{-}71\text{b})$$

The respective variances of the conditional random variables are

$$\text{Var}(X_1 | x_2) = \sigma_1^2(1 - \rho^2) \qquad (4\text{-}72\text{a})$$

$$\text{Var}(X_2 | x_1) = \sigma_2^2(1 - \rho^2) \qquad (4\text{-}72\text{b})$$

An example joint normal is shown in Figure 4-3 using mean values of X_1 and X_2 of 5 and 7, respectively. The corresponding standard deviations are 3 and 3, respectively. The correlation coefficient was varied as shown in the figure. As the correlation coefficient deviates from zero, the surface of the joint density function deviates from the bell shape to show the correlation between the two variables.

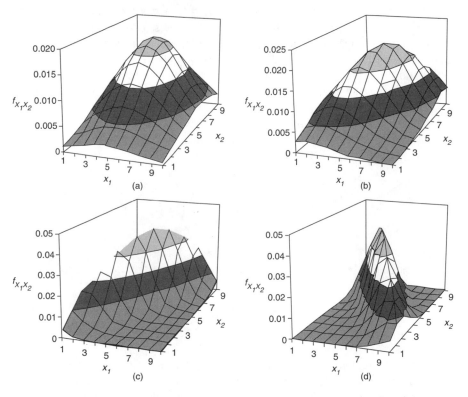

FIGURE 4-3 Joint normal density function. (a) $\rho = 0$; (b) $\rho = 0.5$; (c) $\rho = 0.9$; and (d) $\rho = -0.9$.

4.3. FUNCTIONS OF RANDOM VARIABLES

Many engineering problems deal with a dependent variable that is a function of one or more independent random variables. In this section, analytical tools for determining the probabilistic characteristics of the dependent random variable based on given probabilistic characteristics of independent random variables, and a functional relationship between them, are provided. The discussion in this section is divided into the following cases: (1) probability distributions for dependent random variables, (2) mathematical expectations, and (3) approximate methods. Cases 1 and 2 result in complete distributions and moments without any approximations, respectively, whereas, the third case results in approximate moments.

4.3.1. Probability Distributions for Dependent Random Variables

A random variable X is defined as a mapping from a sample space of an engineering system or experiment to the real line of numbers. This mapping can be one-to-one mapping or many-to-one mapping. If Y is defined to be a dependent variable in terms of a function $Y = g(X)$, then Y is also a random variable. Assuming that both X and Y are discrete random variables and for a given probability mass function of X, i.e., $P_X(x)$, the objective herein is to determine the probability mass function of Y, i.e., $P_Y(y)$. This objective can be achieved by determining the equivalent events of Y in terms of the events of X based on the given relationship between X and Y, i.e., $Y = g(X)$. For each value y_i, all of the values of x that result in y_i should be determined, say, $x_{i_1}, x_{i_2}, ..., x_{i_j}$. Therefore, the probability mass function of Y is given by

$$P_Y(y_i) = \sum_{k=1}^{j} P_X(x_{i_k})$$

(4-73)

If X is continuous but Y is discrete, the probability mass function for Y is given by

$$P_Y(y_i) = \int_{R_e} f_X(x)\, dx$$

(4-74)

where R_e = the region of X that defines an equivalent event to the value $Y = y_i$.

If X is continuous with a given density function $f_X(x)$ and the function $g(X)$ is continuous, then $Y = g(X)$ is a continuous random variable with an unknown density function $f_Y(y)$. The density function of Y can be determined by performing the following four steps:

1. For any event defined by $Y \leq y$, an equivalent event in the space of X needs to be defined.
2. $F_Y(y) = P(Y < y)$ can then be calculated.
3. $f_Y(y)$ can be determined by differentiating $F_Y(y)$ with respect to y.
4. The range of validity of $f_Y(y)$ in the Y space should be determined.

Formally stated, if X is a continuous random variable, $Y = g(X)$ is differentiable for all x, and $g(X)$ is either strictly (i.e., monotonically) increasing or strictly (i.e., monotonically) decreasing for all x, then $Y = g(X)$ is a continuous random variable with the following density function:

$$f_Y(y) = f_X\left[g^{-1}(y)\right]\left|\frac{\partial g^{-1}(y)}{\partial y}\right| \tag{4-75}$$

where $g^{-1}(y) = x$. Equation 4-75 can be derived by developing an expression for the cumulative distribution function of Y as

$$F_Y(y) = \int_{x \leq g^{-1}(y)} f_X(x)\, dx = \int_{-\infty}^{g^{-1}(y)} f_X(x)\, dx \tag{4-76}$$

This result can be expressed as

$$F_Y(y) = \int_{-\infty}^{y} f_X\left(g^{-1}\right)\frac{\partial g^{-1}(y)}{\partial y}\, dy \tag{4-77}$$

By taking the derivative $\dfrac{\partial F_Y(y)}{\partial y}$, Equation 4-75 results. The derivative $\dfrac{\partial g^{-1}(y)}{\partial y}$ is known as the Jacobian of the transformation (or inverse) and can be alternatively determined as follows:

$$\frac{\partial g^{-1}(y)}{\partial y} = \frac{1}{\dfrac{\partial g(x)}{\partial x}}\Bigg|_{\text{evaluated at } x = g^{-1}(y)} \tag{4-78}$$

Then the limits on Y should be determined based on the limits of X and $Y = g(X)$. If the inverse $g^{-1}(y)$ is not unique (say, $= x_1, x_2, ..., x_m$), then the density function of Y is determined as follows:

$$f_Y(y) = \sum_{i=1}^{m} f_X\left[g_i^{-1}(y)\right]\left|\frac{\partial g_i^{-1}(y)}{\partial y}\right| \tag{4-79}$$

where

$$g_i^{-1}(y) = x_i \tag{4-80}$$

The following cases are selected special functions of single and multiple random variables that are commonly used, where the resulting variable (Y) can have known distribution types for some cases:

1. For a multiple independent random variables $X = (X_1, X_2, ..., X_n)$, where the function $g(X)$ is a linear combination as given by

$$Y = g(X) = a_0 + a_1 X_1 + a_2 X_2 + \cdots + a_n X_n \tag{4-81}$$

and $a_0, a_1, a_2, ..., a_n$ are real numbers, the mean value and variance of Y are

$$E(Y) = a_0 + a_1 E(X_1) + a_2 E(X_2) + \cdots + a_n E(X_n) \tag{4-82}$$

and

$$\text{Var}(Y) = \sum_{i=1}^{n}\sum_{j=1}^{n} a_i a_j \, \text{Cov}\left(X_i, X_j\right) \tag{4-83}$$

where $\text{Cov}(X_i, X_j)$ = covariance of X_i and X_j. It should be noted that $\text{Cov}(X_i, X_i) = \text{Var}(X_i) = \sigma_{X_i}^2$. Equation 4-83 can be expressed in terms of the correlation coefficient as follows:

$$\text{Var}(Y) = \sum_{i=1}^{n}\sum_{j=1}^{n} a_i a_j \rho_{X_i X_j} \sigma_{X_i} \sigma_{X_j} \tag{4-84}$$

where $\rho_{X_i X_j}$ = correlation coefficient of X_i and X_j. If the random variables of the vector X are statistically uncorrelated, then the variance of Y is

$$\text{Var}(Y) = \sum_{i=1}^{n} a_i^2 \, \text{Var}\left(X_i\right) \tag{4-85}$$

2. In Equations 4-82 to 4-85, if the random variables $X_1, X_2, X_3, ..., X_n$ have normal probability distributions, then Y has a normal probability distribution with a mean

and variance as given by Equations 4-81 to 4-85. This special case was also described in Equations 3-128 to 3-130.

3. If X has a normal distribution, and $Y = g(X) = \exp(X)$, then Y has a lognormal distribution.

4. If $Y = X_1 X_2 X_3 \ldots X_n$, the arithmetic multiplication of X_1, X_2, X_3, ..., and X_n with lognormal distributions, then Y has a lognormal distribution as given by Equations 3-142 to 3-144.

5. If X_1, X_2, ..., X_n are independent random variables that have Poisson distributions with the parameters, λ_1, λ_2, ..., λ_n, respectively, then $Y = X_1 + X_2 + \ldots + X_n$ has a Poisson distribution with the parameter $\lambda = \lambda_1 + \lambda_2 + \ldots + \lambda_n$.

Example 4-4: Mean and Variance of a Linear Function

Consider the following function:

$$Z = 2X + 5Y + 10 \tag{4-86}$$

where X, Y, and Z are random variables. The means of X and Y are 3 and 5, respectively; and the standard deviations are 1 and 2, respectively. The random variables X and Y are assumed to be uncorrelated. Therefore, the mean of Z, μ_Z, is computed using Equation 4-82 as

$$\mu_Z = 2(3) + 5(5) + 10 = 41 \tag{4-87a}$$

The variance of Z, σ_Z^2, is computed using Equation 4-85 as

$$\sigma_Z^2 = 2^2(1^2) + 5^2(2^2) = 104 \tag{4-87b}$$

The standard deviation of Z is $\sqrt{104}$ = 10.20, and the coefficient of variation is 10.20/41 = 0.25.

Example 4-5: Probability Density Function of a Nonlinear Function

For the following nonlinear function:

$$Y = aX^2 + b \tag{4-88}$$

where a and b are constants and X has an exponential distribution with a parameter λ, determine the probability density function of Y.

In order to use Equation 4-79 to determine $f_Y(y)$, we need to determine X as a function of Y and the derivative of X with respect to Y. The inverse of the function in Equation 4-88 is

$$X = \pm \sqrt{\frac{Y-b}{a}} \tag{4-89}$$

The derivative of X with respect to Y is

$$\frac{dX}{dY} = \mp \frac{1}{2\sqrt{a(Y-b)}} \tag{4-90}$$

Therefore, the density function of Y can be determined based on the density function of the exponential distribution (given by Equation 10-147) and by substituting in Equation 4-79:

$$f_Y(y) = \frac{1}{2\sqrt{a(y-b)}} \left[f_X\left(+\sqrt{\frac{y-b}{a}} \right) + f_X\left(-\sqrt{\frac{y-b}{a}} \right) \right] \tag{4-91a}$$

or

$$f_Y(y) = \frac{\lambda}{2\sqrt{a(y-b)}} \left[\exp\left(-\lambda\sqrt{\frac{y-b}{a}} \right) + \exp\left(+\lambda\sqrt{\frac{y-b}{a}} \right) \right] \tag{4-91b}$$

The resulting function $f_Y(y)$ is not an exponential density function.

4.3.2. Mathematical Expectation

In certain engineering applications, we might be interested in knowing only the moments, specifically the mean and variance, of a random variable Y, based on known probabilistic characteristics of a random variable X and a function $Y = g(X)$. In such cases, mathematical expectation is very effective in achieving this objective.

The mathematical expectation of an arbitrary function $g(X)$ of a continuous random variable X is given by

$$E[g(X)] = \int_{-\infty}^{+\infty} g(x) f_X(x) \, dx \tag{4-92}$$

The corresponding equation for a discrete random variable is

$$E[g(X)] = \sum_{i=1}^{n} g(x_i) P_X(x_i) \tag{4-93}$$

Mathematical expectation can be used to determine the moments of Y for a given probabilistic characteristic of X and the function $Y = g(X)$. The mean (or expected value) can be determined by the direct use of Equations 4-92 and 4-93. The variance can be determined, respectively, for continuous and discrete X random variables as

$$\mathrm{Var}(Y) = \mathrm{Var}[g(X)] = \int_{-\infty}^{+\infty} \{g(x) - \mathrm{E}[g(X)]\}^2 f_X(x)\, dx \qquad (4\text{-}94)$$

and

$$\mathrm{Var}(Y) = \mathrm{Var}[g(X)] = \sum_{i=1}^{n} \{g(x_i) - \mathrm{E}[g(X_i)]\}^2 P_X(x_i) \qquad (4\text{-}95)$$

If we consider a special case where the function, $g(X)$, is the following linear function:

$$Y = g(X) = aX + b \qquad (4\text{-}96)$$

where a and b = real numbers, mathematical expectation can be used to determine, respectively, the mean and variance of Y as follows:

$$\mathrm{E}(Y) = a\mathrm{E}(X) + b \qquad (4\text{-}97)$$

$$\mathrm{Var}(Y) = a^2\, \mathrm{Var}(X) \qquad (4\text{-}98)$$

Equations 4-97 and 4-98 are valid regardless of the distribution type of X. Also, they can be used for both cases of discrete and continuous X. Based on Equation 4-96, it can be stated that mathematical expectation, $\mathrm{E}(.)$, is a linear operator. However, the variance, $\mathrm{Var}(.)$, is not a linear operator.

Equations 4-97 and 4-98 can be generalized to the case of multiple random variables $X = (X_1, X_2, ..., X_n)$, where the function $g(X)$ is given by

$$Y = g(X) = a_0 + a_1 X_1 + a_2 X_2 + ... + a_n X_n \qquad (4\text{-}99)$$

where $a_0, a_1, a_2, ..., a_n$ are real numbers. The mean and variance of Y are

$$\mathrm{E}(Y) = a_0 + a_1 \mathrm{E}(X_1) + a_2 \mathrm{E}(X_2) + ... + a_n \mathrm{E}(X_n) \qquad (4\text{-}100)$$

and

$$\mathrm{Var}(Y) = \sum_{i=1}^{n} \sum_{j=1}^{n} a_i a_j \mathrm{Cov}(X_i, X_j) \qquad (4\text{-}101)$$

where $Cov(X_i,X_j)$ = covariance of X_i and X_j. It should be noted that $Cov(X_i,X_i) = Var(X_i) = \sigma_{X_i}^2$. Equation 4-101 can be expressed in terms of the correlation coefficient as follows:

$$Var(Y) = \sum_{i=1}^{n} \sum_{j=1}^{n} a_i a_j \rho_{X_i X_j} \sigma_{X_i} \sigma_{X_j} \qquad (4\text{-}102)$$

where $\rho_{X_i X_j}$ = correlation coefficient of X_i and X_j. If the random variables of the vector X are statistically uncorrelated, then the variance of Y is expressed as

$$Var(Y) = \sum_{i=1}^{n} a_i^2 \, Var(X_i) \qquad (4\text{-}103)$$

Again, these equations are valid regardless of the distribution types of X, as well as for both cases of discrete and continuous X.

Now, consider the case where the function $g(X) = g(X_1, X_2, ..., X_n)$ is the arithmetic multiplication of all the random variables, and is given by

$$Y = gX = X_1 X_2 X_3 \cdots X_n \qquad (4\text{-}104)$$

For statistically independent random variables, X_1, X_2, X_3, ..., X_n, the mean and variance of Y are

$$E(Y) = E(X_1)E(X_2)\cdots E(X_n) \qquad (4\text{-}105)$$

and

$$Var(Y) = E(X_1^2)E(X_2^2)\cdots E(X_n^2) - \left[E(X_1)E(X_2)\cdots E(X_n)\right]^2 \qquad (4\text{-}106)$$

Again, these equations are valid regardless of the distribution types of X, as well as for both cases of discrete and continuous X.

Example 4-6: Cost of Precast Concrete

The total cost, C, to manufacture a concrete panel in a precast plant is

$$C = 1.5X + 2Y \qquad (4\text{-}107)$$

where X is the cost of materials, Y is the cost of labor, and the constants 1.5 and 2 are overhead cost multipliers for materials and labor, respectively. These multipliers include the costs of storing and handling the materials, management, supervision, and quality control. The costs X and Y are assumed to be uncorrelated random variables with means of \$100/panel and \$250/panel, respectively; and with standard deviations of \$10/panel and \$50/panel, respectively. The means are denoted as μ_X and μ_Y, respectively, and the variances are denoted as σ^2_X and σ^2_Y, respectively. Using mathematical expectation (Equations 4-100 and 4-101), the mean and variance of the total cost are

$$\mu_C = 1.5\mu_X + 2\mu_Y \tag{4-108a}$$

$$= 1.5(100) + 2(250) = \$650/\text{panel} \tag{4-108b}$$

$$\sigma^2_C = 1.5^2 \, \sigma^2_X + 2^2 \, \sigma^2_Y \tag{4-109a}$$

$$= 1.5^2 (10)^2 + 2^2 (50)^2 = 10,225 \; (\$/\text{panel})^2 \tag{4-109b}$$

The standard deviation is $\sqrt{10,225} = \$101.12/\text{panel}$, and the coefficient of variation is $101.12/650 = 0.1556$.

4.3.3. Approximate Methods

The closed-form solutions for the distribution types of dependent random variables, as well as mathematical expectation, provide solutions for the simple cases of functions of random variables. Also, they provide solutions for simple distribution types or a mixture of distribution types for the independent random variables. For cases that involve a more general function $g(X)$ or a mixture of distribution types, these methods are not suitable for obtaining solutions due to the analytical complexity according to these methods. Also, in some engineering applications, precision might not be needed. In such cases, approximate methods based on Taylor-series expansion, with or without numerical solutions of needed derivatives, can be used. The use of Taylor-series expansion, in this section, is divided into the following two headings: (1) single random variable X, and (2) multiple random variables, i.e., a random vector X.

Single random variable X — The Taylor series expansion of a function $Y = g(X)$ about the mean of X, i.e., $E(X)$, is given by

$$Y = g[E(X)] + [X - E(X)]\frac{dg(X)}{dX} + \frac{1}{2}[X - E(X)]^2 \frac{d^2g(X)}{dX^2}$$

$$+ \cdots + \frac{1}{k!}[X - E(X)]^k \frac{d^k g(X)}{dX^k} + \cdots \tag{4-110}$$

in which the derivatives are evaluated at the mean of X. Truncating this series at the linear terms, the *first-order mean and variance* of Y can be obtained by applying the mathematical expectation and variance operators, respectively. The first-order (approximate) mean is

$$E(Y) \approx g[E(X)] \tag{4-111}$$

The first-order (approximate) variance is

$$\text{Var}(Y) \approx \left(\frac{dg(X)}{dX}\right)^2 \text{Var}(X) \tag{4-112}$$

Again the derivative in Equation 4-112 is evaluated at the mean of X.

Random vector X — The Taylor series expansion of a function $Y = g(X)$ about the mean values of X, i.e., $E(X_1)$, $E(X_2)$, ..., $E(X_n)$, is given by

$$Y = g[E(X_1), E(X_2), ..., E(X_n)] + \sum_{i=1}^{n} [X_i - E(X_i)] \frac{\partial g(X)}{\partial X_i}$$

$$+ \sum_{i=1}^{n} \sum_{j=1}^{n} \frac{1}{2} [X_i - E(X_i)][X_j - E(X_j)] \frac{\partial^2 g(X)}{\partial X_i \partial X_j} + ... \tag{4-113}$$

in which the derivatives are evaluated at the mean values of X. Truncating this series at the linear terms, the *first-order mean and variance* of Y can be obtained by applying the mathematical expectation and variance operators, respectively. The first-order (approximate) mean is

$$E(Y) \approx g[E(X_1), E(X_2), ..., E(X_n)] \tag{4-114}$$

The first-order (approximate) variance is

$$\text{Var}(Y) \approx \sum_{i=1}^{n} \sum_{j=1}^{n} \frac{\partial g(X)}{\partial X_i} \frac{\partial g(X)}{\partial X_j} \text{Cov}(X_i, X_j) \tag{4-115}$$

in which the derivatives are evaluated at the mean values of X, i.e., $E(X_1)$, $E(X_2)$, ..., $E(X_n)$.

Example 4-7: Stress in a Beam

The stress, F, in a beam subjected to an external bending moment M is

$$F = \frac{My}{I} \tag{4-116}$$

where y is the distance from the neutral axis of the cross section of the beam to the point where the stress is calculated, and I is the centroidal moment of intertia of the cross section. Assume that the M and I are random variables with means μ_M and μ_I, respectively, and variances σ_M^2 and σ_I^2, respectively. Also assume that y is not a random variable. The mean value of the stress, μ_F, based on first-order approximations (Equation 4-114) is

$$\mu_F \approx \frac{\mu_M \, y}{\mu_I} \qquad (4\text{-}117)$$

The first-order variance of the stress, σ_F^2, is (see Equation 4-115)

$$\sigma_F^2 \approx \left(\frac{y}{\mu_I}\right)^2 \sigma_M^2 + \left(\frac{\mu_M \, y}{\mu_I^2}\right)^2 \sigma_I^2 \qquad (4\text{-}118)$$

4.4. APPLICATIONS

4.4.1. Reactions Due to a Random Load

A random load P is applied at the end of a cantilever beam of length L as shown in Figure 4-4. The vertical force reaction, V, of the beam at the fixed support is

$$V = P \qquad (4\text{-}119)$$

The moment reaction M at the fixed support is

$$M = P L \qquad (4\text{-}120)$$

Assume that the load has a mean value μ_p and variance σ_p^2. The length L is considered to be nonrandom, i.e., a deterministic quantity. Therefore, the mean of the reaction moment based on Equation 4-82 is

$$\mu_M = L \mu_p \qquad (4\text{-}121)$$

FIGURE 4-4 Cantilever beam with one load.

The variance of the moment is computed using Equation 4-83 as follows:

$$\sigma_M^2 = L^2 \sigma_P^2 \tag{4-122}$$

4.4.2. Buckling of Columns

The buckling load-capacity, P, of a column that has an effective length, L, and radius of gyration, r, is given by

$$P = \frac{\pi^2 E}{\left(\dfrac{L}{r}\right)^2} = \frac{\pi^2 r^2 E}{L^2} \tag{4-123}$$

The column is made of a material with a modulus of elasticity, E, and is shown in Figure 4-5. Assume that r and E are deterministic quantities and that L follows a normal distribution, $N(\mu_L, \sigma_L^2)$. Therefore, the buckling load can be expressed as

$$P = \frac{c}{L^2} \tag{4-124}$$

where the constant $c = \pi^2 r^2 E$. Thus, the derivative of L with respect to P can be computed as

$$L = \pm\sqrt{\frac{c}{P}} = \pm\sqrt{c}\, P^{-\frac{1}{2}} \tag{4-125}$$

$$\frac{dL}{dP} = \pm\frac{\sqrt{c}}{2\, P^{1.5}} \tag{4-126}$$

Therefore, the density function of P is given by

$$f_P(p) = \left[f_L\left(\sqrt{\frac{c}{p}}\right) + f_L\left(-\sqrt{\frac{c}{p}}\right) \right] \frac{\sqrt{c}}{2\, P^{1.5}} \tag{4-127}$$

or

$$f_P(p) = \frac{1}{\sigma_L\sqrt{2\pi}} \frac{\sqrt{c}}{2\, P^{1.5}} \left\{ \exp\left[-\frac{1}{2}\left(\frac{\sqrt{\dfrac{c}{p}} - \mu_L}{\sigma_L} \right)^2 \right] + \exp\left[-\frac{1}{2}\left(\frac{-\sqrt{\dfrac{c}{p}} - \mu_L}{\sigma_L} \right)^2 \right] \right\} \tag{4-128}$$

FIGURE 4-5 Buckling of a column.

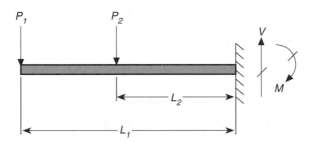

FIGURE 4-6 Cantilever beam with two loads.

4.4.3. Reactions Due to Random Loads

Random concentrated loads P_1 and P_2 are applied at a cantilevered beam of length L_1 as shown in Figure 4-6. The loads are applied at distances L_1 and L_2 from the fixed support, respectively. The vertical force reaction V is

$$V = P_1 + P_2 \tag{4-129}$$

The moment reaction M is

$$M = P_1 L_1 + P_2 L_2 \tag{4-130}$$

The loads P_1 and P_2 have means of μ_{P_1} and μ_{P_2}, respectively, and variances of $\sigma_{P_1}^2$ and $\sigma_{P_2}^2$, respectively. The lengths are considered to be non-random, i.e., deterministic quantities. Assume the loads to be statistically uncorrelated. Using mathematical expectation (Eq. 4-99), the means of the reactions are

$$\mu_V = \mu_{P_1} + \mu_{P_2} \tag{4-131}$$

$$\mu_M = L_1 \mu_{P_1} + L_2 \mu_{P_2} \tag{4-132}$$

The variances of the reactions according to Equation 4-101 are

$$\sigma_V^2 = \sigma_{P_1}^2 + \sigma_{P_2}^2 \tag{4-133}$$

$$\sigma_M^2 = L_1^2 \sigma_{P_1}^2 + L_2^2 \sigma_{P_2}^2 \tag{4-134}$$

Now, assume that the dimensions L_1 and L_2 are random variables with mean values μ_{L_1} and μ_{L_2}, respectively, and variances $\sigma_{L_1}^2$ and $\sigma_{L_2}^2$, respectively. In this case, the mean values of the reactions based on first-order approximations (Equation 4-114) are

$$\mu_V = \mu_{P_1} + \mu_{P_2} \tag{4-135a}$$

$$\mu_M = \mu_{L_1} \mu_{P_1} + \mu_{L_2} \mu_{P_2} \tag{4-135b}$$

The first-order variances of the reactions according to Equation 4-115 are

$$\sigma_V^2 = \sigma_{P_1}^2 + \sigma_{P_2}^2 \tag{4-136a}$$

$$\sigma_M^2 = \mu_{L_1}^2 \sigma_{P_1}^2 + \mu_{P_1}^2 \sigma_{L_1}^2 + \mu_{L_2}^2 \sigma_{P_2}^2 + \mu_{P_2}^2 \sigma_{L_2}^2 \tag{4-136b}$$

4.4.4. Open Channel Flow

In an open channel, water flows because of a difference in elevation between one section of the channel to another. The velocity V of the water flow in an open channel is

$$V = \frac{1}{n} R^{\frac{2}{3}} S^{\frac{1}{2}} \tag{4-137}$$

where R is the hydraulic radius, S is the slope of the hydraulic gradient, and n is a channel roughness factor that depends on the lining of the channel. Equation 4-137 is called the Manning formula. The hydraulic radius, R, can be computed as

$$R = \frac{A}{p} \tag{4-138}$$

where A is the cross-sectional area of flow and p is the wetted perimeter. For a channel near an existing highway, $n = 0.070$, $S = 4\%$, and R is a random variable with the following exponential distribution:

$$f_R(r) = \lambda \exp(-\lambda r) \quad \text{for } r \geq 0 \tag{4-139}$$

where $\lambda = 12.5$ sec/m. Determine the probability density function of the velocity V. Equation 4-137 can be rewritten as

$$V = c R^{\frac{2}{3}} \tag{4-140}$$

where

$$c = \text{constant} = \frac{S^{\frac{1}{2}}}{n} = \frac{\sqrt{0.04}}{0.070} = 0.8571 \tag{4-141}$$

Therefore, from Equation 4-140 r is given by

$$r = \left(\frac{V}{c}\right)^{\frac{3}{2}} \tag{4-142}$$

and

$$\frac{dr}{dV} = \left(\frac{1}{c}\right)^{\frac{3}{2}} \frac{3}{2} \sqrt{V} \tag{4-143}$$

Then, using Equation 4-75,

$$f_V(v) = \frac{3}{2} f_R\left(\left(\frac{v}{c}\right)^{\frac{3}{2}}\right) c^{-\frac{3}{2}} \sqrt{v} \tag{4-144a}$$

$$= \frac{3}{2} \lambda \exp\left[-\lambda\left(\frac{v}{c}\right)^{\frac{3}{2}}\right] \sqrt{v}\, c^{-\frac{3}{2}} \quad \text{for } v \geq 0 \tag{4-144b}$$

The resulting distribution is not of the exponential type. However, the approximate mean of V, $\overline{V}$, can be computed using Equation 4-111. The result is

$$\overline{V} = 2.8571\,\overline{R}^{\frac{2}{3}} \tag{4-145a}$$

where $\overline{R}$ is the mean of R which is $1/\lambda$. Therefore, the mean value is

$$\overline{V} = 2.8571\left(\frac{1}{\lambda}\right)^{\frac{2}{3}} = 0.530 \text{ m/sec} \tag{4-145b}$$

The approximate variance of V, $\text{Var}(V)$, can be computed using Equation 4-112 as

$$\text{Var}(V) = (2.8571)^2\left(\frac{2}{3}\right)^2 \overline{R}^{-0.333}\,\text{Var}(R) \tag{4-146a}$$

where $\text{Var}(R)$ is the variance of R. Therefore, the variance is

$$\text{Var}(V) = (2.8571)^2\left(\frac{2}{3}\right)^2\left(\frac{1}{\lambda}\right)^{-0.333}\frac{1}{\lambda^2} = 0.153887 \text{ (m/sec)}^2 \tag{4-146b}$$

The standard deviation is $\sqrt{0.053887} = 0.2321$ m/sec, and the coefficient of variation is 0.44.

4.4.5. Warehouse Construction

A warehouse is to be constructed from precast concrete elements that are produced by a nearby precast factory. The following construction tasks are identified for building the warehouse:

A: excavation of foundations
B: construction of foundations
C: construction of precast elements at factory
D: transportation of precast elements to construction site
E: assembly of elements at site
F: construction of roof
G: exterior and interior finishing

Figure 4-7 shows the logical network for performing these activities. The figure indicates that tasks C and D can be performed in parallel to tasks A and B, i.e., as excavation and construction of the footings are being performed at the site, the precast elements can be constructed at the factory and then transported to the construction site. Table 4-1 shows the means and standard deviations for the com-

FIGURE 4-7 Construction network.

TABLE 4-1 Moments of Completion Times

Task	Name	Mean (days)	Standard deviation (days)
A	Foundation excavation	3	1
B	Foundation construction	2	0.5
C	Precast-elements construction	4	1
D	Transportation of elements	0.5	0.5
E	Assembly of elements	4	1
F	Roofing	2	1
G	Finishing	3	1

pletion times of these tasks. Normal probability distributions and statistical independence are assumed for these times.

A construction engineer is interested in determining the completion time of the project, i.e., the duration of the project. Figure 4-7 shows the paths that are necessary to complete the project. Therefore, the completion time, T, is a dependent random variable that is given by

$$T = \max\{A+B, C+D\} + E + F + G \tag{4-147}$$

where A, B, C, D, E, F, and G are random variables that model the times for completing the corresponding tasks. The mean duration of the project, $\overline{T}$, can be determined as

$$\overline{T} = \max\{3 + 2, 4 + 0.5\} + 4 + 2 + 3 = 14 \text{ days} \tag{4-148}$$

The variance of the project duration, $\text{Var}(T)$, can be computed approximately as

$$\text{Var}(T) = \text{Var}(A) + \text{Var}(B) + \text{Var}(E) + \text{Var}(F) + \text{Var}(G) \tag{4-149a}$$

$$= (1)^2 + (0.5)^2 + (1)^2 + (1)^2 + (1)^2 = 4.25 \text{ (days)}^2 \tag{4-149b}$$

The computation of the variance was based on the path, from the construction network of Figure 4-7, that resulted in the mean duration of the project. The exact variance of T can be computed using simulation as discussed in Chapter 7. Assuming a normal probability distribution for T, the probability that the project finishes at the most 20 days after the start of the project can be computed as

$$\text{Prob}(T < 20) = \Phi\left(\frac{20-14}{\sqrt{4.25}}\right) = \Phi(2.91) = 0.9986 \qquad (4\text{-}150)$$

4.5. MULTIVARIABLE SIMULATION

Simulation easily handles studies where multiple random variables are involved. Very often, it can provide information about the distributions of random variables that is beyond the ability of theory. Theoretical relationships are often based on restrictive assumptions, such as normal distributions, that may not be valid for a given problem. Simulation would enable the needed statistics to be computed for conditions where the restrictive assumptions are not applicable.

Simulation can provide information beyond that provided by theory. This is especially true in cases that involve multiple random variables. For example, if the distribution of a random variable that represents the output from a function with multiple random variable inputs is needed, simulation can enable the distribution to be empirically approximated.

4.5.1. Simulation of Expected Values

Equations 4-100 and 4-103 provide a method for estimating the mean and variance of uncorrelated multiple random variables. The information from Example 4-6 illustrates the computation of the mean and variance; however, the distribution of the total cost C might be of interest, and it will depend on the distribution of the two random variable inputs X and Y.

Example 4-8: Expectation with Normally Distributed Random Variables

To demonstrate that Equations 4-100 and 4-103 can be applied regardless of the distribution function of the two variables, two simulations were made, each with 50 estimates of the cost C. For the first simulation, the material (X) and labor (Y) costs of Equation 4-107 are assumed to be normally distributed, $X \sim N(\mu = 100, \sigma = 10)$ and $Y \sim N(\mu = 250, \sigma = 50)$. The simulations are given in Table 4-2. The means and standard deviations of the three variables based on the 50 trials are

Variables	Mean	Standard deviation
X	99.07	10.87
Y	231.85	50.24
C	612.31	104.77

TABLE 4-2 Simulated Values of Material (X), Labor (Y), and Cost (C) Assuming Normal Populations

Random number for X	Standard normal variate for X	X	Random number for Y	Standard normal variate for Y	Y	C
0.8443	1.04	110.37	0.4661	−0.09	245.73	657.01
0.1980	−0.85	91.50	0.9932	2.49	374.30	885.85
0.1776	−0.94	90.60	0.1266	−1.16	191.77	519.44
0.1615	−1.01	89.89	0.8295	0.97	298.57	731.98
0.5239	0.06	100.60	0.6777	0.46	273.20	697.30
0.3515	−0.38	96.15	0.0094	−2.36	132.19	408.60
0.8930	1.25	112.51	0.2739	−0.61	219.64	608.04
0.6064	0.27	102.71	0.1731	−0.96	202.00	558.06
0.5811	0.20	102.05	0.9286	1.49	324.47	802.01
0.3413	−0.41	95.88	0.6909	0.50	274.99	693.79
0.0734	−1.47	85.25	0.2223	−0.77	211.46	550.79
0.6733	0.45	104.52	0.0389	−1.77	161.70	480.18
0.1736	−0.96	90.42	0.6915	0.50	275.07	685.77
0.5861	0.22	102.18	0.2933	−0.55	222.72	598.70
0.4850	−0.04	99.62	0.2706	−0.62	219.12	587.67
0.2585	−0.66	93.44	0.1157	−1.21	189.38	518.91
0.3677	−0.34	96.59	0.1653	−0.99	200.29	545.46
0.6824	0.48	104.77	0.6065	0.27	263.55	684.25
0.4980	−0.01	99.95	0.5294	0.07	253.70	657.33
0.6941	0.51	105.08	0.4002	−0.25	237.36	632.34
0.1540	−1.04	89.56	0.3219	−0.47	226.75	587.84
0.5457	0.12	101.15	0.3773	−0.31	234.26	620.24
0.3524	−0.38	96.18	0.7185	0.58	279.15	702.57
0.1393	−1.11	88.91	0.3694	−0.34	233.19	599.74
0.9629	1.79	117.91	0.4530	−0.12	244.08	665.03
0.0512	−1.64	83.64	0.2307	−0.74	212.79	551.03
0.9232	1.45	114.50	0.4632	−0.09	245.37	662.48
0.4701	−0.08	99.25	0.6520	0.39	269.71	688.30
0.6808	0.47	104.72	0.2168	−0.79	210.58	578.25
0.1739	−0.96	90.44	0.0920	−1.34	183.02	501.69
0.9865	2.23	122.31	0.0916	−1.34	182.87	549.21
0.6708	0.45	104.45	0.2004	−0.84	207.98	572.64
0.1648	−1.00	90.04	0.1855	−0.91	204.73	544.51
0.6297	0.33	103.34	0.1087	−1.24	187.84	530.68
0.9849	2.19	121.87	0.9313	1.51	325.45	833.71
0.7241	0.60	106.01	0.9626	1.79	339.35	837.71
0.2591	−0.65	93.46	0.8913	1.24	312.16	764.52
0.0157	−2.17	78.29	0.0723	−1.48	175.86	469.15
0.2928	−0.55	94.53	0.1062	−1.25	187.29	516.36
0.0017	−2.94	70.58	0.0542	−1.61	169.28	444.44
0.3103	−0.50	95.04	0.3740	−0.32	233.81	610.17
0.4511	−0.12	98.77	0.3404	−0.41	229.26	606.66
0.9829	2.13	121.33	0.8335	0.99	299.45	780.89
0.1553	−1.04	89.62	0.4301	−0.18	241.20	616.82
0.1830	−0.92	90.84	0.2275	−0.75	212.28	560.82
0.7176	0.58	105.80	0.2582	−0.66	217.15	593.00
0.0907	−1.35	86.51	0.2206	−0.78	211.19	552.14
0.5069	0.02	100.17	0.0563	−1.6	170.05	490.35
0.7756	0.76	107.64	0.1468	−1.08	196.22	553.90
0.9003	1.28	112.84	0.0806	−1.42	178.87	527.00

Using Equations 4-100 and 4-103, the computed moments are

$$\overline{C} = 1.5(99.07) + 2.0(231.85) = \$612.3$$

$$S_c^2 = (1.5)^2 (10.87)^2 + (2.0)^2 (50.24)^2 = 10,362$$

$$S_c = \$101.8$$

The difference between 101.8 and 104.77 is due to sampling variation.

Example 4-9: Expectation with Exponential Variables

For this second simulation, X and Y of Example 4-8 are assumed to be exponentially distributed, $X \sim \text{Exp}$(with $\lambda = 100$) and $Y \sim \text{Exp}$(with $\lambda = 250$). The exponential distribution assumes that the mean equals the standard deviation, which was not the case for the simulation based on the normal distribution. The simulations for the exponential distribution are given in Table 4-3. The means and standard deviations of the three variables based on the 50 trials are

Variables	Mean	Standard deviation
X	111.98	115.31
Y	333.48	245.35
C	834.93	559.16

Using Equations 4-100 and 4-103, the computed moments are

$$\overline{C} = 1.5(111.98) + 2.0(333.48) = \$834.93$$

$$S_c^2 = (1.5)^2 (115.31)^2 + (2.0)^2 (245.35)^2 = 270,703$$

$$S_c = \$520.3$$

The difference between 520.3 and 559.16 is due to sampling variation.

While the relationships for the two moments, Equations 4-100 and 4-103, do not depend on the distributions of X and Y, the distribution of C depends on the distributions of X and Y. Histograms for the 50 simulated values of C from Tables 4-2 and 4-3 are shown in Figure 4-8. Since the means and standard deviations are different, the two histograms can only be assessed using only their shapes, which are quite different. The distribution of C depends on the distributions of X and Y, the inputs. For the histogram of Figure 4-8a, which is based on normally distributed

TABLE 4-3 Simulated Values of Material (*X*), Labor (*Y*),
and Cost (*C*) Assuming Exponential Populations

Random number for X	X	Random number for Y	Y	C
0.8443	16.92	0.4661	190.84	407.06
0.1980	161.95	0.9932	1.71	246.33
0.1776	172.82	0.1266	516.68	1292.59
0.1615	182.33	0.8295	46.73	366.95
0.5239	64.65	0.6777	97.26	291.49
0.3515	104.55	0.0094	1166.76	2490.35
0.8930	11.32	0.2739	323.75	664.47
0.6064	50.02	0.1731	438.47	951.98
0.5811	54.28	0.9286	18.52	118.46
0.3413	107.50	0.6909	92.44	346.13
0.0734	261.18	0.2223	375.93	1143.64
0.6733	39.56	0.0389	811.69	1682.72
0.1736	175.10	0.6915	92.22	447.10
0.5861	53.43	0.2933	306.64	693.42
0.4850	72.36	0.2706	326.78	762.10
0.2585	135.29	0.1157	539.19	1281.31
0.3677	100.05	0.1653	450.00	1050.07
0.6824	38.21	0.6065	125.01	307.35
0.4980	69.72	0.5294	159.00	422.58
0.6941	36.51	0.4002	228.95	512.67
0.1540	187.08	0.3219	283.38	847.38
0.5457	60.57	0.3773	243.68	578.21
0.3524	104.30	0.7185	82.65	321.74
0.1393	197.11	0.3694	248.97	793.61
0.9629	3.78	0.4530	197.97	401.60
0.0512	297.20	0.2307	366.66	1179.12
0.9232	7.99	0.4632	192.40	396.78
0.4701	75.48	0.6520	106.93	327.08
0.6808	38.45	0.2168	382.20	822.06
0.1739	174.93	0.0920	596.49	1455.37
0.9865	1.36	0.0916	597.58	1197.20
0.6708	39.93	0.2004	401.86	863.61
0.1648	180.30	0.1855	421.18	1112.80
0.6297	46.25	0.1087	554.79	1178.96
0.9849	1.52	0.9313	17.79	37.87
0.7241	32.28	0.9626	9.53	67.48
0.2591	135.05	0.8913	28.77	260.12
0.0157	415.41	0.0723	656.73	1936.58
0.2928	122.83	0.1062	560.61	1305.46
0.0017	637.71	0.0542	728.77	2414.11
0.3103	117.02	0.3740	245.87	667.28
0.4511	79.61	0.3404	269.41	658.23
0.9829	1.72	0.8335	45.53	93.65
0.1553	186.24	0.4301	210.93	701.23
0.1830	169.83	0.2275	370.15	995.04
0.7176	33.18	0.2582	338.51	726.79
0.0907	240.02	0.2206	377.85	1115.73
0.5069	67.94	0.0563	719.27	1540.45
0.7756	25.41	0.1468	479.67	997.46
0.9003	10.50	0.0806	629.56	1274.88

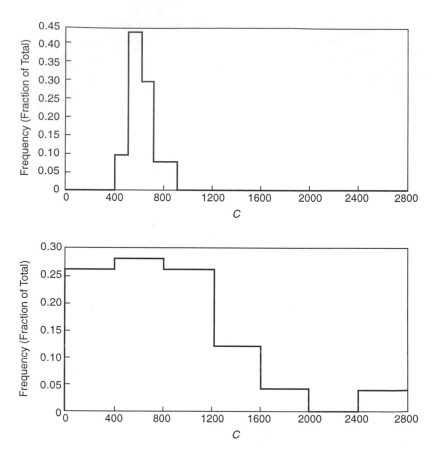

FIGURE 4-8 Distributions of total cost (C) for 50 simulation trials assuming cost components X and Y are (a) normally distributed and (b) exponentially distributed.

inputs, the histogram has a more centered peak than the histogram based on exponentially distributed inputs (Figure 4-8b). The latter histogram is relatively skewed, much as the exponential distribution is.

4.5.2. Simulation and Correlation

Correlation was introduced in Section 4.2.3. It is an important topic in statistical analysis. Equation 4-43 provides the general concept of correlation based on probability theory. In practice, the numerical value depends on the calculation of the covariance. Several coefficients have been developed, with the Pearson product-moment correlation coefficient the most widely used as discussed in Chapter 6. For continuous random variables, the covariance is defined as the average of the products of the deviations of the two random variables (Equation 4-39); thus, the correlation coefficient for sample values of X_1 and X_2 becomes:

$$\rho_{X_1 X_2} = \frac{\text{Cov}(X_1, X_2)}{\sigma_{X_1} \sigma_{X_2}} = \frac{1}{n} \sum_{i=1}^{n} \frac{(X_{1i} - \overline{X}_1)(X_{2i} - \overline{X}_2)}{S_{X_1} S_{X_2}} \tag{4-151}$$

or expressed as standardized variates

$$\rho_{X_1 X_2} = \frac{1}{n} \sum_{i=1}^{n} \left(\frac{X_{1i} - \overline{X}_1}{S_{X_1}} \right) \left(\frac{X_{2i} - \overline{X}_2}{S_{X_2}} \right) = \frac{1}{n} \sum_{i=1}^{n} z_{1i} \, z_{2i} \tag{4-152}$$

Of special interest in many statistical analyses is the distribution of $\rho_{X_1 X_2}$. While a sample estimate of the value is easily computed with the above equation, knowledge of the distribution of the sample value is important when deciding how much weight to place on the sample value when making a decision about a causal relationship between X_1 and X_2. The distribution depends on the sample size and the true value $\rho_{X_1 X_2}$ from which the sample values of X_1 and X_2 were drawn.

In order to simulate random samples that can be used to determine the distribution of the correlation coefficient, a model is needed. Since the distribution of sample values of the correlation coefficient depends on the population value, ρ, the model should be a function of ρ. The following relationship can be used to generate sample values of the random variable Y for given values of the second random variable X and a population value of the correlation coefficient ρ:

$$Y = \beta_0 + \beta_1 X + Z \sigma_Y \sqrt{1 - \rho^2} \tag{4-153}$$

where β_0 and β_1 are the population values of a linear relationship between X and Y, σ_Y is the standard deviation of Y, and Z is the standard normal deviate. The slope coefficient β_1 is related to the population correlation by $\beta_1 = \rho \sigma_Y / \sigma_X$. Thus, the above equation becomes

$$Y = \mu_Y + \rho \left(\frac{\sigma_Y}{\sigma_X} \right) (x - \mu_X) + Z \sigma_Y \sqrt{1 - \rho^2} \tag{4-154}$$

The first two terms on the right-hand side of Equation 4-154 represent the correlation between Y and X, with μ_Y, the mean of Y, being a vertical location parameter and σ_Y and σ_X being scaling parameters. The third term on the right-hand side of the above equation represents the extent to which Y and X are uncorrelated. If ρ equals 1 or -1, then the third term drops out and the relationship between Y and X is defined entirely by the deterministic linear relationship $Y = \beta_0 + \beta_1 X$. If ρ equals 0, then the random component and the vertical location parameter defines the relationship between Y and X.

To generate random samples that are from a population with correlation ρ, values of σ_Y, σ_X, μ_Y, μ_X, and the sample size n must be set. Then values of X with set characteristics μ_x and σ_x can be generated randomly. Using random values of Z, where Z is normally distributed with mean of 0 and variance of 1, the above equation is used to generate values of Y. Other distributions can be used; however, the normal distribution is almost always assumed.

Example 4-10: Correlation Between Functions

Equations 4-45 and 4-46 provide two functions that depend on the value of a random variable X and parameter n. Figure 4-2 shows the relationship between the correlation coefficient ρ and parameter n. For $n = 2$, Equation 4-62 shows that the population correlation coefficient is 0.87. For $n = 3$, the population value equals 0.71. The development of Equation 4-62 applies to the population, but not to sample estimates of the correlation coefficient. Furthermore, the theoretical development does not provide knowledge of the distribution of the sample correlation coefficient (R), which is a function of the size of the random sample (N).

Simulation can provide the distribution of sample estimates of the correlation between the two functions of Equations 4-45 and 4-46. Using uniform random numbers, samples of 5, 10, and 25 were generated and the correlation between the two functions computed, with the following means:

N	Mean R
5	0.891
10	0.832
25	0.803

As the sample size N increases, the mean would approach the population value of 0.71.

In addition to means, the distributions of the sample R were obtained for each N. Figure 4-9 shows the three simulated distributions with the values listed in Table 4-4. As expected, as the sample size increases, the spread of the distribution decreases, which reflects the increase in the accuracy of the statistic as N increases. The centers of the distributions move toward the population value of 0.71.

Example 4-11: Multivariable Simulation of Stage and Discharge

Example 4-10 evaluated the correlation between two functions of one random variable, X. In this example, values of two random variables were simulated and the correlation between the simulated data was assessed. The simulation assumes values of the population parameters of Equation 4-154 and the distribution of X, which along with the distribution of Z determine the distribution of Y. In this example, both X and Z are assumed to be normally distributed. The following statistics were assumed: $\mu_y = 55$; $\rho = 0.921$; $\sigma_y = 54$; $\sigma_x = 0.71$; $\mu_x = 2.84$. These are approximately the sample values for the data of Table 2-2.

FIGURE 4-9 Sample distributions of the correlation coefficients of the functions of Equations 4-45 and 4-46 for sample sizes (N) of 5, 10, and 25.

TABLE 4-4 Sample Distributions of the Correlation of the Functions of Equations 4-45 and 4-46 for Sample Sizes (N) of 5, 10, and 25

Correlation coefficient	N = 5	N = 10	N = 25
0.575	0.0000	0.0002	
0.600	0.0002	0.0007	
0.625	0.0016	0.0036	0.0003
0.650	0.0022	0.0099	0.0028
0.675	0.0065	0.0179	0.0104
0.700	0.0106	0.0352	0.0269
0.725	0.0195	0.0591	0.0819
0.750	0.0325	0.0759	0.1468
0.775	0.0484	0.1016	0.1960
0.800	0.0672	0.1328	0.2068
0.825	0.0824	0.1376	0.1785
0.850	0.1022	0.1401	0.0989
0.875	0.1123	0.1199	0.0396
0.900	0.1374	0.0877	0.0097
0.925	0.1378	0.0529	0.0014
0.950	0.1371	0.0213	
0.975	0.1021	0.0036	
1.000			

Table 4-5 gives 50 simulated values of the stage and discharge. The values of stage X were simulated to have a normal distribution and Equation 4-154 was used to simulate the discharges Y. Figure 4-10 shows the distributions of the simulated values.

Simulated data should always be examined to ensure that the values are reasonable. Certainly, the distributions of Figure 4-10 are an important means of assessing the data. In this case, both appear to be normally distributed. It is also useful to

TABLE 4-5 Simulated Stages and Discharges

Observation number	Simulated stage	Random number	Simulated discharge
1	3.342	−0.257	84.8
2	2.927	0.546	72.6
3	2.477	−0.100	27.5
4	1.384	2.931	14.7
5	1.908	1.111	13.1
6	2.544	−0.735	18.8
7	2.546	1.711	70.4
8	3.182	0.005	79.1
9	3.564	0.150	108.9
10	3.305	−0.188	83.6
11	3.602	−0.466	98.6
12	2.237	−0.523	1.8
13	1.404	2.790	13.1
14	2.753	0.618	61.9
15	2.522	−0.307	26.3
16	2.529	0.174	36.8
17	2.734	0.052	48.7
18	2.135	0.705	20.5
19	3.219	−0.293	75.4
20	4.285	−0.223	151.5
21	3.132	−0.674	61.3
22	2.959	−1.042	41.4
23	3.415	−0.235	90.3
24	2.519	−0.553	20.9
25	3.459	−0.714	83.3
26	2.111	0.752	19.7
27	2.414	0.467	35.0
28	3.039	0.112	71.3
29	3.371	0.629	105.4
30	2.761	−1.198	24.2
31	3.008	0.056	67.9
32	3.417	−1.633	61.1
33	2.194	0.323	16.5
34	2.031	0.262	3.8
35	2.838	0.160	58.2
36	3.718	−0.585	104.2
37	3.112	−0.302	67.7
38	2.739	−0.739	32.4
39	2.639	−0.639	27.5
40	1.810	1.103	6.1
41	2.663	0.314	49.2
42	2.030	0.735	13.7
43	2.353	0.144	23.9
44	2.030	0.667	12.3
45	2.554	−0.434	25.8
46	3.758	−0.193	115.2
47	2.702	0.600	57.9
48	2.927	0.711	76.1
49	3.075	1.424	101.4
50	3.644	1.426	141.3

(a) Stage (X)

(b) Normal Variate (z)

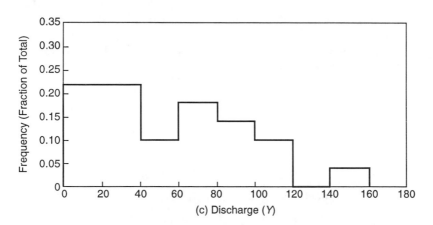

(c) Discharge (Y)

FIGURE 4-10 Distributions of (a) simulated stages; (b) random normal variate used to compute simulated discharges; (c) simulated discharges.

compare the sample statistics of the simulated data with the assumed population. The following tabular summary compares the statistics:

Statistic	Assumed population	Simulated data
Stage: mean	2.84	2.780
Stage: standard deviation	0.71	0.619
Discharge: mean	55.	54.5
Discharge: standard deviation	54.	37.3

While the standard deviation of the simulated discharges is smaller than that of the population, the simulated statistics are within the expected sampling variation of the population values.

4.6. PROBLEMS

4-1. A shipping company records the conditions of cargoes at the delivery points as not damaged (ND), damaged (D), and partly damaged (PD). The following probability mass function was established for three shipping methods:

X = condition of cargo

		ND	D	PD
$Y =$	**Sea**	0.20	0.04	0.02
shipping	**Air**	0.40	0.02	0.03
method	**Ground**	0.27	0.01	0.01

Determine the marginal probability mass functions $P_X(x)$ and $P_Y(y)$. If a cargo is damaged, what is the probability that it was shipped by sea? by air? by ground?

4-2. An electronics company tests for the performance quality of its products by performing nondestructive thermal and magnetic-field tests. The products are classified as adequate (A), not adequate (NA), and partly adequate (PA). A sample of 100 produced the following results:

X = thermal test

		A	NA	PA
$Y =$	**A**	80	2	0
magnetic - field	**NA**	4	2	2
	PA	4	2	4

Determine the joint probability mass function $P_{XY}(x,y)$, and the marginal probability mass functions $P_X(x)$ and $P_Y(y)$.

4-3. Are the random variables in Problem 4-1 statistically uncorrelated? Why?

4-4. Are the random variables in Problem 4-2 statistically uncorrelated? Why?

4-5. The following is a joint probability density function for two continuous random variables, X and Y:

$$f_{XY}(x,y) = c\,xy \quad \text{for } 0 < x < 1 \text{ and } 0 < y < 1$$

Determine the constant c such that $f_{XY}(x,y)$ is a legitimate joint density function. Evaluate the marginal density functions $f_X(x)$ and $f_Y(y)$. Evaluate the probability that $0 < x < 0.5$ and $0 < y < 0.25$.

4-6. The following is a joint probability density function for two continuous random variables, X and Y:

$$f_{XY}(x,y) = c\,x\sqrt{y} \quad \text{for } 0 < x < 1 \text{ and } 0 < y < 1$$

Determine the constant c such that $f_{XY}(x,y)$ is a legitimate joint density function. Evaluate the marginal density functions $f_X(x)$ and $f_Y(y)$. Evaluate the probability that $0 < x < 0.5$ and $0 < y < 0.25$.

4-7. Are the random variables in Problem 4-5 statistically uncorrelated? Why?

4-8. Are the random variables in Problem 4-6 statistically uncorrelated? Why?

4-9. Determine the conditional probability mass function $P_{X|Y}(X = \text{ND})$ for all y values for the random variables in Problem 4-1.

4-10. Determine the conditional probability mass function $P_{Y|X}(Y = \text{sea})$ for all x values for the random variables in Problem 4-1.

4-11. Determine the conditional probability mass function $P_{X|Y}(X = \text{PA})$ for all y values for the random variables in Problem 4-2.

4-12. Determine the conditional probability mass function $P_{Y|X}(Y = \text{NA})$ for all x values for the random variables in Problem 4-2.

4-13. Determine the conditional expected values $E(X|y)$ and $E(Y|x)$ for the random variables in Problem 4-5.

4-14. Determine the conditional expected values $E(X|y)$ and $E(Y|x)$ for the random variables in Problem 4-6.

4-15. Determine the correlation coefficient ρ_{XY} for the random variables in Problem 4-5.

4-16. Determine the correlation coefficient ρ_{XY} for the random variables in Problem 4-6.

4-17. The random variable X is normally distributed with mean and standard deviation $\mu_X = 0$ and $\sigma_X = 1$, respectively. A dependent random variable Y is defined as

$$Y = X^2$$

Show that the probability density function of Y is a chi-square distribution (as given by Equation 3-162).

4-18. The random variable Y is normally distributed with mean and standard deviation μ_Y and σ_Y, respectively. The dependent random variable Y is defined as

$$Y = \ln(X)$$

Show that the probability density function of X is a lognormal distribution (as given by Equation 3-134).

4-19. For a simply supported beam with a concentrated load P at midspan and uniformly distributed load of intensity w over the entire span L, the maximum shear force and the maximum bending moment are given by

$$V = \frac{wL}{2} + \frac{P}{2}$$

$$M = \frac{wL^2}{8} + \frac{PL}{4}$$

Find the covariance of V and M, i.e., $\text{Cov}(V,M)$, using first-order approximations. Assume P and w to be random variables with known mean values and variances, i.e., μ_P, σ_P, μ_w, and σ_w, respectively.

PROBLEM 4-20. Deformation of a Rod

4-20. The change in the length of a rod shown in the figure due to an axial force P is given by

$$\Delta L = \frac{PL}{AE}$$

where L = length of rod, P = applied axial force, A = cross-sectional area of rod, and E = modulus of elasticity. If P and E are normally distributed with μ_P and σ_P, and μ_E and σ_E, respectively, determine the first-order mean and variance of ΔL for the following two cases: (1) P and E are uncorrelated, and (2) P and E are correlated with a correlation coefficient ρ.

4-21. The ultimate moment capacity, M, of an under-reinforced concrete rectangular section is given by

$$M = A_s f_y \left(d - \frac{a}{2} \right)$$

where

$$a = \frac{A_s f_y}{0.85 b f_c'}$$

in which the following are random variables: A_s is the cross-sectional area of the reinforcing steel, f_y is the yield stress (strength) of the steel, d is the distance from the reinforcing steel to the top fibers of the beam, b is the width of the beam, and f_c' is the ultimate stress (strength) of the concrete. If the random variables are assumed to be statistically uncorrelated, determine the first-order mean and variance of the moment capacity in terms of the means and variances of the random variables.

4-22. For the reinforced concrete section described in Problem 4-21, determine the second-order mean and variance of the moment capacity by including the second-order term of Taylor series expansion (Equation 4-113) in the approximate expression for the mean and variance.

4-23. Using the midsquare method with a seed of 2941, generate 30 uniform variates. Use these to compute the material costs X assuming $X \sim N(\mu = 150, \sigma = 20)$. Using the midsquare method with a seed of 7676, generate 30 uniform variates and use them to compute the labor costs (Y) assuming $Y \sim N(\mu = 225, \sigma = 25)$. Generate the 30 total cost (C) values assuming that $C = 2.5X + 1.7Y$. Compute the mean and standard deviation of C and plot histograms of X, Y, and C. Discuss the results.

4-24. Using the midsquare method with a seed of 9507, generate a sample of ten values (X) that follow $X \sim N(\mu = 10, \sigma = 2)$. Assume $\sigma_Y = 3$, $\mu_Y = 20$, and $\rho = 0.5$. Generate ten values of Y using Equation 4-154 with values of Z generated using the midsquare method with a seed of 2782. Compute the correlation coefficient between the X and Y values. Compute the sample and population correlation coefficient.

4-25. Assume the river stage X is $N(\mu = 4, \sigma = 0.7)$. Assume $\mu_Y = 650$, $\sigma_Y = 60$, and $\rho = 0.82$. Using the midsquare method with a seed of 1073, generate 15 values of the stage and generate values of the discharge Y using Equation 4-154 with values of Z generated using the midsquare method and a seed of 6639. Compute the mean, standard deviation, and frequency histograms of the simulated data. Make comparisons with the assumed populations.

Chapter 5
Fundamentals of Statistical Analysis

Table of Contents

5.1. INTRODUCTION

Values of random variables obtained from sample measurements are commonly used in making important engineering decisions. For example, samples of river water are collected to estimate the average level of a pollutant in the entire river at that location. Samples of stopping distances are used to develop a relationship between the speed of a car at the time the brakes are applied and the distance traveled before the car comes to a complete halt. The average of sample measurements of the compressive strength of concrete collected during the pouring of a large concrete slab, such as the deck of a parking garage, is used to help decide whether or not the deck has the strength specified in the design specifications. It is important to recognize the random variables involved in these cases. In each case, the individual measurements or samples are values of a random variable, and the computed mean is also the value of a random variable. For example, the transportation engineer measures the stopping distance; each measurement is a sample value of the random variable. If ten measurements are made for a car stopping from a speed of 50 mph, then the sample consists of ten values of the random variable. Thus, there are two random variables in this example: the stopping distance and the estimated mean of the stopping distance; this is also true for the water-quality-pollutant and compressive-strength examples.

The estimated mean for a random variable is considered by itself to be a random variable, because different samples about the random variable can produce different estimated mean values; hence the randomness in the estimated mean. When a sample of n measurements of a random variable is collected, the n values are not necessarily identical. The sample is characterized by variation. For example, let's assume that five independent estimates of the compressive strength of the concrete in a parking garage deck are obtained from samples of the concrete obtained when the concrete was poured. For illustration purposes, let's assume that the five compressive strength measurements are 3250, 3610, 3460, 3380, and 3510 psi. This produces a mean of 3442 psi and a standard deviation of 135.9 psi. Assume that another sample of five measurements of concrete strength was obtained from the same concrete pour; however, the values were 3650, 3360, 3328, 3420, and 3260 psi. In this case, the estimated mean and standard deviation are 3404 and 149.3 psi, respectively. Therefore, the individual measurements and the mean are values of two different random variables, i.e., X and $\overline{X}$.

It would greatly simplify decision making if the sample measurements were identical, i.e., there was no sampling variation so the standard deviation was zero. Unfortunately, that is never the case, so decisions must be made in spite of the uncertainty. For example, let's assume in the parking garage example that the building code requires a mean compressive strength of 3500 psi. Since the mean of 3442 psi based on the first sample is less than the required 3500 psi, should we conclude that the garage deck does not meet the design specifications? Unfortunately, decision making is not that simple. If a third sample of five measurements had been randomly collected from other locations on the garage deck, the following values are just as likely to have been obtained: 3720, 3440, 3590, 3270, and 3610 psi. This sample of five produces a mean of 3526 psi and a standard deviation of 174.4 psi. In this case, the mean exceeds the design standard of 3500 psi. Since the sample mean is greater than the specified value of 3500 psi, can we conclude that the concrete is of adequate strength? Unfortunately, we cannot conclude with certainty that the strength is adequate any more than we could conclude with the first sample that the strength was inadequate. The fact that different samples lead to different means is an indication that we cannot conclude that the design specification is not met just because the sample mean is less than the design standard. We need to have more assurance.

The need then is for a systematic decision process that takes into account the variation that can be expected from one sample to another. The decision process must also be able to reflect the risk of making an incorrect decision. This decision making can be made using, for example, hypothesis testing as described in Section 5.4.

5.1.1. Samples and Population

The data that are collected represent sample information but it is not complete by itself, and predictions are not made directly from the sample. The intermediate step between sampling and prediction is the identification of the underlying population. The sample is used to identify the population and then the population is used to make predictions or decisions. This sample-to-population-to-prediction sequence is true for the univariate methods of this chapter or for the bivariate and multivariate methods that follow.

A known function or model is most often used to represent the population. The normal and lognormal distributions are commonly used to model the population for a univariate problem. For bivariate and multivariate prediction, linear ($\hat{Y} = a + bX$) and power ($\hat{Y} = aX^b$) models are commonly assumed functions for representing the population, where $\hat{Y}$ = predicted value of dependent variable Y, X = independent random variable, and a and b = model parameters. When using a probability function to represent the population, it is necessary to estimate the parameters. For example, for the normal distribution, the location and scale parameters need to be estimated, which are the mean and standard deviation, respectively. For the exponential distribution, the rate (λ) is a distribution parameter that needs to be estimated, which is not equal to the mean or standard deviation. When using the linear or power models as the population, it is necessary to estimate the coefficients a and b. In both the univariate and multivariate cases, they are called sample estimators of the population parameters.

5.2. ESTIMATION OF PARAMETERS

In developing models for populations, models can be classified as univariate, bivariate, or multivariate, with parameters that provide the needed complete definition of a model. Models can have one, two, or more parameters. For example, the normal distribution as a univariate model has two parameters, the exponential distribution has one parameter, and the bivariate power model ($\hat{Y} = aX^b$) has two parameters. Samples are used to develop a model that can adequately represent the population and to estimate the parameters of the population model. The parameters can be estimated in the form of point estimates (single values) or interval estimates (ranges of values) using the samples. The equations or methods used to estimate the parameters are called estimators. In this section, estimators and their properties are introduced. The statistical uncertainty associated with the estimators is considered in the next sections for statistical decision making using hypothesis testing (Sections 5.4 to 5.6). Then interval estimation is discussed in Section 5.7.

5.2.1. Properties of Estimators

Several properties of estimators are of interest to engineers. The concepts that are widely used, and sometimes misunderstood, include accuracy, bias, precision, consistency, efficiency, and sufficiency. Because of the lack of uniform terminology, definitions that represent the most common usage are given in the following paragraphs.

5.2.1.1. Bias

An estimate of a parameter θ made from the sample statistic is said to be an unbiased estimate if the expected value of the sample quantity $\hat{\theta}$ is θ; that is,

$$E(\hat{\theta}) = \theta \tag{5-1}$$

where $E(\cdot)$ denotes the mathematical expectation. The bias is defined as $[E(\hat{\theta}) - \theta]$. Therefore, bias deprives a statistic result of representativeness by systematically distorting it. It is not the same as a random error that may distort at any one occasion but balances out on the average. It is important to note that bias is a systematic error.

Figure 5-1 shows the results of four experiments, where each experiment is repeated six times. It is known that the true (i.e., population) mean of each experiment is 15. The sample means for the four experiments are 15, 24, 7, and 14. Thus experiment A is unbiased because its expected value equals the true value. Experiments B, C, and D show varying degrees of bias. Whereas experiment B has a positive bias of 9, the bias of experiments C and D are negative; that is, experiment B tends to overestimate θ and experiments C and D underestimate θ. A question that is discussed in Sections 5.5 and 5.6 is whether or not the biases are significant.

5.2.1.2. Precision

In addition to the systematic variation, variation in an observation may be due to random error. Random error is measured by fluctuations in the value of a variable that occur when the same experiment is performed more than once. It may be the result of (1) errors in judgment, (2) uncontrollable variation in environmental conditions, (3) differences due to deficiencies in defining the quantity being measured, and (4) intrinsically random process.

Precision can be defined as the ability of an estimator to provide repeated estimates that are very close together. Therefore, precision may be expressed in terms of the variance of an estimator, with a large variance signifying lack of precision and a small variance signifying high precision. Obviously, absolute precision implies a zero variance, a level that practically cannot be achieved.

Using the variance of the observations, as measured about the sample mean, the precision of each experiment is shown in Figure 5-1. Experiments A and B show considerably more precision (i.e., they have lower variances among the observations) than do experiments C and D. Experiment C has the largest variation (i.e., it is the least precise experiment). In comparing experiments A and B, they have the same level of precision but experiment A is unbiased, whereas B is highly biased.

5.2.1.3. Accuracy

It can be defined as the closeness or nearness of the measurements to the true or actual value of the quantity being measured. The concept of accuracy can be considered to comprise both the precision and bias.

FIGURE 5-1 Bias, precision, and accuracy of experimental observations.

Definitions of accuracy often imply that the term should only be used in a dichotomous manner; that is, a measurement or a statistic is either accurate or inaccurate. In practice, the term "accuracy" is used for comparisons, and quantitative estimates of accuracy are used to describe a measurement or statistic. That is, accuracy is a characteristic that should be measured on a continuous scale, rather than as a discrete variable.

The variance of the observations in Figure 5-1, as measured about the true value of the statistic (i.e., 15), is used as a measure of the accuracy. The values of the four experiments are shown in Figure 5-1. Experiment B is the least accurate; in addition to its high bias, it is imprecise. Experiment D is reasonably accurate, in spite of its noticeable lack of precision. Experiment A is the most accurate.

5.2.1.4. Comparison of Bias, Precision, and Accuracy

It is interesting to distinguish among accuracy, bias (systematic error), and precision (random error). The measurements using experiment A are all close to the true value, which indicates no significant bias and high precision; thus experiment A is highly accurate. There is little variation among the six observations in experiment B, which indicates that it provides precise measurements; however, because the values are consistently higher than the true value, the experiment is biased. The significant bias contributes to its poor accuracy. Experiment C is both biased and imprecise because the systematic and random variations are both significant; thus it is also highly inaccurate. Experiment D is imprecise but relatively unbiased. In summary, inaccuracy can result from either a bias or lack of precision. Although accuracy can be of greatest overall concern, one should examine the bias and precision components to identify the most likely sources of the inaccuracy.

5.2.1.5. Mean Square Error (MSE)

One objective of data analysis is to explain variation in a set of measurements. Accuracy reflects both systematic and random errors, and both sources of variation must be assessed when attempting to select one estimator from among many; specifically, a biased estimator may be preferable to an unbiased estimator if the precision of the biased estimator is significantly better than that of the unbiased estimator. The variance of an estimator may be viewed as a measure of its precision. The MSE can be defined as the expected value of the square of the deviation of the estimate from the true value of the quantity being estimated. It is equal to the variance of the estimate plus the square of the bias. Thus, the MSE is a measure of accuracy.

5.2.1.6. Consistency, Sufficiency, and Efficiency

In addition to unbiasness, there are other properties that a statistic should have in order to be a good estimator. The estimator should be both consistent and sufficient, and the method of estimation should be relatively efficient.

An estimator is consistent if the probability (P) that $\hat{\theta}$ will deviate from θ more than any fixed amount $\varepsilon > 0$, no matter how small, approaches zero as the sample size (n) becomes larger and larger. Mathematically, consistency is given by

$$P\left(\left|\hat{\theta} - \theta\right| \leq \varepsilon\right) \to 1 \quad \text{as } n \to \infty \tag{5-2}$$

in which n is the sample size. In contrast with biasedness, consistency is a "sample size" property of an estimator. It can be shown that unbiased estimators are consistent estimators. However, a consistent estimator is not necessarily unbiased; for example, the sample variance is consistent but must be corrected for its biasedness by multiplying it by the factor $n/(n-1)$.

An estimate $\hat{\theta}$ is said to be sufficient if it exhausts all possible information on θ in a sample of any size. Sufficiency implies that no other estimator computed from the same sample can provide additional information about θ. Sufficiency is the most desirable property of an estimator, and such estimators are said to be optimum.

Efficiency is an important criterion for evaluating the quality of an estimator. Because it is desirable to have an estimate that is close to the true value and the variance is a measure of closeness, the efficiency of an estimator is inversely proportional to its variance. A consistent estimate, $\hat{\theta}_1$, is said to be more efficient than another estimate, $\hat{\theta}_2$, if Var($\hat{\theta}_1$) < Var($\hat{\theta}_2$), where Var($\cdot$) is the variance. An estimate is said to be efficient if it has the smallest variance of all available estimators.

5.2.2. Estimation of Moments

In engineering, it is common to characterize a sample of interest by certain estimated moments. These measures, which take numerical values, can be easily communicated to others and quantify the main characteristics of the sample. The concept of moments was introduced in Chapter 3. The first moment is a commonly used quantity that provides a measure of central tendency, or mean. The second central moment or variance is a common measure of dispersion. In this section, methods for estimating these two moments are defined and discussed. A formal discussion of these measures is provided in Chapter 3.

The mean or average value of n observations, if all observations are given equal weights, is given by

$$\overline{X} = \frac{1}{n}\sum_{i=1}^{n} x_i \tag{5-3}$$

where x_i = a sample point, and $i = 1, 2, ..., n$. Although this moment conveys certain information about the underlying sample, it does not completely characterize the underlying variable. Two variables can have the same mean, but different histograms. Other characteristics are also important and necessary. The estimator of the mean as given by Equation 5-3 was discussed in Chapter 2.

The dispersion measure (variance) describes the level of scatter in the data about the central tendency location. For n observations in a sample that are given equal weight, the variance (S^2) is given by

$$S^2 = \frac{1}{n-1}\sum_{i=1}^{n} \left(x_i - \overline{X}\right)^2 \tag{5-4}$$

The units of the variance are the square of the units of the parameter or variable x; for example, if the variable is measured in pounds per square inch (psi), the variance has units of (psi)^2. Computationally, the variance of a sample can be determined using the following alternative equation:

$$S^2 = \frac{1}{n-1}\left[\sum_{i=1}^{n} x_i^2 - \frac{1}{n}\left(\sum_{i=1}^{n} x_i\right)^2\right] \tag{5-5}$$

Equation 5-5 provides the same answer as Equation 5-4 when computations are made using an appropriate number of significant digits. Equations 5-4 and 5-5 provide an estimate of the variance which is an average of the squared difference between the x values and their average value. In these equations, $(n - 1)$ is used to compute the average, instead of (n), in order to obtain an unbiased estimate of the variance. The concept of variance was introduced in Chapter 2.

There are two commonly used measures based on the variance, the standard deviation and coefficient of variation. By definition the standard deviation (S) is the square root of the variance as follows:

$$S = \sqrt{\frac{1}{n-1}\left[\sum_{i=1}^{n} x_i^2 - \frac{1}{n}\left(\sum_{i=1}^{n} x_i\right)^2\right]} \qquad (5\text{-}6)$$

The coefficient of variation (COV or δ) is a normalized quantity based on the standard deviation and mean value. Therefore, it is dimensionless. The COV is defined as

$$\text{COV} = \frac{S}{\overline{X}} \qquad (5\text{-}7)$$

The standard deviation and coefficient of variation were also introduced in Chapter 2.

5.2.3. Method-of-Moments Estimation

The method of moments is one method of estimating population parameters using the moments of samples. The method of moments is frequently used to provide estimates of the parameters of a distribution, primarily because of its computational simplicity. Equations that relate the moments of a sample to the parameters of a distribution can be derived. Estimates obtained by the method of moments are always consistent, but they may not be efficient in a statistical sense.

Because of its structural simplicity, the uniform distribution is used to illustrate the method of moments. The uniform distribution is a function of two parameters, α and β. The location parameter α defines the lower limit of the distribution, while the scale parameter β indicates the spatial extent of the density function which is given by

$$f_X(x) = \begin{cases} \dfrac{1}{b-a} & a \leq x \leq b \\ 0 & \text{otherwise} \end{cases} \qquad (5\text{-}8)$$

where a and b are the estimators of the parameters α and β, respectively. The sample mean can be related to the definition of the mean of the population (Chapter 3) as follows:

$$\overline{X} = \int_{-\infty}^{\infty} x f_X(x)\,dx = \int_a^b x\left(\frac{1}{b-a}\right) dx = \frac{1}{b-a}\int_a^b x\,dx$$

$$= \frac{1}{b-a}\left(\frac{x^2}{2}\right)\Big|_a^b = \frac{b^2 - a^2}{2(b-a)} = \frac{(b-a)(b+a)}{2(b-a)} = \frac{b+a}{2}$$

(5-9)

The sample variance can be related to the variance of the population. The following relationship simplifies the computation:

$$\mu = \mu'_2 - \left(\mu_1\right)^2$$

(5-10)

where μ_1 is the mean and μ'_2 is the second moment about the origin. Thus Equation 5-10 becomes

$$S^2 = \int_a^b x^2\left(\frac{1}{b-a}\right) dx - \left(\frac{b+a}{2}\right)^2 = \frac{1}{b-a}\left(\frac{x^3}{3}\right)\Big|_a^b - \left(\frac{b+a}{2}\right)^2$$

$$= \frac{b^3 - a^3}{3(b-a)} - \left(\frac{b+a}{2}\right)^2 = \left(\frac{b-a}{12}\right)^2$$

(5-11)

Equations 5-10 and 5-11 provide the means for evaluating the parameters of the uniform density function. This provides two equations with two unknowns, b and a, as follows:

$$\overline{X} = \frac{b+a}{2}$$

(5-12)

$$S^2 = \frac{(b-a)^2}{12}$$

(5-13)

Solving Equations 5-12 and 5-13 for a and b provides

$$a = \overline{X} - S\sqrt{3}$$

(5-14)

and

$$b = \overline{X} + S\sqrt{3}$$

(5-15)

Table 5-1 provides a summary of the relationships between the parameters of commonly used distributions, and the mean and variance. These relationships can be developed using a similar approach to the method used for the uniform distribution and using the concepts in Chapter 3.

TABLE 5-1 Relationships for the Method of Moments: (a) Discrete Distributions, (b) Continuous Distributions

Distribution type	Probability mass or density function	Parameters	Relationships
	(a) Discrete Distributions		
Bernoulli	$$P_X(x) = \begin{cases} p & x = 1 \\ 1-p & x = 0 \\ 0 & \text{otherwise} \end{cases}$$	p	$\bar{X} = p$ $S^2 = p(1-p)$
Binomial	$$P_X(x) = \begin{cases} \binom{N}{x} p^x (1-p)^{N-x} & \text{for } x = 0,\ 1,\ 2, ...,\ N \\ 0 & \text{otherwise} \end{cases}$$	p	$\bar{X} = Np$ $S^2 = Np(1-p)$
Geometric	$$P_X(x) = \begin{cases} p(1-p)^{x-1} & x = 0,\ 1,\ 2,\ 3, ... \\ 0 & \text{otherwise} \end{cases}$$	p	$\bar{X} = 1/p$ $S^2 = (1-p)/p^2$
Poisson	$$P_{X_t}(x) = \begin{cases} \dfrac{(\lambda t)^x \exp(-\lambda t)}{x!} & x = 0,\ 1,\ 2,\ 3, ... \\ 0 & \text{otherwise} \end{cases}$$	λ	$\bar{X} = \lambda t$ $S^2 = \lambda t$

TABLE 5-1 Relationships for the Method of Moments: (a) Discrete Distributions, (b) Continuous Distributions (continued)

(b) Continuous Distributions

Distribution type	Probability mass or density function	Parameters	Relationships
Uniform	$f_X(x) = \begin{cases} \dfrac{1}{b-a} & \text{for } a \le x \le b \\ 0 & \text{otherwise} \end{cases}$	a, b	$\bar{X} = (a+b)/2$ $S^2 = \dfrac{1}{12}(b-a)^2$
Normal	$f_X(x) = \dfrac{1}{\sigma\sqrt{2\pi}}\exp\left[-\dfrac{1}{2}\left(\dfrac{x-\mu}{\sigma}\right)^2\right]$ for $-\infty < x < \infty$	μ, σ	$\bar{X} = \mu$ $S^2 = \sigma^2$
Lognormal	$f_X(x) = \dfrac{1}{x\sigma_Y\sqrt{2\pi}}\exp\left[-\dfrac{1}{2}\left(\dfrac{\ln x - \mu_Y}{\sigma_Y}\right)^2\right]$ for $0 < x < \infty$	μ_Y, σ_Y	$\bar{X} = \exp(\mu_Y + 0.5\sigma_Y^2)$ $S^2 = \mu_Y^2\,[\exp(\sigma_Y^2) - 1]$
Exponential	$f_X(x) = \begin{cases} \lambda\exp(-\lambda x) & \text{for } x \ge 0 \\ 0 & \text{otherwise} \end{cases}$	λ	$\bar{X} = 1/\lambda$ "$S^2 = 1/\lambda^2$

FIGURE 5-2 Uniform probability function fit using the method of moments.

Example 5-1: Method of Moments for the" Uniform Distribution

To illustrate the method of moments, consider the sample histogram of Figure 5-2. This sample was introduced as sample number 2 in Section 2.6.1 (in Table 2-1 and Figure 2-17). It is shown that the sample mean and standard deviation are 10 and 3.0625, respectively. Using Equations 5-14 and 5-15, the parameters are estimated to be $a = 4.7$ and $b = 15.3$. The density function $f_X(x)$ is thus given by

$$f_X(x)\begin{cases} 0.0943 & 4.7 \le x \le 153 \\ 0 & \text{otherwise} \end{cases} \tag{5-16}$$

The population density function of Equation 5-16 is shown in Figure 5-2. Although the population density function differs from the sample probability histogram, the population should be used to make probability statements about the random variables X. For example, the probability that X is between 7 and 10 equals $0.0943(10 - 7) = 0.283$.

Example 5-2: Method of Moments and the Normal Distribution

For the normal distribution, an estimator for the location parameter (μ) can be found by equating the population and the sample means:

$$\bar{X} = \int_{-\infty}^{\infty} x f_X(x)\, dx = \int_{-\infty}^{\infty} x \left[\frac{1}{\sigma\sqrt{2\pi}} \exp\left(-\frac{1}{2}\left(\frac{x-\mu}{\sigma}\right)^2 \right) \right] dx \tag{5-17}$$

$$= \mu$$

Thus, the location parameter (μ) equals the mean ($\overline{X}$). To derive an estimator for the scale parameter (σ) the sample variance can be equated to the definition of the variance:

$$S^2 = \int_{-\infty}^{\infty} (x-\mu)^2 \, f_X(x) \, dx = \int_{-\infty}^{\infty} (x-\mu)^2 \left[\frac{1}{\sigma\sqrt{2\pi}} \exp\left(-\frac{1}{2}\left(\frac{x-\mu}{\sigma}\right)^2 \right) \right] dx \tag{5-18}$$

$$= \sigma^2$$

Thus, the scale parameter (σ) equals the standard deviation of the sample (S).

5.2.4. Maximum-Likelihood Estimation

The most common statistical method of parameter estimation is the method of maximum likelihood. This method is based on the principle of calculating values of parameters that maximize the probability of obtaining the particular sample.

The likelihood of the sample is the total probability of drawing each item of the sample. The total probability is the product of all the individual item probabilities. This product is differentiated with respect to the parameters, and the resulting derivatives are set to zero to achieve the maximum.

Maximum likelihood solutions for model parameters are statistically efficient solutions, meaning that parameter values have minimum variance. This definition of a best method, however, is theoretical. Maximum likelihood solutions do not always produce solvable equations for the parameters. The following examples illustrate easy to moderately difficult solutions. For some distributions, including notably the normal distribution, the method of moments and maximum likelihood estimation produce identical solutions for the parameters.

Example 5-3: Maximum-Likelihood Estimation for the Exponential Distribution

For this example we will find the maximum likelihood estimate of the parameter λ in the density function $\lambda \exp(-\lambda x)$. Consider a sample of n items: $x_1, x_2, x_3, ..., x_n$. By definition the likelihood function, L, is

$$L = \prod_{i=1}^{n} \lambda \exp\left(-\lambda x_i\right) \tag{5-19}$$

The product form of the function in Equation 5-19 is difficult to differentiate. We make use of the fact that the logarithm of a variate must have its maximum at the same place as the maximum of the variate. Taking logarithms of Equation 5-19 gives

$$\ln(L) = n \ln(\lambda) - \lambda \sum_{i=1}^{n} x_i \qquad (5\text{-}20)$$

The differential of $\ln(L)$ with respect to λ, set to zero, produces the value of the parameter that maximizes the likelihood function. The derivative is given by

$$\frac{d \ln(L)}{d\lambda} = \frac{n}{\lambda} - \sum_{i=1}^{n} x_i = 0 \qquad (5\text{-}21)$$

Equation 5-21 yields the following:

$$\frac{1}{\lambda} = \frac{1}{n} \sum_{i=1}^{n} x_i = \overline{X} \qquad (5\text{-}22)$$

Thus, the maximum likelihood value of $1/\lambda$ is the mean of the sample of x's.

Example 5-4: Maximum-Likelihood Estimation of an Arbitrary Density Function

Consider the problem of finding the maximum likelihood value of parameter A in the density function:

$$f_X(x) = c\,x \exp(-Ax) \quad \text{for } x \geq 0 \qquad (5\text{-}23)$$

where c = constant. To use this equation as a probability density function, we must first find c from the condition that the total probability equals 1 as follows:

$$c \int_{0}^{\infty} x \exp(-Ax)\, dx = 1 \qquad (5\text{-}24)$$

Solution of this equation gives $c = A^2$. Thus the likelihood function is

$$L = \prod_{i=1}^{n} A^2 x_i \exp(-Ax_i) \qquad (5\text{-}25)$$

The logarithm of this function is

$$\ln(L) = 2n \ln(A) + \sum_{i=1}^{n} \ln(x_i) - A \sum_{i=1}^{n} x_i \qquad (5\text{-}26)$$

and

$$\frac{d\ln(L)}{dA} = \frac{2n}{A} - \sum_{i=1}^{n} x_i = 0 \qquad (5\text{-}27)$$

We find that the maximum likelihood value of $1/A$ is one half the mean of the sample.

Example 5-5: Maximum-Likelihood Estimation and the Gamma Distribution

The problem here is to find the maximum likelihood expressions for the parameters of the gamma distribution. The density function of the gamma distribution can be written as

$$f_X(x) = \frac{b^{a+1}}{a!} \exp(-bx) x^a \qquad (5\text{-}28)$$

where $a!$ is the factorial of a and is related to the gamma integral by $a! = \Gamma(a + 1)$ and $\Gamma(a + 1) = a\Gamma(a)$. Solving for the likelihood expressions:

$$L = \prod_{i=1}^{n} \frac{b^{a+1}}{a!} \exp(-bx_i) x_i^a \qquad (5\text{-}29)$$

and

$$\ln(L) = n[(a+1)\ln(b) - \ln(a!)] - b\sum_{i=1}^{n} x_i + a\sum_{i=1}^{n} \ln(x_i) \qquad (5\text{-}30)$$

Therefore,

$$\frac{\partial\ln(L)}{\partial a} = n\ln(b) - \frac{n\partial\ln(a!)}{\partial a} + \sum_{i=1}^{n} \ln(x_i) \qquad (5\text{-}31)$$

and

$$\frac{\partial\ln(L)}{\partial b} = \frac{n(a+1)}{b} - \sum_{i=1}^{n} x_i \qquad (5\text{-}32)$$

Setting the two partial differentials of Equations 5-31 and 5-32 to zero for their maxima and solving simultaneously for a and b give

$$b = \frac{n(a+1)}{\displaystyle\sum_{i=1}^{n} x_i} \qquad (5\text{-}33a)$$

and

$$\ln(a+1) - \frac{\partial \ln(a!)}{\partial a} = \ln\left(\frac{\displaystyle\sum_{i=1}^{n} x_i}{n}\right) - \frac{\displaystyle\sum_{i=1}^{n} \ln(x_i)}{n} \qquad (5\text{-}33b)$$

Equations 5-33a and 5-33b illustrate the potential complexity of the maximum likelihood solutions. Equation 5-33b can be solved by a trial-and-error process for a. With a evaluated, Equation 5-33a can be solved for b. The expression $\partial\ln(a!)/\partial a$ in Equation 5-33b is called the gamma derivative function of a and designated $\Psi(a)$. Tables of values of the gamma function can be found in mathematical handbooks as shown in Appendix B.

A sample of 36 synthetic numbers is shown as a histogram in Figure 5-3. Values for the sample-summation terms in Equation 5-33b are

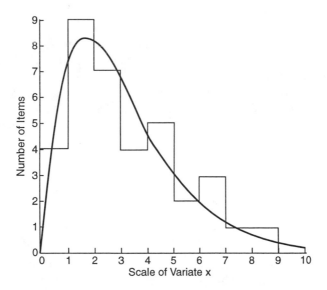

FIGURE 5-3 Gamma function fitted by maximum likelihood.

$$\ln\left(\frac{\sum\limits_{i=1}^{n} x_i}{n}\right) = \ln\left(\frac{116}{36}\right) = 1.17 \tag{5-34a}$$

and

$$\frac{\sum\limits_{i=1}^{n} \ln(x_i)}{n} = \frac{33.002}{36} = 0.9167 \tag{5-34b}$$

Substituting in Equation 5-33b and rearranging gives

$$\ln(a+1) - \left(\frac{\partial \ln(a!)}{\partial a} + 0.2533\right) = 0 \tag{5-35}$$

We must now search for values of a in Equation 5-35. One systematic method is to try values of a at the midpoints of intervals that are known to contain a root of the equation. This method is illustrated in Table 5-2. In this table, A denotes the first term on the left side of Equation 5-35, B is the second expression in brackets in this equation. Trial values of a of 0, 1, and 2 were assumed. The quantity $(A - B)$ changed from positive to negative between 1 and 2; therefore, a zero point of the equation falls in this interval. The successive midinterval values of 1.5, 1.25, and 1.125 were assumed, yielding the approximate values of a and b as shown. The function, $\psi(a)$, is $\partial \ln(a!)/a$ which can be related to the gamma function. The values for the gamma function were obtained from Appendix B. The resulting value of a of 1.12 was then substituted in Equation 5-33a to obtain a value for b of 0.658.

The gamma distribution (Equation 5-28), using the estimated values of a and b, is superimposed on the histogram in Figure 5-3. Probability estimates for the occurrence of various values of X could now be made, using the assumed population and not the sample histogram.

TABLE 5-2 Maximum Likelihood Solution for Gamma Parameter a

Trial value for a	$\psi(a)$	B	A	$A - B$
0	−0.5772	−0.3239	0	0.3239
1.0	0.4228	0.6761	0.6931	0.0170
2.0	0.9228	1.1761	1.0986	−0.0775
1.5	0.7032	0.9565	0.9163	−0.0402
1.25	0.5725	0.8258	0.8109	−0.0149
1.125	0.5034	0.7567	0.7538	−0.0029

5.2.5. Other Estimation Methods

The estimation methods discussed in the previous sections are commonly used in engineering to obtain point estimates of parameters in univariate analysis. Other methods of point estimation are available. They include, for example, Bayes estimators and least-squares estimators.

The Bayes estimators treat the unknown parameter as a random variable with a known prior probability distribution. The sample provides additional information that is used to modify the prior distribution to obtain a posterior probability distribution for the parameter. The modification of the distribution from prior to posterior is performed using Bayes theorem as discussed in Chapter 3. The Bayes estimator is, therefore, updated as additional information is obtained. This concept can be used for both univariate and multivariate analyses. The process of Bayesian updating of prior quantities constitutes the basis of Bayesian techniques discussed in Chapter 9.

The least-squares estimators are commonly used in bivariate and multivariate analysis. The parameter estimation is performed such that the mean-squared error is minimized. The MSE was defined as the expected value of the square of the deviation of the estimate $\hat{\theta}$ from the parameter being estimated θ, i.e.,

$$MSE = E\left[\left(\hat{\theta} - \theta\right)^2\right] \tag{5-36}$$

This method constitutes the basis of least-squares regression in bivariate and multivariate analyses. Regression analysis is discussed in Chapter 6.

5.3. SAMPLING DISTRIBUTIONS

In Section 5.1, two samples of five measurements of the compressive strength of concrete were given. The means differed (3442 vs. 3526) and the standard deviations differed (135.9 and 174.4). If we took many samples of five and computed the mean and standard deviation for each sample, we could form two histograms, one for the sample means and one for the sample standard deviations. What would the histograms look like? The histogram for the sample means would approximate the sampling distribution of the estimated mean $\overline{X}$, where $\overline{X}$ is treated as a random variable. The histogram of the sample standard deviations would approximate the sampling distribution of the estimated standard deviation S, where S is treated as a random variable.

5.3.1. Sampling Distribution of the Mean

The sampling distribution of the mean depends on whether or not the population variance σ^2 is known. If it is known, then the mean of a random sample of size n

from a population with mean μ and variance σ^2 has a normal distribution with mean μ and variance σ^2/n. The statistic Z has a standard normal distribution (i.e., mean = 0 and variance = 1) as follows:

$$Z = \frac{\overline{X} - \mu}{\sigma/\sqrt{n}} \qquad (5\text{-}37)$$

If the population variance is not known, then the distribution of the mean depends on the distribution of the random variable. For a random variable with a normal distribution with mean μ, the distribution of the mean has a mean μ and a standard deviation $S/\sqrt{n}$. The statistic t has a t distribution with $n - 1$ degrees of freedom:

$$t = \frac{\overline{X} - \mu}{S/\sqrt{n}} \qquad (5\text{-}38)$$

To illustrate the distribution of the sample mean, consider the 40 samples of five measurements from the standard normal population, i.e., N(0,1), that are given in Table 5-3a. A histogram of the 200 sample values of the random variable and the density function for the underlying population are given in Figure 5-4a. The 200 values give a reasonably good approximation of the population N(0,1). For each of the 40 samples, the mean was computed as shown in Table 5-3b. A sample histogram is shown in Figure 5-4b, with the density function N(0,1/5) also shown. This density function of the sample mean has a variance of σ^2/n, where n is the size of each sample (i.e., $n = 5$). Thus the standard error of the mean is $\sigma/\sqrt{n}$ or $1/\sqrt{5}$. The two population distributions (X and $\overline{X}$) are given in Figure 5-4c for the purpose of comparison. It is important to note that, while both have identical means of 1, the sampling distribution of the mean has a much smaller spread. This should be expected because there is much less variation in means than in the values that are used to compute the means.

If two independent samples of sizes n_1 and n_2 are drawn from populations with means μ_1 and μ_2 and variances σ_1^2 and σ_2^2, respectively, then the difference of the sample means $\overline{X}_1 - \overline{X}_2$ has a sampling distribution that is approximately normal with a mean $\mu_1 - \mu_2$ and variance $(\sigma_1^2/n_1 + \sigma_2^2/n_2)$. Thus, the statistic Z has a standard normal distribution:

$$Z = \frac{\left(\overline{X}_1 - \overline{X}_2\right) - \left(\mu_1 - \mu_2\right)}{\left(\dfrac{\sigma_1^2}{n_1} + \dfrac{\sigma_2^2}{n_2}\right)^{0.5}} \qquad (5\text{-}39)$$

TABLE 5-3 Sampling Distributions for the Mean and Variance for Samples Taken from N(0,1) Population

(a) Forty Samples from N(0,1), Each with a Sample Size of 5

0.048	1.040	−0.111	−0.120	1.396	−0.393	−0.220	0.422	0.233	0.197
−0.521	−0.563	−0.116	−0.512	−0.518	−2.194	2.261	0.461	−1.533	−1.836
−1.407	−0.213	0.948	−0.073	−1.474	−0.236	−0.649	1.555	1.285	−0.747
1.822	0.898	−0.691	0.972	−0.011	0.517	0.808	2.651	−0.650	0.592
1.346	−0.137	0.952	1.467	−0.352	0.309	0.578	−1.881	−0.488	−0.329
0.420	−1.085	−1.578	−0.125	1.337	0.169	0.551	−0.745	−0.588	1.810
−1.760	−1.868	0.677	0.545	1.465	0.572	−0.770	0.655	−0.574	1.262
−0.959	0.061	−1.260	−0.573	−0.646	−0.697	−0.026	−1.115	3.591	−0.519
0.561	−0.534	−1.730	−1.172	−0.261	−0.049	0.173	0.027	1.138	0.524
−0.717	0.254	0.421	−1.891	2.592	−1.443	−0.061	−2.520	−0.497	0.909
−2.097	−0.180	−1.298	−0.647	0.159	0.769	−0.735	−0.343	0.966	0.595
0.443	−0.191	0.705	0.420	−0.486	−1.038	−0.396	1.406	0.327	1.198
0.481	0.161	−0.044	−0.864	−0.587	−0.037	−1.304	−1.544	0.946	−0.344
−2.219	−0.123	−0.260	0.680	0.224	−1.217	0.052	0.174	0.692	−1.068
1.723	−0.215	−0.158	0.369	1.073	−2.442	−0.472	2.060	−3.246	−1.020
−0.937	1.253	0.321	−0.541	−0.648	0.265	1.487	−0.554	1.890	0.499
−0.568	−0.146	0.285	1.337	−0.840	0.361	−0.468	0.746	0.470	0.171
−1.717	−1.293	−0.556	−0.545	1.344	0.320	−0.087	0.418	1.076	1.669
−0.151	−0.266	0.920	−2.370	0.484	−1.915	−0.268	0.718	2.075	−0.975
2.278	−1.819	0.245	−0.163	0.980	−1.629	−0.094	−0.573	1.548	−0.896

(b) Sample Means

0.258	0.205	0.196	0.347	−0.246	−0.399	0.556	0.642	−0.231	−0.425
−0.491	−0.634	−0.694	−0.643	0.897	−0.290	−0.027	−0.740	0.641	0.797
−0.334	−0.110	−0.211	−0.008	0.077	−0.793	−0.571	0.351	−0.063	−1.28
−0.219	−0.454	0.243	−0.456	0.264	−0.520	0.114	0.151	1.412	0.094

(c) Sample Variances

1.764	0.514	0.529	0.694	1.272	1.147	1.257	2.829	1.113	0.882
0.955	0.752	1.325	0.880	1.777	0.627	0.231	1.462	3.305	0.766
3.042	0.024	0.514	0.484	0.445	1.493	0.248	2.033	3.233	1.002
2.280	1.402	0.276	1.745	0.945	1.318	0.613	0.442	0.421	1.192

(d) Sample Standard Deviations

1.328	0.717	0.727	0.833	1.128	1.071	1.121	1.682	1.055	0.939
0.977	0.867	1.151	0.938	1.333	0.792	0.481	1.209	1.818	0.875
1.744	0.155	0.717	0.696	0.667	1.222	0.498	1.426	1.798	1.001
1.510	1.184	0.525	1.321	0.972	1.148	0.783	0.665	0.649	1.092

If the population means and variances are equal, then the Z statistic of Equation 5-39 is

$$Z = \frac{\overline{X}_1 - \overline{X}_2}{\sigma^2 \left(\dfrac{1}{n_1} + \dfrac{1}{n_2} \right)^{0.5}} \tag{5-40}$$

Equations 5-37 to 5-40 can be used to test hypotheses about the means and to form confidence intervals.

FIGURE 5-4 Sample distribution of (a) random variable from N(0,1) population; (b) sample means; (c) the population of the random variable and the mean; and (d) variances.

5.3.2. Sampling Distribution of the Variance

The estimated variance of a sample is a random variable, and so it has a distribution. The distribution depends on the characteristics of the underlying population from which the sample is derived. If the population is normal, then it can be shown that for the unbiased estimate of the variance S^2, the quantity $(n-1)S^2/\sigma^2$ is a random variable distributed as chi-square (χ^2, called C in Chapter 3) with $n-1$ degrees of freedom. Thus, inferences about the variance of a single normally distributed population are made with:

$$\chi^2 = \frac{(n-1)\,S^2}{\sigma^2} \qquad (5\text{-}41)$$

The chi-square statistic of Equation 5-41 can be used to test hypotheses about the variance of a single random variable and to form confidence intervals.

The 40 samples of 5 were used to compute the sample variances as shown in Table 5-3c from which the standard deviations were also computed as shown in Table 5-3d. Since each sample was based on a size of 5, the number of degrees of freedom is four. The histogram of sample variance is shown in Figure 5-4d and the underlying chi-square distribution is also shown. The sample does not match the population as well for the variance as for the mean because the sampling variation of the variance is relatively greater than that of the mean. It would take considerably more than 40 samples to provide a reasonable approximation of the sampling distribution of the variance. However, it is evident from Figure 5-4d that the variance has a chi-square distribution.

5.3.3. Sampling Distributions for Other Parameters

Any estimated quantity using a sample can be treated as a random variable, and so it has a distribution. The distribution depends on the characteristics of the underlying population from which the sample is derived. For example, in Chapter 6 correlation analysis and regression analysis are introduced. The estimated correlation coefficient and the estimated parameters (or coefficients) in the regression models are treated as random variables. Therefore, they are random variables and have probability distributions. Their distributions are discussed after their introduction in Chapter 6.

5.4. HYPOTHESIS TESTING: PROCEDURE

Hypothesis testing is the formal procedure for using statistical concepts and measures in performing decision making. The following six steps can be used to make a statistical analysis of a hypothesis:

1. Formulate hypotheses.
2. Select the appropriate statistical model (theorem) that identifies the test statistic.
3. Specify the level of significance, which is a measure of risk.
4. Collect a sample of data and compute an estimate of the test statistic.
5. Define the region of rejection for the test statistic.
6. Select the appropriate hypothesis.

These six steps are discussed in detail in the following sections.

5.4.1. Step 1: Formulation of Hypotheses

Hypothesis testing represents a class of statistical techniques that are designed to extrapolate information from samples of data to make inferences about populations for the purpose of decision making. The first step is to formulate two or more hypotheses for testing. The hypotheses are usually statements indicating that a random variable has some specific distribution or that a population parameter has some specific value. If the objective is to compare two or more specific parameters, such as the means of two populations, the hypotheses are then statements formulated to indicate the absence or presence of differences. It is important to note that the hypotheses are composed of statements involving either population distributions or parameters; hypotheses should not be expressed in terms of sample statistics.

The first hypothesis is called the *null hypothesis*, is denoted by H_o, and is formulated as an equality, thus indicating that a difference does not exist. The second hypothesis, which is called the *alternative hypothesis*, is formulated to indicate that a difference does exist. The alternative hypothesis is denoted by either H_1 or H_A. The null and alternative hypotheses should be expressed both grammatically and in mathematical terms and should represent mutually exclusive conditions. Thus, when a statistical analysis of sampled data suggests that the null hypothesis should be rejected, the alternative hypothesis must be accepted. In some cases, the users may hope that the experiment leads to a difference; thus, there may be a desire to present the null hypothesis in a way that would suggest that a difference exists. This practice should not be followed. The null hypothesis must be a statement of equality, not inequality.

5.4.2. Step 2: The Test Statistic and Its Sampling Distribution

The alternative hypothesis of step 1 provides for a difference between specified populations or parameters. To test the two hypotheses, it is necessary to develop a test statistic that reflects the difference suggested by the alternative hypothesis. The computed value of a test statistic varies from one sample to the next. Therefore, the test statistic is a random variable and has a sampling distribution. A hypothesis test should be based on a theoretical model that defines the distribution function of the test statistic and the parameters of the sampling distribution. Based on the distributions of the test statistic, probability statements about computed values may be made.

Theoretical models are available for all of the more frequently used hypothesis tests. In cases where theoretical models are not available, approximations have usually been developed. In any case, a model or theorem that specifies the test statistic, its distribution, and its parameters must be found. The test statistic reflects the hypotheses and the data that are usually available. Also, the test statistic is a random variable, and thus it has a distribution function that is defined by a functional form and one or more parameters.

5.4.3. Step 3: The Level of Significance

A set of hypotheses were formulated in step 1. In step 2, a test statistic and its distribution were selected to reflect the problem for which the hypotheses were formulated. In step 4, data will be collected to test the hypotheses. Therefore, before the data are collected, it is necessary to provide a probabilistic framework for accepting or rejecting the null hypothesis and, subsequently, making a decision. The framework should reflect the allowance to be made for the chance variation that can be expected in a sample of data. This chance variation is referred to as sampling variation.

TABLE 5–4 Decision Table for Hypothesis Testing

	Situation	
Decision	H_0 is true	H_0 is not true
Accept H_0	Correct decision	Incorrect decision: type II error
Reject H_0	Incorrect decision: type I error	Correct decision

Table 5-4 shows the available situations and the potential decisions involved in a hypothesis test. The decision table suggests two types of error:

1. Type I error: reject H_0 when, in fact, H_0 is true.
2. Type II error: accept H_0 when, in fact, H_0 is false.

These two types of incorrect decisions are not independent; however, the decision process is most often discussed with reference only to the type I error.

The level of significance, which is a primary element of the decision process in hypothesis testing, represents the probability of making a type I error and is denoted by α. The probability of a type II error is denoted by β. The two possible incorrect decisions are not independent. The level of significance should not be made exceptionally small, because the probability of making a type II error will then be increased. Selection of the level of significance should, therefore, be based on a rational analysis of the effect on decisions and should be selected prior to the collection and analysis of the sample data. Specifically, one would expect the level of significance to be different when considering a case involving the loss of human life and a case involving minor property damage. However, the value chosen for α is often based on convention and the availability of statistical tables, with values for α of 0.05 and 0.01 being selected frequently. However, this traditional means of specifying α should be understood.

5.4.4. Step 4: Data Analysis

Given the significance level α, it is possible to determine the sample size required to meet any rejection criterion. The selection of a sample size is discussed in Section 5.8. After obtaining the necessary data, the sample is used to compute an estimate of the test statistic. In most cases, the data are also used to provide estimates of other parameters required to define the sampling distribution of the test statistic.

5.4.5. Step 5: The Region of Rejection

The region of rejection consists of those values of the test statistic that would be unlikely to occur when the null hypothesis is, in fact, true. Conversely, the region of acceptance consists of those values of the test statistic that would be expected when the null hypothesis is, in fact, true. Extreme values of the test statistic are least likely to occur when the null hypothesis is true. Thus the region of rejection is usually represented by one or both tails of the distribution of the test statistic.

The critical value of the test statistic is defined as that value which separates the region of rejection from the region of acceptance. The critical value of the test statistic depends on (1) the statement of the alternative hypothesis, (2) the distribution of the test statistic, (3) the level of significance, and (4) characteristic of the sample or data. These four components represent the first four steps of a hypothesis test.

Depending on the statement of the alternative hypothesis, the region of rejection may consist of values associated with either one or both tails of the distribution of the test statistic. This may best be illustrated with examples. Consider the case of a manufacturer of cold drinks. It must be recognized that the accuracy of the bottling process is not sufficient to assure that every bottle will contain exactly 12 oz; some bottles will contain less, whereas others will contain more. If the label on the bottle indicates that the bottle contains 12 oz, the manufacturer wants to be assured that, on the average, each bottle contains at least that amount; otherwise, the manufacturer may be subject to a lawsuit for false advertising. But if the bottles contain, on the average, more than 12 oz, the manufacturer may be losing money. Thus the manufacturer is interested in both extremely large and extremely small deviations from 12 oz and would be interested in the following two-tailed test:

FIGURE 5-5 Representation of the region of rejection (cross-hatched area), region of acceptance, and critical value (z_α): (a) two-sided test; (b) lower tail one-sided test; and (c) upper tail one-sided test.

$$H_o: \mu = \mu_o \qquad \text{(5-42a)}$$

$$H_A: \mu \neq \mu_o \qquad \text{(5-42b)}$$

where μ is the population mean and μ_o is the standard of comparison (i.e., 12 oz). In this case, the region of rejection will consist of values in both tails. This is illustrated in Figure 5-5a, where Z is the test statistic that has the continuous probability density function shown in the figure. In this case, the test is called a two-tailed test.

In other cases, the region of rejection may consist of values in only one tail of the distribution. For example, if the mean concentration of a pollutant from ten samples is used as an indicator of water quality, a regulatory agency would only be interested in whether or not the state water quality standard for the pollutant is met. Specifically, if a manufacturing plant is discharging waste that contains the pollutant, a state may fine the manufacturer or close down the plant if the effluent causes the concentration to exceed the state standard. Thus the regulatory agency is only interested in whether or not the quality is greater than the standard limit. It is not concerned with the degree to which the concentration is below the standard. This situation represents a one-tailed hypothesis test because the test is directional. The following hypotheses reflect this problem:

$$H_o: \mu = \mu_o \qquad \text{(5-43a)}$$

$$H_A: \mu > \mu_o \qquad (5\text{-}43b)$$

In this case, the region of rejection is associated with values in only the upper tail of the distribution, as illustrated in Figure 5-5c. The test is called a one-tailed test.

Other engineering situations involve concern over the lower tail. Engineers design concrete structures such that a minimum concrete strength is achieved. Strengths higher than a standard or specified value are acceptable assuming the ductility of the structure is adequate. During the construction of the structure, samples of the concrete are collected and then tested after they have cured for a required period of time. The samples are tested and the mean strength is computed. This mean is compared with the standard or specified values (μ_o). This situation leads to the following hypothesis:

$$H_o: \mu = \mu_o \qquad (5\text{-}43c)$$

$$H_A: \mu < \mu_o \qquad (5\text{-}43d)$$

The test is called a one-tailed, lower test because the interest is in strengths significantly below the standard. In this case, the region of rejection is associated with values in only the lower tail of the distribution, as illustrated in Figure 5-5b. Rejection of the null hypothesis would suggest that the strength is not adequate. This may necessitate a decision to retrofit the structure to ensure adequate strength.

Although the region of rejection should be defined in terms of values of the test statistic, it is often pictorially associated with an area of the sampling distribution that is equal to the level of significance. The region of rejection, region of acceptance, and the critical value are shown in Figure 5-5 for the two-tailed and both one-tailed tests.

5.4.6. Step 6: Select the Appropriate Hypothesis

A decision on whether or not to accept the null hypothesis depends on a comparison of the computed value of the test statistic and the critical value. The null hypothesis is rejected when the computed value lies in the region of rejection. Rejection of the null hypothesis implies acceptance of the alternative hypothesis. When a computed value of the test statistic lies in the region of rejection, there are two possible explanations. First, the sampling procedure may have resulted in an extreme value purely by chance; although this is a very unlikely event, this corresponds to the type I error of Table 5-4. Because the probability of this event is relatively small, this explanation is most often rejected. Second, the extreme value of the test statistic may have occurred because the null hypothesis is false; this is the explanation most often accepted and forms the basis for statistical inference.

5.5. HYPOTHESIS TESTS OF MEANS

Since the mean of a population is often a primary element of statistical decision making, hypothesis tests of the mean are used frequently. The six-step procedure for testing hypotheses is applicable for testing hypotheses concerning the population mean μ.

5.5.1. Test of the Mean with Known Population Variance

When the standard deviation of the population is known, the procedure for testing the mean is as follows:

Step 1: Formulate hypotheses — The null and alternative hypotheses must be stated in terms of the population parameter μ and the value selected for comparison, which may be denoted as μ_o. The null hypothesis should state that the mean of the population equals a preselected standard value. Acceptance of the null hypothesis implies that is not significantly different from μ_o. Mathematically, the null hypothesis could be stated as:

$$H_o: \mu = \mu_o \tag{5-44}$$

One of three alternative hypotheses may be selected:

$$H_{A1}: \mu < \mu_o \quad \text{one-tailed test} \tag{5-45a}$$

$$H_{A2}: \mu > \mu_o \quad \text{one-tailed test} \tag{5-45b}$$

$$H_{A3}: \mu \neq \mu_o \quad \text{two-tailed tests} \tag{5-45c}$$

Each of the alternative hypotheses indicates that a significant difference exists between the population mean and the standard value. The selected alternative hypothesis depends on the statement of the problem.

Consider, for example, the problem of defining the yield strength of a shipment of steel. A study may be proposed to decide whether or not a shipment of steel meets the required yield strength of 3250 kgf (kilogram force). In this case, the shipment would be rejected only if the mean yield strength of the specimens in the sample was significantly less than 3250 kgf. Whether or not the strength is greater than 3250 kgf is not important since the strength of the structure is assumed to be adequate as long as it is at least 3250 kgf. Thus a one-sided hypothesis test would be appropriate with the following hypotheses:

$$H_o: \mu = 3250 \text{ kgf} \tag{5-46a}$$

$$H_A: \mu < 3250 \text{ kgf} \tag{5-46b}$$

This represents the one-tailed (lower) test.

Step 2: Select the appropriate model — The mean $\overline{X}$ of a random sample is used in testing hypotheses about the population mean μ; $\overline{X}$ is itself a random variable. If the population from which the random sample is drawn has mean μ and variance σ^2, the distribution of the random variable $\overline{X}$ has mean μ and variance σ^2/n for samples from infinite populations. For samples from finite populations of size N, the variance is $[\sigma^2(N - n)]/[n(N - 1)]$.

For a random sample of size n, the sample mean $\overline{X}$ can be used in calculating the value of the test statistic z as

$$z = \frac{\overline{X} - \mu}{\sigma/\sqrt{n}} \tag{5-47}$$

in which z is the value of a random variable whose distribution function is a standard normal.

Step 3: Select the level of significance — A level of significance of 1% is selected for demonstration of this hypothesis test. However, in actual practice the level selected for use should vary with the problem being studied and the impact of making an incorrect decision.

Step 4: Compute estimate of the test statistic — A random sample consisting of 100 specimens is selected, with a computed mean of 3190 kgf. The standard deviation of the population is 160 kgf. The value of the test statistic of Equation 5-47 is

$$z = \frac{3190 - 3250}{160/\sqrt{100}} = -3.750 \tag{5-48}$$

Step 5: Define the region of rejection — For the standard normal distribution, the level of significance is the only characteristic required to determine the critical value of the test statistic. The region of rejection depends on the statement of the alternative hypothesis:

if H_A is	then the region of rejection is	
$\mu < \mu_o$	all $z < -z_\alpha$	
$\mu > \mu_o$	all $z > z_\alpha$	(5-49)
$\mu \neq \mu_o$	all $z < -z_{\alpha/2}$ or $z > z_{\alpha/2}$	

For the case study, because the statement of the alternative hypothesis indicates a one-tailed test, the critical value of z for a 1% level of significance (α) can be obtained from Table A-1 as

$$-z_\alpha = -\Phi^{-1}(1-\alpha) = -2.326 \qquad (5\text{-}50)$$

Thus the region of rejection consists of all values of Z less than -2.326.

Step 6: Select the appropriate hypothesis — Since the computed statistic lies in the region of rejection, the null hypothesis must be rejected. That is, there is no statistical basis for assuming the population mean to be equal to, or greater than, the selected value of 3250 kgf. If steel having a yield strength of at least 3250 kgf was required, then the sampled data suggest that the shipment from which the sample was obtained is inadequate.

Summary — The decision criterion specified in step 3 was limited to the specification of the level of significance. Since the null hypothesis was rejected for a 1% level of significance, there is a 1% chance of making a type I error; that is, there is a chance of 1 in 100 of returning the shipment of steel when, in fact, it was of adequate strength. The decision criterion of step 3 did not take into account the possibility of a type II error (β). The results of a type II error would be the use of steel of inadequate strength. Obviously, the consequences of a type II error (e.g., structural failure) are probably more severe than the consequences of a type I error (e.g., economic loss). However, it is easier and more direct to specify a value for α than to specify a value for β.

Example 5-6: Hypothesis Testing for the Mean

Assume that the population mean is 7, with a variance of 5.833. Also, assume that a sample size of $n = 15$ has a mean of 7.2. Based on this, the test statistic of Equation 5-47 is

$$z = \frac{7.2 - 7.0}{2.416/\sqrt{15}} = 0.321 \qquad (5\text{-}51)$$

For a level of significance of 5% and a two-tailed test, the critical value of the standard normal distribution is ± 1.96. Since the computed value of 0.321 is within the region of acceptance, the null hypothesis H_o: $\mu = 7.0$ is accepted.

Example 5-7: Daily Evaporation Measurements

A mean pan evaporation of 0.1387 in./day is computed from a sample of 354 measurements. Is the mean estimate a biased estimate if the long-term records indicate an average value of 0.1575 in./day and a standard deviation of $\sigma = 0.1$ in./day? The hypothesis test for the mean with σ known can be used to test the significance of the bias. The null and alternative hypotheses are

$$H_o: \mu = 0.1575 \text{ in./day} \qquad (5\text{-}52a)$$

$$H_A: \mu \neq 0.1575 \text{ in./day} \qquad (5\text{-}52b)$$

The computed value of the test statistic is

$$z = \frac{0.1387 - 0.1575}{0.1/\sqrt{354}} = -3.537 \qquad (5\text{-}53)$$

For a level of significance of 5%, the region of rejection consists of all values of Z less than -1.96 and all values of Z greater than 1.96. Thus the null hypothesis must be rejected. This suggests that the sample is biased. Since the sample mean is less than the population mean, one might expect that the sample included more days during cold months.

5.5.2. Test of the Mean with Unknown Population Variance

When the population variance is unknown, the theorem used in the preceding section is not applicable, even though the null and alternative hypotheses and the steps are the same. In such cases, a different theorem is used for testing a hypothesis about a mean. Specifically, for a random sample of size n, the sample mean $\overline{X}$ and standard deviation S can be used in calculating the value of the test statistic t:

$$t = \frac{\overline{X} - \mu}{S/\sqrt{n}} \qquad (5\text{-}54)$$

The test statistic, t, is the value of a random variable having the Student-t distribution with the degree of freedom $\nu = n - 1$. This statistic requires that the sample be drawn from a normal population. The region of rejection depends on the level of significance, the degrees of freedom, and the statement of the alternative hypothesis:

if H_A is	then reject H_0 if	
$\mu < \mu_0$	$t < -t_\alpha$	
$\mu > \mu_0$	$t > t_\alpha$	$(5\text{-}55)$
$\mu \neq \mu_0$	$t < -t_{\alpha/2}$ or $t > t_{\alpha/2}$	

When the standard deviation of the population is unknown but the sample is large, say, greater than 30, the Z statistic of the preceding section can be used as an approximation of the t statistic. For a sample size of 30, the Z values for 5 and 1% are obtained from Table A-1 to be 1.645 and 2.324, respectively; the corresponding t values from Table A-2 are 1.699 and 2.462, respectively. The differences are 3.3 and 5.8%, respectively. The decision on whether or not to use the Z statistic in place of the t statistic should be based on the consequences of erroneous decisions.

Example 5-8: Dissolved Oxygen Concentration

The case of sampling the dissolved oxygen concentration of effluent from a manufacturing plant can be used to illustrate the use of this test statistic. If a water quality regulation required a minimum dissolved oxygen concentration of three parts per million (3 ppm), the following procedure may be used in testing the quality of the effluent:

1. H_o: $\mu = 3$ ppm
   H_A: $\mu < 3$ ppm
2. Theorem: t statistic of Equation 5-54 with $\nu = n - 1$
3. $\alpha = 5\%$
4. $n = 5$
   $\overline{X} = 2.8$ ppm
   $S = 0.32$ ppm

$$t = \frac{\overline{X} - \mu_o}{S/\sqrt{n}} = \frac{2.8 - 3.0}{0.32/\sqrt{5}} = -1.398$$

   $\nu = n - 1 = 4$
5. For $\nu = 4$, $\alpha = 5\%$, a one-tailed lower test, and using Table A-2, $t_\alpha = -2.132$.
6. Since $t > t_\alpha$, then accept H_o.

Because there is no statistical basis for rejecting the null hypothesis, one must assume that the effluent meets the water quality regulations of 3 ppm, even though the sample mean is less than the standard of 3 ppm. The fact that H_o cannot be rejected even though the sample mean is less than the standard of 3 ppm reflects the tolerance that must be permitted, because decision making based only on the sample represents decision making under conditions of uncertainty.

5.5.3. Hypothesis Test of Two Means

The previous hypothesis tests have involved a single random variable. In some cases, samples are obtained from two different populations, and it is of interest to determine if the population means are equal. For example, two manufacturers may advertise that their tires will last for 30,000 mi; samples of tires may be used to test whether or not the means of the two populations are equal. Similarly, tests could be conducted on engineering products to determine whether or not the means are equal. The fraction of downtime for two computer types could be tested to decide whether or not the mean times differ.

A number of tests can be used to test a pair of means. The method presented here should be used to test the means of two independent samples. This test is frequently of interest in engineering research when the investigator is interested in comparing an experimental group to a control group. For example, an environmental engineer might be interested in comparing the mean growth rates of microorganisms in a polluted and natural environment. The procedure presented in this section can be used to make the test.

Step 1: Formulate hypotheses — If we denote the means of two populations as μ_1 and μ_2, the null hypothesis for a test on two independent means would be

$$H_0: \text{the means of two populations are equal} \qquad (5\text{-}56a)$$

Mathematically, this is

$$H_0: \mu_1 = \mu_2 \qquad (5\text{-}56b)$$

Both one-sided and two-sided alternatives can be used:

$$H_{A1}: \mu_1 < \mu_2 \qquad (5\text{-}57a)$$

$$H_{A2}: \mu_1 > \mu_2 \qquad (5\text{-}57b)$$

$$H_{A3}: \mu_1 \neq \mu_2 \qquad (5\text{-}57c)$$

The selection of the alternative hypotheses should depend on the statement of the problem.

Step 2: Select the appropriate model — For the case of two independent samples, the hypotheses of step 1 can be tested using the following test statistic:

$$t = \frac{\overline{X}_1 - \overline{X}_2}{S_p \left(\dfrac{1}{n_1} + \dfrac{1}{n_2} \right)^{0.5}} \qquad (5\text{-}58)$$

in which $\overline{X}_1$ and $\overline{X}_2$ are the means of the samples from populations 1 and 2, respectively; n_1 and n_2 are the samples sizes from populations 1 and 2, respectively; t is the value of a random variable having a t distribution with degrees of freedom of $(\nu = n_1 + n_2 - 2)$; and S_p is the square root of the pooled variance given by

$$S_p^2 = \frac{(n_1 - 1)S_1^2 + (n_2 - 1)S_2^2}{n_1 + n_2 - 2} \qquad (5\text{-}59)$$

in which S_1^2 and S_2^2 are the variances of the samples from population 1 and 2, respectively. This test statistic assumes that the variances of the two populations are equal, but unknown.

Step 3: Select the level of significance — As usual, the level of significance should be selected on the basis of the problem. However, values of either 5% or 1% are used most frequently.

Step 4: Compute an estimate of test statistic — Samples are drawn from the two populations, and the sample means and variances computed. Equation 5-58 can be computed to test the null hypothesis of Equation 5-56.

Step 5: Define the region of rejection — The region of rejection is a function of the degrees of freedom ($v = n_1 + n_2 - 2$), the level of significance (α), and the statement of the alternative hypothesis. The regions of rejection for the alternative hypotheses are as follows:

$$
\begin{array}{ll}
\text{if } H_A \text{ is} & \text{then reject } H_o \text{ if} \\[6pt]
\mu_1 < \mu_2 & t < -t_\alpha \\[6pt]
\mu_1 > \mu_2 & t > t_\alpha \\[6pt]
\mu_1 \neq \mu_2 & t < -t_{\alpha/2} \text{ or } t > t_{\alpha/2}
\end{array}
\tag{5-60}
$$

Step 6: Select the appropriate hypothesis — The sample estimate of the t statistic from step 4 can be compared with the critical value, which is based on either t_α or $t_{\alpha/2}$ obtained from step 5. If the sample value lies in the region of rejection, then the null hypothesis should be rejected.

Example 5-9: Effect of Development on Total Nitrogen

A study was made to measure the effect of suburban development on total nitrogen levels in small streams. A decision was made to use the mean concentrations before and after the development as the criterion. Eleven measurements of the total nitrogen (mg/l) were taken prior to the development with a mean of 0.78 mg/l and a standard deviation of 0.36 mg/l. Fourteen measurements were taken after the development with a mean of 1.37 mg/l and a standard deviation of 0.87 mg/l. The data are assumed to test the null hypothesis that the population means are equal against the alternative hypothesis that the urban development increased total nitrogen levels; this requires the following one-tailed test:

$$
H_o : \mu_b = \mu_a \tag{5-61a}
$$

$$
H_A : \mu_b < \mu_a \tag{5-61b}
$$

where μ_b and μ_a are the before and after development means, respectively. Rejection of H_o would show that the nitrogen levels after development exceed the nitrogen levels before development, with the implication that the development might have caused the increase.

Based on the sample data, the pooled variance of Equation 5-59 is

$$S_p^2 = \frac{(11-1)(0.36)^2 + (14-1)(0.87)^2}{11+14-2} = 0.4842 \qquad (5\text{-}62)$$

The computed value of the test statistic is

$$t = \frac{0.78 - 1.37}{\sqrt{0.4842\left(\dfrac{1}{11} + \dfrac{1}{14}\right)}} = -2.104 \qquad (5\text{-}63)$$

which has $v = 11 + 14 - 2 = 23$ degrees of freedom. From Table A-2, with a 5% level of significance and 23 degrees of freedom, the critical value of t is -1.714. Thus, the null hypothesis is rejected, with the implication that the development caused a significant change in nitrogen level. However, for a 1% significance level, the critical value is -2.500, which leads to the decision that the increase is not significant. This shows the importance of selecting the level of significance before analyzing the data.

5.5.4. Summary

Three hypothesis tests were introduced. Each test can be conducted using the six steps that were identified in Section 5.4. In applying a hypothesis test, the important ingredients are the test statistic, the level of significance, the degrees of freedom, and the critical value of a test statistic. Table 5-5 includes a convenient summary of the three statistical tests introduced in this section and other important tests.

5.6. HYPOTHESIS TESTS OF VARIANCES

The variance of a random sample is a measure of the dispersion of the observations about the sample mean. Although the variance is used to indicate the degree of variation about the mean, it is an important statistic in its own right. Large variation in engineering systems reflects instability or nonuniformity, both of which can be considered nonoptimal.

5.6.1. One-Sample Chi-Square Test

Consider, for example, the case of water distribution systems used for irrigation. They should be designed to distribute water uniformly over an area, such as a lawn or an agricultural field. Failure to provide a uniform application of water over the

TABLE 5-5 Summary of Hypothesis Tests

H_o	Test statistic	H_A	Region of rejection
$\mu = \mu_o$ (σ known)	$Z = \dfrac{\overline{X} - \mu}{\sigma / \sqrt{n}}$	$\mu < \mu_o$ $\mu > \mu_o$ $\mu \neq \mu_o$	$z < -z_\alpha$ $z > z_\alpha$ $z < -z_{\alpha/2}$ or $z > z_{\alpha/2}$
$\mu = \mu_o$ (σ unknown)	$t = \dfrac{\overline{X} - \mu}{S / \sqrt{n}}$ $v = n - 1$	$\mu < \mu_o$ $\mu > \mu_o$ $\mu \neq \mu_o$	$t < -t_\alpha$ $t > t_\alpha$ $t < -t_{\alpha/2}$ or $t > t_{\alpha/2}$
$\mu_1 = \mu_2$ ($\sigma_1^2 = \sigma_2^2$, but unknown)	$t = \dfrac{\overline{X}_1 - \overline{X}_2}{S_p \left(\dfrac{1}{n_1} + \dfrac{1}{n_2} \right)^{0.5}}$ $v = n_1 + n_2 - 2$ $S_p^2 = \dfrac{(n_1 - 1)S_1^2 + (n_2 - 1)S_2^2}{n_1 + n_2 - 2}$	$\mu < \mu_o$ $\mu > \mu_o$ $\mu \neq \mu_o$	$t < -t_\alpha$ $t > t_\alpha$ $t < -t_{\alpha/2}$ or $t > t_{\alpha/2}$
$\sigma^2 = \sigma_o^2$	$\chi^2 = \dfrac{(n-1)S^2}{\sigma_o^2}$ $v = n - 1$	$\sigma^2 < \sigma_o^2$ $\sigma^2 > \sigma_o^2$ $\sigma^2 \neq \sigma_o^2$	$\chi^2 < \chi^2_{\alpha - 1}$ $\chi^2 < \chi^2_\alpha$ $\chi^2 < \chi^2_{1-\alpha/2}$ or $\chi^2 < \chi^2_{\alpha/2}$
$\sigma_1^2 = \sigma_2^2$ (assuming $\sigma_1^2 > \sigma_2^2$)	$F = \dfrac{S_1^2}{S_2^2}$ $v_1 = n_1 - 1$ $v_2 = n_2 - 1$	$\sigma_1^2 \neq \sigma_2^2$	$F > F_{\alpha/2}$
Goodness of fit $X \sim f_X(x,p)$	$\chi^2 = \displaystyle\sum_{i=1}^{k} \dfrac{(O_i - E_i)^2}{E_i}$ $v = k - 1 - $ (*number of parameters estimated using sample*)	$X \not\sim f_X(x,p)$	$\chi^2 > \chi^2_\alpha$

area may lead to nonoptimum grass or crop output; thus equipment that does not apply water uniformly would probably not be purchased. A company that manufactures irrigation distribution systems wishes to check whether or not a new system increases the uniformity of water application in comparison with existing models. The variance of depths of water measured at different locations in a field would serve as a measure of uniformity of water application. The following procedure is used to test for a statistical difference in the uniformity of application rates, i.e., a test on the variance of a random variable.

Step 1: Formulate hypotheses — To investigate the possibility of a significant difference existing between the variance of a population σ^2 and the preselected standard variance value (σ_o^2), the following null hypothesis can be used:

H_o: there is no significant difference between the variance of the population

σ^2 and the preselected standard value for the variance (σ_o^2) (5-64a)

Mathematically, the null hypothesis is

$$H_o: \sigma^2 = \sigma_o^2 \tag{5-64b}$$

The null hypothesis can be tested against either a one-tailed or two-tailed alternative hypothesis as follows:

$$H_{A1}: \sigma^2 < \sigma_o^2 \tag{5-65a}$$

$$H_{A2}: \sigma^2 > \sigma_o^2 \tag{5-65b}$$

$$H_{A3}: \sigma^2 \neq \sigma_o^2 \tag{5-65c}$$

For the irrigation distribution system case, assume that the standard deviation of an existing system is 0.1 cm/hr. The following hypotheses are examined by a manufacturing company wishing to determine if their new system provides a more uniform distribution of water:

$$H_o: \sigma^2 = (0.1 \text{ cm/hr})^2 \tag{5-66a}$$

$$H_A: \sigma^2 < (0.1 \text{ cm/hr})^2 \tag{5-66b}$$

Step 2: Select the appropriate model — The variance S^2 of a random sample is a random variable itself and is used in testing the hypotheses about the variance of a population, σ^2. The sampling distribution of the estimated variance of a random sample that is drawn from a normal population has a chi-square distribution. The test statistic for testing the hypotheses is

$$\chi^2 = \frac{(n-1)S^2}{\sigma_o^2} \tag{5-67}$$

in which χ^2 is the value of a random variable that has a chi-square distribution with $v = n - 1$ degrees of freedom, and n is the sample size used in computing the sample variance S^2.

Step 3: Select the level of significance — For the case study, a level of significance of 2.5% is selected.

Step 4: Compute estimate of test statistic — To test the uniformity of application of water for the new irrigation system, the amount of water in each of 25 randomly placed recording devices was observed after 1 h. The mean and standard deviation of the random sample were 0.31 and 0.063 cm/hr, respectively. The computed test statistic is

$$\chi^2 = \frac{(25-1)(0.063)^2}{(0.1)^2} = 9.526 \tag{5-68}$$

Step 5: Define the region of rejection — The region of rejection for a test statistic having a chi-square distribution is a function of the level of significance, the statement of the alternative hypothesis, and the degrees of freedom. The regions of rejection for the alternative hypotheses are as follows:

if H_A is	then reject H_o if	
$H_{A1}: \sigma^2 < \sigma_o^2$	$\chi^2 < \chi_{\alpha-1}^2$	
$H_{A2}: \sigma^2 > \sigma_o^2$	$\chi^2 > \chi_\alpha^2$	(5-69)
$H_{A3}: \sigma^2 \neq \sigma_o^2$	$\chi^2 < \chi_{1-\alpha/2}^2$ or $\chi^2 > \chi_{\alpha/2}^2$	

For values of $\alpha = 0.025$ and $v = k = 24$, the critical value was obtained from Table A-3 to be 12.401. Thus the region of rejection consists of all values of the test statistic χ^2 that are less than 12.401.

Step 6: Select the appropriate hypothesis — The computed value of the test statistic is less than the critical value, and thus the null hypothesis must be rejected. This implies that the new irrigation equipment applies water with more uniformity than existing irrigation systems. It is unlikely that the difference between the sample variance (0.004) and the variance of existing irrigation equipment (0.01) is due solely to chance variation within the sample; it is probably due to a real difference in the equipment.

Rationale of test statistic — If the sample variance was exactly equal to the population variance specified in the null hypothesis (i.e., $S^2 = \sigma_o^2$), then the value of the test statistic χ^2 will equal $(n-1)$. For the two-tailed alternative a significant difference exists if the computed value of χ^2 is either significantly larger or smaller than $(n-1)$. For the one-tailed alternative, a significant difference exists only when the computed value of χ^2 is much less than $(n-1)$; in the case of the irrigation system with $(n-1) = 24$, the computed value of 9.526 is considered to be significant because it is less than the critical value. For the upper one-tailed alternative, a significant difference exists only when the computed value of χ^2 is much greater than $(n-1)$.

5.6.2. Two-Sample F Test

For comparing the variances of two random samples, several strategies have been recommended, with each strategy valid when the underlying assumptions hold. One of these strategies is presented here.

For a two-tailed test, an F-ratio is formed as the ratio of the larger sample variance to the smaller sample variance as follows:

$$F = \frac{S_1^2}{S_2^2} \qquad (5\text{-}70)$$

with degrees of freedom of $v_1 = n_1 - 1$ for the numerator and $v_2 = n_2 - 1$ for the denominator, where n_1 and n_2 are the sample sizes for the samples used to compute S_1^2 and S_2^2, respectively. The computed F is compared with the tabulated values for the F probability distribution provided in Table A-4, and the null hypothesis of equal variances (H_o: $\sigma_1^2 = \sigma_2^2$) is accepted if the computed F is less than the tabulated F value for $k = v_1$, $u = v_2$, and α. If the computed F is greater than the tabulated F value, then the null hypothesis is rejected in favor of the alternative hypothesis (H_A: $\sigma_1^2 \neq \sigma_2^2$). An important note for this two-tailed test is the level of significance is twice the value from which the tabulated F value was obtained; for example, if the 5% F table is used to obtain the critical F statistic, then the decision to accept or reject the null hypothesis is being made at a 10% level of significance. This is the price paid for using the sample knowledge that one sample has the larger variance.

For a one-tailed test, it is necessary to specify which of the two samples is expected to have the larger population variance. This must be specified prior to collecting the data. The computed F statistic is the ratio of the sample variance of the group expected to have the larger population variance to the sample variance from the second group. If it turns out that the sample variance of the group expected to have the larger variance is smaller than that of the group expected to have the smaller variance, then the computed F statistic will be less than 1. For a test with the level of significance equal to that shown on the table, the null hypothesis is rejected if the computed F is greater than the critical F. Because the direction is specified, the null hypothesis is accepted when the computed F is less than the critical F; the null hypothesis is rejected when the computed F is greater than the critical F.

5.6.3. Summary

Two hypothesis tests for the variance were introduced. Table 5-5 includes a summary of these tests.

5.7. CONFIDENCE INTERVALS

From a sample we obtain single-valued estimates such as the mean, variance, a correlation coefficient, or a regression coefficient. These single-valued estimates represent our best estimate of the population values. But they are only estimates of random variables, and we know that they probably do not equal the corresponding true values. Thus we should be interested in the accuracy of these sample estimates.

If we are only interested in whether or not an estimate of a random variable is significantly different from a standard of comparison, we can use a hypothesis test. However, the hypothesis test only gives us a "yes" or "no" answer, and not a statement of the accuracy of an estimate of a random variable, which may be the object of our attention.

In Example 5-8, a water quality standard of 3 ppm was introduced for illustration purposes. The hypothesis test showed that the sample mean of 2.8 ppm was not significantly different from 3 ppm. The question arises: Just what is the true mean? Although the best estimate (i.e., expected value) is 2.8 ppm, values of 2.75 or 3.25 ppm could not be ruled out. Is the true value between 2 and 4 ppm, or is it within the range 2.75 to 3.25 ppm? The smaller range would suggest that we are more sure of the population value; that is, the smaller range indicates a higher level of accuracy. The higher level of accuracy makes for better decision making; this is the reason for examining confidence intervals as a statistical tool.

Confidence intervals represent a means of providing a range of values in which the true value can be expected to lie. Confidence intervals have the additional advantage, compared with hypothesis tests, of providing a probability statement about the likelihood of correctness.

5.7.1. General Form of a Confidence Interval

The theoretical basis of a confidence interval is the same as that for the theorem for the corresponding hypothesis test. That is, the theorem used in making a statistical hypothesis test is also used to derive the confidence interval. Just as hypothesis tests can be one- or two-sided, confidence intervals can be computed as either one- or two-sided.

Many two-sided confidence intervals have the following general form:

$$K_{est} \pm F_d D \qquad (5\text{-}71)$$

in which K_{est} is the estimated value of the statistic K, F_d is a distribution factor, and D is a measure of the dispersion of the sampling distribution of K_{est}. For one-sided confidence intervals the $\pm$ is replaced by either $+$ or $-$, depending on the intent, i.e., upper or lower interval, respectively.

5.7.2. General Procedure

The following general procedure can be used to compute a confidence interval:

1. Specify whether a one-sided or two-sided confidence interval is of interest.
2. Identify the theorem that specifies the sampling distribution and distribution factor F_d of the statistic of interest.
3. State the desired level of confidence, $\gamma = 1 - \alpha$, where α is the level of significance used in hypothesis tests.
4. Collect the sample and compute the necessary statistics (e.g., K_{est} and D of Equation 5-71).
5. Determine the value of the distribution factor F based on the confidence level and possibly the sample size.
6. Compute the confidence interval.

The six steps of this procedure correspond directly to the six steps of the hypothesis test. It is important to notice that the theoretical model is selected prior to obtaining the sample. This corresponds to making the statement of the hypothesis prior to obtaining the data to test the hypothesis.

5.7.3. Confidence Interval for the Mean

The same theorems that were used for testing hypotheses on the mean are used in computing confidence intervals. In testing a hypothesis for the mean, the choice of test statistic depends on whether or not the standard deviation of the population, σ, is known; that is also true in computing confidence intervals. The theorem for the case where σ is known specifies a Z statistic, whereas the t statistic is used when σ is unknown; the theorems are not repeated here.

For the case where σ is known, the theorem is given in Section 5.5.1. Confidence intervals on the population mean are given by

$$\overline{X} - Z_{\alpha/2}\left(\frac{\sigma}{\sqrt{n}}\right) \leq \mu \leq \overline{X} + Z_{\alpha/2}\left(\frac{\sigma}{\sqrt{n}}\right) \quad \text{Two-sided interval} \quad (5\text{-}72)$$

$$\overline{X} - Z_{\alpha}\left(\frac{\sigma}{\sqrt{n}}\right) \leq \mu \leq \infty \quad \text{Lower one-sided interval} \quad (5\text{-}73)$$

$$-\infty \leq \mu \leq \overline{X} + Z_{\alpha}\left(\frac{\sigma}{\sqrt{n}}\right) \quad \text{Upper one-sided interval} \quad (5\text{-}74)$$

in which $\overline{X}$ is the sample mean; n is the sample size; Z_{α} and $Z_{\alpha/2}$ are values of random variables having the standard normal distribution and cutting off $(1 - \gamma)$ or $(1 - \gamma/2)$

percent in the tail of the distribution, respectively; and α is the level of significance and equals $1 - \gamma$. The value of $\overline{X}$ corresponds to K_{est} of Equation 5-71, while Z_α and $Z_{\alpha/2}$ correspond to the distribution factor F_d. The measure of dispersion is given by $\sigma/\sqrt{n}$ since $\sigma/\sqrt{n}$ is the standard error of the mean. Equation 5-72 is a two-sided confidence interval, while Equations 5-73 and 5-74 are one-sided. Equation 5-73 gives a lower confidence limit, with no limit on the upper side of the mean; similarly, Equation 5-74 gives an upper limit, with no lower limit.

For the case where σ is unknown, the theorem is given in Section 5.5.2. Confidence intervals on the population mean are given by

$$\overline{X} - t_{\alpha/2}\left(\frac{S}{\sqrt{n}}\right) \le \mu \le \overline{X} + t_{\alpha/2}\left(\frac{S}{\sqrt{n}}\right) \quad \text{Two-sided interval} \qquad (5\text{-}75)$$

$$\overline{X} - t_\alpha\left(\frac{S}{\sqrt{n}}\right) \le \mu \le \infty \qquad\qquad \text{Lower one-sided interval} \qquad (5\text{-}76)$$

$$-\infty \le \mu \le \overline{X} + t_\alpha\left(\frac{S}{\sqrt{n}}\right) \qquad\qquad \text{Upper one-sided interval} \qquad (5\text{-}77)$$

in which S is the sample standard deviation, and t_α and $t_{\alpha/2}$ are values of random variables having a t distribution with $v = n - 1$. The significance level (α) is used for one-sided confidence interval, and $\alpha/2$ is used for a two-sided confidence interval. The value of $\overline{X}$ corresponds to K_{est} of Equation 5-71, while t_α and $t_{\alpha/2}$ are the distribution factors (F_d). Since $S/\sqrt{n}$ is the sample estimate of the standard error of the mean, it is the measure of dispersion (D). Equation 5-75 gives a two-sided confidence interval on the mean, while Equations 5-76 and 5-77 give one-sided lower and upper confidence intervals, respectively. For sample sizes greater than 30, Equations 5-72, 5-73, and 5-74 can be used in place of Equations 5-75, 5-76, and 5-77, respectively.

Example 5-10: Daily Evaporation Measurements

Using a sample of 354 daily evaporation measurements, the mean evaporation was computed to be 0.1387 in./day. The long-term mean and standard deviation for that location were 0.1575 and 0.1 in./day, respectively. The test of hypothesis in Example 5-7 indicated that the sample mean was significantly different from the long-term mean value. The confidence interval for would give a better indication of the difference between $\overline{X}$ and the long-term average. The confidence interval can be computed using Equation 5-72 because the sample size is greater than 30. The two-sided, 95% confidence interval is given by

$$0.1387 \pm \frac{1.96(0.1)}{\sqrt{354}} \tag{5-78}$$

or

$$0.1283 \le \mu \le 0.1491 \text{ in./day} \tag{5-79}$$

Because the upper limit of the 95% confidence interval is not near the long-term mean, one must conclude that the sample is definitely biased; that is, the sample of evaporation data is not representative of the long-term evaporation characteristics at the site.

Example 5-11: Water Quality

Example 5-8 showed that the mean water quality parameter of 2.8 ppm was not significantly different from the standard of 3 ppm. The accuracy of the mean value is usually more important than a simple "yes-no" decision; therefore, a confidence interval on the mean would be useful in decision making. Since the standard deviation of the population is not known, the confidence interval can be computed using Equation 5-76; the one-sided lower interval is selected because the upper limit is of no physical importance. For a 95% confidence (i.e., $\alpha = 0.05$), the distribution factor was obtained from Table A-2 to be -1.833 for 9 degrees of freedom. The confidence interval is

$$2.80 - 1.833(0.4)/\sqrt{10} \le \mu \le \infty \tag{5-80a}$$

$$2.568 \text{ ppm} \le \mu \le \infty \tag{5-80b}$$

Thus, we are 95% confident that the true mean is at least 2.568 ppm. Note that if the sample statistics (i.e., $\overline{X}$ and S) were based on a sample size of 20 rather than 10, the lower limit would be

$$2.80 - 1.729(0.4)/\sqrt{20} \le \mu \le \infty \tag{5-81a}$$

$$2.645 \text{ ppm} \le \mu \le \infty \tag{5-81b}$$

The higher value of the lower limit reflects the increased confidence that results from the larger sample size; as n is increased, both the standard error of the mean $S/\sqrt{n}$ and the distribution factor decrease.

5.7.4. Factors Affecting a Confidence Interval

It is of interest to notice that a confidence interval is a function of the level of confidence, the theoretical sampling distribution, and the sample characteristic. The level of confidence, γ, is a user-supplied measure of the desired accuracy. It equals 1 minus the level of significance (i.e., $\gamma = 1 - \alpha$). For a given set of sample statistics, the width of the confidence interval increases as the confidence level is increased. The increased width is the price that must be paid for the increased level of confidence (i.e., increased accuracy).

The size of each sample is an important factor that affects the width of the confidence interval. It was shown that the standard error of the mean is inversely proportional to the square root of the sample size. Because D of Equation 5-71 is a measure of the sampling variation of the statistic, one would expect the width of the confidence interval to decrease as the sample size is increased.

5.7.5. Confidence Interval for the Variance

The confidence interval on the population variance (σ^2) can be computed using the same theorem that was used in testing a hypothesis for the variance (see Section 5.6.1). The two-sided and one-sided confidence intervals are

$$\frac{(n-1)S^2}{\chi^2_{\alpha/2}} \le \sigma^2 \le \frac{(n-1)S^2}{\chi^2_{1-\alpha/2}} \qquad \text{Two-sided interval} \qquad (5\text{-}82)$$

$$\frac{(n-1)S^2}{\chi^2_{\alpha}} \le \sigma^2 \le \infty \qquad \text{Lower one-sided interval} \qquad (5\text{-}83)$$

$$0 \le \sigma^2 \le \frac{(n-1)S^2}{\chi^2_{1-\alpha}} \qquad \text{Upper one-sided interval} \qquad (5\text{-}84)$$

in which $\chi^2_{\alpha/2}$ and χ^2_{α} are values of a random variable having a chi-square distribution that cuts $\alpha/2$ and α percent of right tail of the distribution, respectively; and similarly, $\chi^2_{1-\alpha/2}$ and $\chi^2_{1-\alpha}$ are values of a random variable having a chi-square distribution with cuts at $1 - \alpha/2$ and $1 - \alpha$, respectively. The confidence interval provides an interval in which we are $100(1 - \alpha)$ percent confident that the population value lies within the interval. Although Equations 5-82, 5-83, and 5-84 do not have the same form as Equation 5-71, S^2 is the estimated value of the statistic and the χ^2 value represents the distribution factor. Although the general form is not the same, the six-step procedure for confidence intervals is directly applicable.

Example 5-12: Uniformity of Irrigation

In Section 5.6.1, the example centered on the variance of water depths resulting from irrigation. Twenty-five measurements produced a variance of 0.003969 $(\text{cm/hr})^2$, or $S = 0.063$ cm/hr. In this case, the greater the variance, the poorer the equipment, so an upper limit would be of interest. Thus, an upper confidence limit is computed using Equation 5-84. For a 95% level of confidence ($\alpha = 0.05$) and 24 degrees of freedom, the critical χ^2 value is 13.848. Thus, the upper limit on the interval is

$$\frac{(n-1)S^2}{\chi^2_{1-\alpha}} = \frac{24(0.003969)}{13.848} = 0.006879 \qquad (5\text{-}85a)$$

The confidence interval on the standard deviation is

$$0 \leq \sigma \leq 0.0829 \text{ cm/hr} \qquad (5\text{-}85b)$$

Since the standard deviation of 0.01 cm/hr in Section 5.6.1 falls outside this limit, the null hypothesis of equality was rejected.

It is important to note that $\chi^2_{1-\alpha}$, which equals 13.848, is used for computing the upper limit. If χ^2_{α} were used, which equals 36.415, then the upper limit would actually be smaller than the sample value; thus, using χ^2_{α} would be incorrect.

5.8. SAMPLE-SIZE DETERMINATION

The selection of a sample size is a first step in performing statistical analysis. The selection of the sample size should be based on a clearly stated objective. For example, if we are interested in estimating the mean value of concrete strength that is received at a construction site, then the sample size should be determined based on the objective of estimating the mean concrete strength with specified confidence level (α) and desired width for the corresponding confidence interval. A different sample size would be required if our objective was the development of a regression model with concrete strength as one of the variables. The development of methods for determining the sample size is based on the theorems and methods presented for hypothesis testing and confidence intervals.

5.8.1. Sample Size for Estimating a Mean Value with Known Population Variance

The methods for determining the sample size for the purpose of estimating the mean value depend on whether or not the variance of the population is known or unknown. If the population variance is known, Equation 5-72 can be used to solve for half the width of the confidence interval (H) as

$$H = Z_{\alpha/2} \frac{\sigma}{\sqrt{n}} \qquad (5\text{-}86)$$

Equation 5-86 can be written as

$$\frac{H}{\sigma} = \frac{Z_{\alpha/2}}{\sqrt{n}} \tag{5-87}$$

The right-hand size of Equation 5-87 can be viewed as the half-width of the desired confidence interval for the parameter of interest expressed in units of standard deviations. This quantity is called the normalized half-width H_n, i.e.,

$$H_n = \frac{H}{\sigma} \tag{5-88}$$

Values for H_n in the range of 0.01 to 1 are commonly used. This range will result in half-width for the desired confidence interval of 0.01 to 1 standard deviation, which is sufficient for most engineering applications. Therefore, a decision needs to be made about the value of H_n; this should be based on the importance of the decision and practical aspects of the engineering problem under consideration. The smaller the value of H_n, the larger the sample size, which would reflect our interest in a greater certainty level in the mean and, thus, a smaller confidence interval. Substituting Equation 5-88 into Equation 5-87 and solving for the sample size n produces

$$n = \left(\frac{Z_{\alpha/2}}{H_n} \right)^2 \tag{5-89a}$$

or

$$n = \left(\frac{Z_{\alpha/2}}{\dfrac{H}{\sigma}} \right)^2 = \left(\frac{Z_{\alpha/2}\, \sigma}{H} \right)^2 \tag{5-89b}$$

Example 5-13: Sample Size for Estimating Daily Evaporation

In Examples 5-7 and 5-10, daily evaporation measurements were discussed. The information in these examples is used to illustrate the determination of the sample size in the case of known population variance. Assuming that we are interested in a normalized half-width of 0.1 and α of 5%, the sample size using Equation 5-89a is

$$n = \left(\frac{1.96}{0.1} \right)^2 = 384.2 \tag{5-90}$$

FIGURE 5-6 Sample size with known variance.

Therefore, a sample size of 385 measurements is selected. Figure 5-6 shows the sample size as a function of the normalized half-width of the confidence interval for two α values of 5 and 1%.

5.8.2. Sample Size for Estimating a Mean Value with Unknown Population Variance

If the population variance is unknown, Equation 5-75 can be used to solve for the half-width of the confidence interval (H) as

$$H = t_{\alpha/2} \frac{S}{\sqrt{n}} \tag{5-91}$$

Equation 5-91 can be written as

$$\frac{H}{S} = \frac{t_{\alpha/2}}{\sqrt{n}} \tag{5-92}$$

The normalized half-width H_n is

$$H_n = \frac{H}{S} \tag{5-93}$$

Therefore, the sample size n is

$$n = \left(\frac{t_{\alpha/2}}{H_n}\right)^2 \tag{5-94a}$$

or

$$n = \left(\frac{t_{\alpha/2}}{\frac{H}{S}}\right)^2 = \left(\frac{t_{\alpha/2}S}{H}\right)^2 \tag{5-94b}$$

In Examples 5-8 and 5-11, water quality data were used to illustrate one-sided hypothesis testing and confidence interval estimation. Assuming that we are interested in a normalized half-width of 0.5 and α of 5% for one-sided interval, the sample size using Equation 5-94a can be utilized with t_α instead of $t_{\alpha/2}$. In order to look up a value for t_α, a sample size is needed. Therefore, the solution procedure is iterative starting by assuming a sample size, then looking up t_α from Table A-2, then computing the sample size (using Equation 5-94b), and then repeating the process.

Example 5-14: Sample Size for Water Quality

This example uses the same problem presented in Examples 5-8 and 5-11 in which water quality data were used to illustrate one-sided hypothesis testing and confidence interval estimation. For a normalized half-width of 0.5 and α of 5% for one-sided interval, a sample size is assumed to be 9; therefore, from Table A-2, $t_\alpha = 1.833$. The resulting sample size is

$$n = \left(\frac{1.833}{0.5}\right)^2 = 13.4 \tag{5-95a}$$

Therefore, a sample size of 14 measurements is obtained. A revised estimate of t_α is obtained from Table A-2 to be 1.771. Therefore, the revised sample size is

$$n = \left(\frac{1.771}{0.5}\right)^2 = 12.55 \tag{5-95b}$$

Therefore, a sample size of 13 can be used.

5.9. SELECTION OF MODEL PROBABILITY DISTRIBUTIONS

In engineering analysis, univariate statistical methods, which are techniques that use a single random variable, are commonly employed. Even the two-sample tests for the mean and variance are univariate methods since there is just one random variable. In addition to making tests of hypotheses on means and variances and computing confidence intervals, it is often necessary to (1) select a probability function based on sample data, (2) assess the validity of an assumption that underlies a statistical method, or (3) make predictions of either a probability for a selected value of the random variable or the value of the random variable for a selected probability. Sample data can be used to identify a probability function that can be a valid representation of the population, which can then be used to make the predictions. Two of the many cases where it is necessary to test the validity of an assumption are (1) the t test on a mean assumes the data are from a normal population; and (2) least-squares regression analysis assumes that the residuals are normally distributed (see Chapter 6).

The purpose of this section is to introduce a method that can be used to select a probability distribution function to represent data. Once the probability function has been selected, then it can be used to make predictions. The frequency analysis method is introduced for use with the normal and lognormal distributions. To statistically validate whether or not the sample data could realistically be sampled from an assumed population, the chi-square goodness-of-fit test and the Kolmogorov-Smirnov one-sample tests are introduced in this section.

5.9.1. Frequency Analysis Using the Normal Distribution

The input for a frequency analysis consists of a sample of data and either probability paper or a mathematical model for the probability distribution. While many probability papers are available, only normal probability paper is used here. A normal probability paper uses the ordinate for the value of the random variable and the abscissa for the cumulative probability; Figure 5-7 is an example of the probability paper. Lognormal probability papers, which are also available, use a logarithmic scale for the ordinate.

In a broad sense, frequency analysis can be divided into two parts, deriving the population and comparing the sample data with the assumed population as a means of validating the population. The procedure for fitting the normal population on the probability paper is as follows:

1. Compute the mean $(\overline{X})$ standard deviation (S) of the sample.
2. For the graphical approach, compute $\overline{X} + S$ and $\overline{X} - S$.
3. Plot the following two points: $\overline{X} + S$ vs. a cumulative probability of 84.1% and $\overline{X} - S$ vs. a cumulative probability of 15.9% (these probabilities correspond to the cumulative values of the standard normal curve for Z equal to 1 and -1, respectively). The value of $\overline{X} + S$ and $\overline{X} - S$ should be plotted on the ordinate, and the cumulative probability values should be plotted on the abscissa in Figure 5-7.
4. Connect a straight line between the two points plotted in step 3; the straight line is the assumed normal population.

The computed frequency curve, which represents the population, can then be used to make probability statements about the likelihood of occurrence of values of the random variable.

5.9.2. Construction of Probability Paper

It is important to recognize that it is not necessary to plot the data points in order to make probability statements about the random variable. While the four steps given above lead to an estimate of the population frequency curve, the data

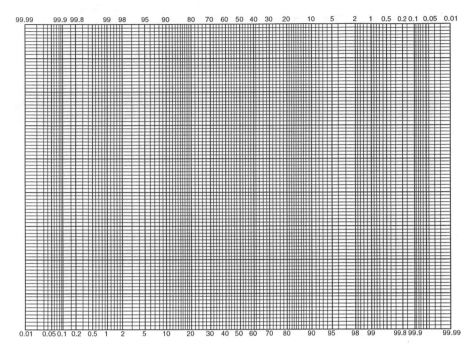

FIGURE 5-7 Normal-distribution probability paper.

should be plotted to ensure that the population curve is a good representation of the data. The plotting of the data is, therefore, a somewhat separate part of a frequency analysis; it is for assessing the quality of the fit rather than part of the estimation process.

Probability papers can be constructed for other probability distributions. The procedure for constructing a probability paper for a probability distribution $f_X(x)$ is given by

1. Using a regular graphing paper with arithmetic scales on both the ordinate and abscissa, draw a scale for the X random variable on both axes.
2. Convert the abscissa to an nonarithmetic scale by placing cumulative values $F_X(x)$ at the locations of the corresponding values of x.
3. The resulting paper will have an arithmetic scale on the ordinate for the random variable, and a nonarithmetic abscissa for the cumulative probability. The paper is designed for use with the density function $f_X(x)$.

A cumulative distribution of a population plotted on paper designed for that distribution will always result in a straight line. Other populations that do not have the same distribution results in lines with curvature. For some probability distributions, it might be convenient to transform the ordinate to nonarithmetic scale, such as logarithmic scale, to obtain the desired probability papers, such as a lognormal paper.

5.9.3. Plotting the Sample Data

Plotting the sample data on a probability paper with the assumed population of the paper is used to verify if the sample is likely to have come from the assumed population. A rank-order method is used to plot the data. This analysis involves ordering the data from the largest event to the smallest and using the rank (i) of the event to obtain a probability plotting position; numerous plotting position formulas are available. The most commonly used formula is the Weibull formula given by

$$p_i = \frac{i}{n+1} \qquad\qquad (5\text{-}96\text{a})$$

where p_i = exceedence probability for the ith value. However, the Weibull formula can give biased positions for the probability distributions except for data from a uniform distribution. The following Hazen formula gives reliable positions for the probability distributions and is also used in engineering:

$$p_i = \frac{i - 0.5}{n} \qquad\qquad (5\text{-}96\text{b})$$

The resulting p_i is used as the cumulative measure for the ith sample value and is called the expected exceedence probability. The exceedence probability is one minus the cumulative probability. The sampled values and their corresponding p_i values should be plotted on a probability paper.

In practice, the measured values should then be plotted to decide whether or not the sample is representative of the population. If the data provide a reasonable fit to the line, one can assume that the underlying population is the assumed distribution and the parameters derived from the sample are reasonable estimates of the population parameters. A poor fit would indicate that either the assumed distribution is not the population or that the sample statistics are poor estimators of the population parameters, or both.

5.9.4. Prediction

The graph of the population curve can be used to make estimates of either a value of the random variable for a given probability or an estimate of the probability for a given value of the random variable. Graphical estimation is made by entering with the value of either the random variable (X) or the cumulative value (p), and finding the value of the other at the point where the population line is intersected.

While the frequency curve is useful for obtaining approximate values, design predictions should be made using a mathematical formula suitable for the assumed population distribution. For a normal distribution, the predicted value can be made with the standard normal transformation as

$$x = \overline{X} + zS \qquad (5\text{-}97a)$$

where $\overline{X}$ and S are the sample mean and standard deviation, respectively, and z is the standard normal deviate for a specified probability. If the magnitude x is known and the exceedence probability that corresponds to the magnitude is wanted, then Equation 5-97a can be used as follows:

$$z = \frac{x - \overline{X}}{S} \qquad (5\text{-}97b)$$

Using the computed value of z, the exceedence probability can be obtained from Table A-1 for the standard normal distribution.

5.9.5. Return Period

When using a frequency curve, it is common to discuss the likelihood of events in terms of either the exceedence probability, the cumulative probability, or the return period. The exceedence probability is defined as one minus the cumulative probability, and it is the probability of the random variable exceeding the corresponding magnitude in one period of time. The return period (T) is related to the exceedence probability (p) by $p = 1/T$, or $T = 1/p$. Thus an event with an exceedence probability of 0.01 should be expected to occur 1 time in 100. In many cases, a time unit is attached to the return period. For example, if the data represent annual floods at a location, the basic time unit is 1 year. In such a case, the return period for an event with an exceedence probability of 0.01 would be the 100-year event (i.e., $T = 1/0.01 = 100$); similarly, the 25-year event has an exceedence probability of 0.04 (i.e., $p = 1/25 = 0.04$). It is important to emphasize that two T-year events will not necessarily occur exactly T years apart. Two T-year events could occur in successive years; then, again, two T-year events may be spaced three times T years apart. But on the average the events will be spaced T years apart. Thus in a long period of time, say 10,000 years, we would expect $10,000/T$ such events to occur. In any one 10,000-year period, we may have more or less occurrences than the mean number of occurrences (i.e., $10,000/T$).

Example 5-15: Peak Discharge for a River Using the Normal Distribution

Figure 5-8 shows a frequency histogram of the annual maximum discharge data (Table 5-6) for the Piscataquis River. The sample consists of 58 annual maximum instantaneous flood discharges and has a mean of 8620 ft³/sec and a standard deviation of 4128 ft³/sec. The normal frequency curve was fitted using the procedure of the preceding section; Figure 5-9 shows the cumulative normal distribution using

FIGURE 5-8 Histogram for annual maximum discharge in cubic feet per second for the Piscataquis River.

the sample mean and the standard deviation as estimates of the location and scale parameters. The data were plotted using the Weibull plotting position formula (Equation 5-96a). It is evident that the data do not provide a reasonable fit to the population; the data show a significant skew with an especially poor fit in the tails of the distribution (e.g., high and low exceedence probabilities). Because of the poor fit, we would conclude that the measured floods do not follow a normal distribution and that the line shown in Figure 5-9 should not be used to make probability statements about future occurrences of floods; for example, the assumed normal population (i.e., the line) would suggest a 1% chance flood magnitude of slightly more than 18,000 ft³/sec. However, if a line were drawn subjectively through the points, the 1% chance flood would be considerably larger, say about 23,000 ft³/sec.

5.9.6. Frequency Analysis Using the Lognormal Distribution

In Example 5-15, the observed data resulted in a poor fit to the assumed; therefore, a different distribution function should be considered. In this case, the data demonstrate a concave-upward curve as shown in Figure 5-9. Therefore, it is reasonable to try a lognormal distribution or an extreme-value distribution. In some cases, it may be preferable to fit with a distribution that requires an estimate of the skew coefficient, such as a log-Pearson type III distribution. However, sample estimates of the skew coefficient may be inaccurate for small samples, and thus they should be used with caution.

The same procedure that is used for fitting the normal distribution can be used to fit the lognormal distribution. The assumption is made that the underlying population is lognormal. The data are first transformed to logarithms, using either $Y = \log(X)$ or $Y = \ln(X)$. This transformation creates a new random variable Y and

TABLE 5-6 Annual Maximum Discharge (q_p) and Expected Exceedence Probability (p_i)

Rank i	q_p (ft³/sec)	p_i	Rank i	q_p (ft³/sec)	p_i
1	21,500	0.0169	30	7,600	0.5085
2	19,300	0.0339	31	7,420	0.5254
3	17,400	0.0508	32	7,380	0.5424
4	17,400	0.0678	33	7,190	0.5593
5	15,200	0.0847	34	7,190	0.5763
6	14,600	0.1017	35	7,130	0.5932
7	13,700	0.1186	36	6,970	0.6102
8	13,500	0.1356	37	6,930	0.6271
9	13,300	0.1525	38	6,870	0.6441
10	13,200	0.1695	39	6,750	0.6610
11	12,900	0.1864	40	6,350	0.6780
12	11,600	0.2034	41	6,240	0.6949
13	11,100	0.2203	42	6,200	0.7119
14	10,400	0.2373	43	6,100	0.7288
15	10,400	0.2542	44	5,960	0.7458
16	10,100	0.2712	45	5,590	0.7627
17	9,640	0.2881	46	5,300	0.7797
18	9,560	0.3051	47	5,250	0.7966
19	9,310	0.3220	48	5,150	0.8136
20	8,850	0.3390	49	5,140	0.8305
21	8,690	0.3559	50	4,710	0.8475
22	8,600	0.3729	51	4,680	0.8644
23	8,350	0.3898	52	4,570	0.8814
24	8,110	0.4068	53	4,110	0.8983
25	8,040	0.4237	54	4,010	0.9153
26	8,040	0.4407	55	4,010	0.9322
27	8,040	0.4576	56	3,100	0.9492
28	8,040	0.4746	57	2,990	0.9661
29	7,780	0.4915	58	2,410	0.9831

the analysis assumes that the values of Y have a normal distribution. The mean and standard deviation of the logarithms (i.e., $\overline{Y}$ and S_Y) are computed and used as the parameters of the population; it is important to recognize that the logarithm of the mean does not equal the mean of the logarithms, which is also true for the standard deviation. Thus one should not use the logarithms of the mean and standard deviation as parameters; the mean and standard deviation of the logarithms should be computed and used as the parameters. Either natural or base 10 logarithms may be used, although the latter is more common in engineering. The population line is defined by plotting the straight line between the points $(\overline{Y} + S_Y, 0.8413)$ and $(\overline{Y} - S_Y, 0.1587)$ using cumulative probabilities, where $\overline{Y}$ and S_Y are the mean and standard deviation of the logarithms. In plotting the data, either logarithms can be plotted on an arithmetic scale or the untransformed data can be plotted on a logarithmic scale.

Predictions with the lognormal distribution are made in almost the same way as for the normal distribution. Graphical estimates can be made with the plot of the frequency curve. However, final estimates should be made using the standard normal

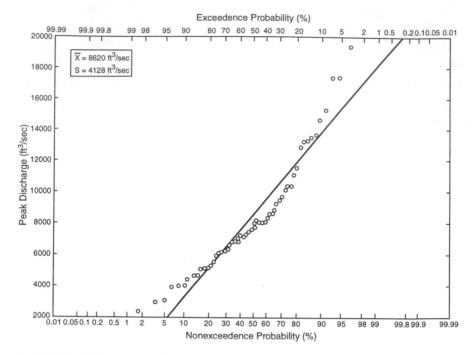

FIGURE 5-9 Normal probability paper for the Piscataquis River annual maximum discharge.

transformation using the log mean ($\overline{Y}$) and log standard deviation (S_Y). To estimate a magnitude (x), its logarithm [$y = \log(x)$ or $y = \ln(x)$] is computed using the standard normal deviate corresponding to a desired exceedence probability as follows:

$$y = \overline{Y} + zS_Y \qquad (5\text{-}98a)$$

The value of x can then be computed using the transformation $x = 10^y$ or $x = e^y$, for base 10 and natural logarithms, respectively. Similarly, if a probability is wanted for a specified magnitude (x), the value of z is computed for the natural logarithmic transformation as

$$z = \frac{\ln(x) - \overline{Y}}{S_Y} \qquad (5\text{-}98b)$$

or for the base-10 logarithmic transformation as

$$z = \frac{\log(x) - \overline{Y}}{S_Y} \qquad (5\text{-}98c)$$

The exceedence probability is then obtained by entering the standard normal table (Table A-1) with z and locating the probability.

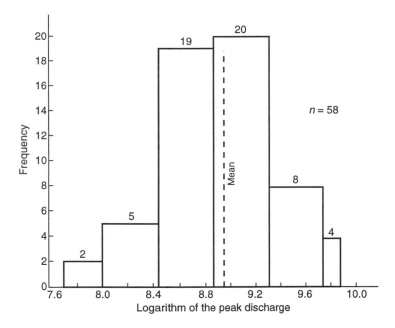

FIGURE 5-10 Histogram for the logarithm of the annual maximum discharge
in cubic feet per second for the Piscataquis River.

Example 5-16: Peak Discharge for a River Using the Lognormal Distribution

The peak discharge data for the Piscataquis River were transformed by taking the natural logarithm of each of the 58 values. The resulting statistics for $Y = \ln(X)$ transformation are $\overline{Y} = 8.9572$, $S_Y = 0.4641$. A histogram of the logarithms is shown in Figure 5-10. In comparison to the histogram of Figure 5-8, the logarithms of the sample data are less skewed. The standardized skew is practically zero. Thus the discharges can be represented with a lognormal distribution. The frequency curve is shown in Figure 5-11. The data points show a much closer agreement with the population line in comparison to the points in Figure 5-9. Therefore, there is no reason to believe that the measured peak discharge rates do not come from a lognormal distribution. Future flood behavior of the watershed can be described statistically using a lognormal distribution.

If one were interested in the probability that a flood of 20,000 ft³/sec would be exceeded in any 1 year, the natural logarithm of 20,000, which is 9.903, would be entered on the discharge axis and followed to the assumed population line; reading the exceedence probability corresponding to that point on the frequency curve, one concludes that there is about a 1.7% chance of a flood of 20,000 ft³/sec being exceeded in any 1 year. The value can also be computed using Equation 5-98b as follows:

$$z = \frac{\ln(20,000) - \overline{Y}}{S_Y} = \frac{9.903 - 8.9572}{0.4641} = 2.0379$$

FIGURE 5-11 Lognormal probability paper for the Piscataquis River annual maximum discharge.

This corresponds to a probability according to Table A-1 of 0.0208 or 2.1%. The difference in the results between the graphical and the analytical methods can be attributed to the inaccuracy of the graphical method. It can also be interpreted that over the span of 1000 years, one could expect the flood discharge to exceed 20,000 ft³/sec in 17 of those years; it is important to understand that this is an average value. In any period of 1000 years a value of 20,000 ft³/sec may be exceeded more or less than 17, but on the average, there would be 17 exceedences in a period of 1000 years.

The frequency curve can also be used to estimate flood magnitudes for selected probabilities. The flood magnitude is found by entering the figure with the exceedence probability, moving vertically downward to the frequency curve, and finally moving horizontally to the flood magnitude. For example, the 100-year flood for Piscataquis River can be found by entering with an exceedence probability of 1% moving to the curve of Figure 5-11 and then to the ordinate, which indicates a logarithm of about 10.036 or a discharge of 22,800 ft³/sec. Discharges for other exceedence probabilities can be found in the same way. They can also be found using Equation 5-98a. For a 1% exceedence flood the z value is 2.327. Thus, Equation 5-98a gives

$$y = \overline{Y} + zS_Y = 8.9572 + 2.327(0.4641) = 10.037$$

Using the inverse of the transformation, the discharge is $x = e^{10.037} = 22,857$ ft³/sec, which agrees with the graphical solution.

5.9.7. Chi-Square Test for Goodness of Fit

Data analysts are often interested in identifying the density function of a random variable so that the population can be used to make probability statements about the likelihood of occurrence of values of the random variable. Very often, a graphical analysis such as a histogram of the data is used to suggest a likely candidate for the population density function. However, the graphical approach involves a subjective assessment and lacks the validity accorded to systematic analysis such as hypothesis tests. This has led to several tests that can be applied to evaluate the likelihood that a sample is from an assumed population. The chi-square test for goodness of fit is used to assess whether or not a sample of measurements from a random variable can be represented by a selected theoretical probability density function. The following paragraphs provide the six-step framework for performing the chi-square test. To demonstrate the quantitative evaluation, the distribution of arrival times for automobiles at an intersection during a nonpeak-volume period is used to illustrate these steps.

Step 1: Formulate hypothesis — The first step is to formulate both the null and the alternative hypotheses that reflect the theoretical density function (continuous random variables) or mass function (discrete random variables). Because a function is not completely defined without the specification of its parameters, the hypothesis must also include specific values for the parameters of the function. For example, if the population is hypothesized to be normal, values for μ and σ must be specified; if the hypotheses deal with the uniform distribution, values for a and b must be specified. Estimates of the parameters may be obtained either empirically or from external conditions. If estimates of the parameters are obtained from the data set used in testing the hypotheses, the degrees of freedom in this test must be modified to reflect this.

The following null and alternative hypotheses are typical examples:

H_o: the time between arrivals of automobiles at the specific intersection under investigation (X) can be characterized by a uniform density function with a location parameter of zero and a scale parameter of 40 sec.

H_A: the time between arrivals of automobiles at a specific intersection is not uniformly distributed with parameters of 0 and 40 sec.

Mathematically, these hypotheses are

$$H_o: X \sim U(a = 0, b = 40)$$

$$H_A: X \neq U(0, 40)$$

Rejection of the null hypothesis would not necessarily imply that the arrival time is not uniformly distributed. While rejection of the null hypothesis may occur because the random variable is not uniformly distributed, it may also be rejected because one or both of the parameters, in this case 0 and 40, are incorrect.

The chi-square goodness-of-fit test is always a one-tailed test because the structure of the hypotheses is unidirectional; that is, the random variable is either distributed as specified or it is not.

Step 2: Select the appropriate model — To test the hypotheses formulated in step 1, the chi-square test is based on a comparison of the observed frequencies of sample values with frequencies expected from the population density function that is specified in the null hypothesis. The test statistic, which is a random variable, is a function of the observed and expected frequencies, which are also random variables. The test statistic is

$$\chi^2 = \sum_{i=1}^{k} \frac{(O_i - E_i)^2}{E_i} \tag{5-99}$$

where χ^2 is the computed value of a random variable having a chi-square distribution with v degrees of freedom; O_i and E_i are the observed and expected frequencies in cell (or interval) i, respectively; and k is the number of discrete cells (intervals) into which the data were separated. The random variable χ^2 has a sampling distribution that can be approximated by the chi-square distribution with $(k - j)$ degrees of freedom, where j is the number of quantities that are estimated from the sample of data for use in calculating the expected frequencies. Specifically, since the total number of observations n is used to compute the expected frequencies, one degree of freedom is lost. If the mean and standard deviation of the sample are needed to compute the expected frequencies, then two additional degrees of freedom are subtracted (i.e., $k - 3$). However, if the mean and standard deviation are obtained from past experience or other data sources, then the number of degrees of freedom for the test statistic remains $(k - 1)$.

Step 3: Select the level of significance — If the decision is not considered critical, a level of significance of 5% may be considered appropriate, because of convention. For the test of the hypotheses given previously, a value of 5% is used for illustration purposes.

Step 4: Compute estimate of test statistic — The observed data are usually used to form a histogram that shows the observed frequencies in a series of k cells (or intervals). The cell bounds are usually selected such that the width for each cell is the same; however, unequal cell widths are a practical alternative to ensure more uniform values of the observed and expected frequencies across cells. Having selected the cell bounds and counted the observed frequencies (O_i) in the k cells, the expected frequencies (E_i) for the cells can be computed using the density function of the population specified in the hypotheses of step 1. To compute the expected frequencies, the expected probability for each cell is determined and multiplied by the sample size. The expected probability for cell i, p_i, is either the area under the density function between the cell bounds for that cell for a continuous random variable or the sum of the probabilities from the probability mass function for the values within the cell bounds for a discrete random variable. The sum of the expected

Cell bound	$-\infty$				...		∞
Cell number i	1	2	3		...	k	
Observed frequency (O_i)	O_1	O_2	O_3		...	O_k	
Expected frequency (E_i)	E_1	E_2	E_3		...	E_k	
$(O_i - E_i)^2/E_i$	$\dfrac{(O_1 - E_1)^2}{E_1}$	$\dfrac{(O_2 - E_2)^2}{E_2}$	$\dfrac{(O_3 - E_3)^2}{E_3}$		...	$\dfrac{(O_k - E_k)^2}{E_k}$	

FIGURE 5-12 Cell structure of arrival times for chi-square goodness-of-fit test.

Cell bound	0	10 sec	20 sec	30 sec	40 sec
Cell number		1	2	3	4
Observed frequency (O_i)		18	19	25	18
Expected frequency (E_i)		20	20	20	20
$(O_i - E_i)^2/E_i$		0.20	0.05	1.25	0.20

FIGURE 5-13 Cell structure of arrival times for chi-square goodness-of-fit test.

frequencies must equal the sample size n. The problem is best summarized using the general structure shown in Figure 5-12.

The value of the test statistic of Equation 5-99 is obtained from the cell frequencies of Figure 5-13. The range of the random variable was separated into four equal intervals of 10 sec. Thus, the expected probability for each cell is 0.25 (because the random variable is assumed to have a uniform distribution and the width of each of the four cells is the same). For a sample size of 80, the expected frequency for each of the four cells is 20 (i.e., the expected probability times the total number of observations). If observed frequencies of 18, 19, 25, and 18 are determined from the sample, then the cell structure would appear as shown in Figure 5-13. The computed c statistic is, therefore, 1.70, which is obtained by adding up 0.2, 0.05, 1.25, and 0.2. Because the sample size was used in computing the expected frequencies, the number of degrees of freedom is given by $(k - 1)$, or $4 - 1 = 3$.

Step 5: Define the region of rejection — According to the underlying theorem of step 2, the test statistic has a chi-square distribution with degrees of freedom that depend on the number of cells and the number of pieces of information obtained from the sample. In this case, only the sample size was used. Thus, there are 3 degrees of freedom. For level of significance of 5%, the critical value of the test statistic from Table A-3 is 7.81. Thus the region of rejection consists of all values of the test statistic greater than 7.81.

Step 6: Select the appropriate hypothesis — Because the computed value of the test statistic (1.70) is less than the critical value, i.e., it is not within the region of rejection; thus there is no statistical basis for rejecting the null hypothesis based on the location and scale parameters of 0 and 40, respectively; thus, U(0,40) may be used to represent the distribution of arrival times.

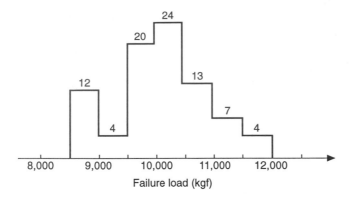

FIGURE 5-14 Frequency distribution of failure loads.

Summary — In summary, the chi-square test for goodness of fit provides the means for comparing the observed distribution of a random variable with a theoretical probability mass or density function. An additional point concerning the use of the chi-square test should be noted. The effectiveness of the test is diminished if the expected frequency in any cell is less than five. When this condition occurs, both the expected and observed frequencies of the appropriate cell should be combined with the values of an adjacent cell; the value of k should be reduced to reflect the number of cells in computing the test statistic. Also, it is desirable to have a value for k larger than three. The chi-square test is included in the summary of Table 5-5.

Example 5-17: Chi-Square Test for a Normal Distribution

To illustrate the use of the chi-square test for the normal distribution, a sample of 84 beams was tested to failure. The histogram is shown in Figure 5-14. In general, the normal distribution has a bell shape, but in this case there is one large cell in the lower tail. The sample mean and standard deviation of the failure loads were 10,100 and 780 lb, respectively. A null hypothesis is proposed that the failure loads are normally distributed with a mean and standard deviation of 10,100 and 780 lb, respectively. Table 5-7 gives the standardized variates z_i for the bounds of each interval, the probability that the variate Z is less than z_i, the expected probabilities for each interval, the expected and observed frequencies, and the chi-square values. The test statistic has a computed value of 10.209. Note that because the expected frequency for the seventh interval was less than 5, both the observed and expected frequencies were combined with those of the sixth cell. Thus, $k = 6$ is used to compute the degrees of freedom. Because the sample mean and standard deviation were used to define the hypothesis and compute the expected frequencies, 2 degrees of freedom are lost. Also, 1 degree of freedom is lost because of the sample size, Thus, the remaining degrees of freedom are 3 for the test statistic. If past evidence had indicated a mean of 10,000 lb and a standard deviation of 1000 lb, and these

TABLE 5-7 Computations for Failure Load Example

Cell i	z_i	$P(Z < z_i)$	Expected probability	Expected frequency, E_i	Observed frequency, O_i	$\dfrac{(O_i - E_i)^2}{E_i}$
Cell 1: $X \leq 9{,}000$	$\dfrac{9000-10100}{780} = -1.41$	0.0793	0.0793	6.66	12	$\dfrac{(12-6.66)^2}{6.66} = 4.282$
Cell 2: $9{,}000 < X \leq 9{,}500$	$\dfrac{9500-10100}{780} = -0.77$	0.2206	0.1413	11.87	4	$\dfrac{(4-11.87)^2}{11.87} = 5.218$
Cell 3: $9{,}500 < X \leq 10{,}000$	$\dfrac{10000-10100}{780} = -0.13$	0.4483	0.2277	19.13	20	$\dfrac{(20-19.13)^2}{19.13} = 0.040$
Cell 4: $10{,}000 < X \leq 10{,}500$	$\dfrac{10500-10100}{780} = 0.52$	0.6985	0.2502	21.02	24	$\dfrac{(24-21.02)^2}{21.02} = 0.422$
Cell 5: $10{,}500 < X \leq 11{,}000$	$\dfrac{11000-10100}{780} = 1.15$	0.8749	0.1764	14.82	13	$\dfrac{(13-14.82)^2}{14.82} = 0.224$
Cell 6: $11{,}000 < X \leq 11{,}500$	$\dfrac{11500-10100}{780} = 1.79$	0.9633	0.0884	$\left.\begin{array}{c}7.42\\[6pt]3.08\end{array}\right\} 10.50$	11	$\dfrac{(11-10.50)^2}{10.50} = 0.024$
Cell 7: $11{,}500 < X$	∞	1.0000	0.0367			
Total		1.0000		84	84	10.209

statistics were used in Table 5-7 for computing the expected probabilities, then 5 degrees of freedom would be used. For a level of significance of 5% and 3 degrees of freedom, the critical χ_α^2 value is 7.817. The null hypothesis is, therefore, rejected because the computed value is greater than the critical value. One may conclude that failure loading is not normally distributed with $\mu = 10,100$ lb and $\sigma = 780$ lb. The reason for the rejection of the null hypothesis may be due to one or more of the following: (1) failure loading is not normally distributed, (2) $\mu \neq 10,100$ lb, or (3) $\sigma \neq 780$ lb.

Example 5-18: Chi-Square Test for an Exponential Distribution — Case 1

Large volumes of sediment yield are less likely than small volumes. Thus, it is not unreasonable to assume that sediment yields might follow an exponential distribution, which has the following density $f_X(x)$ and cumulative functions $F_X(x)$, respectively:

$$f_X(x) = \lambda e^{-\lambda x} \tag{5-100}$$

and

$$F_X(x) = 1 - e^{-\lambda x} \tag{5-101}$$

in which λ is a scale parameter. The method of moments yields the reciprocal of a sample mean as an estimator of λ, i.e., $\lambda = 1/\overline{X}$.

The sample shown in Table 5-8 includes measured rates of sediment yield in acre-feet per square mile per year. Since the sample is small ($n = 37$) and five or more expected frequencies are needed per cell, the bounds of the cells will be set to get the maximum number of cells possible, which is seven cells. Using an increment of 0.14 for the cumulative probability provides an expected frequency of 5.18, with one cell having a probability of 0.16 and $E_i = 5.92$, as shown in Table 5-9. The value of λ is computed using the mean of the 37 sample values, which is 0.6579; thus, the estimate of λ is 1.52. Therefore, the hypotheses are

H_o: sediment yield has the following density function: $f_X(x) = 1.52 \exp(-1.52x)$.
H_A: $f_X(x)$ cannot be used to represent the sediment yield.

The calculation of the computed chi-square value is shown in Table 5-9. The computed value is 15.32. There are seven cells, but 2 degrees of freedom were lost, one for using n and the second for using the mean. Thus, there is a total of 5 degrees of freedom. Critical values of the χ^2 statistic are obtained from Table A-3. For 5, 1, and 0.5%, the critical values are 11.070, 15.086, and 16.75, respectively. Thus, the null hypothesis would be rejected for 5 and 1%, but not for 0.5%.

TABLE 5-8 Data for Sediment Yield for Example 5-18
(in Acre-Feet per Square Mile per Year)

2.67	1.55	0.61	0.21	0.16	0.076
2.65	1.42	0.51	0.21	0.15	0.070
2.37	0.99	0.37	0.20	0.14	0.040
2.31	0.69	0.35	0.18	0.14	0.036
2.20	0.66	0.25	0.17	0.12	0.030
1.65	0.64	0.22	0.17	0.09	0.020
					0.020

TABLE 5-9 Goodness-of-Fit Test for Exponential Density Function — Case 1

Cell i (1)	$F_X(x)$ (2)	x (3)	Expected frequency, E_i (4)	Observed frequency, O_i (5)	$\dfrac{(O_i - E_i)^2}{E_i}$ (6)
1	0.14	0.099	5.18	8	1.53
2	0.28	0.216	5.18	11	6.54
3	0.42	0.358	5.18	3	0.92
4	0.56	0.540	5.18	2	1.95
5	0.70	0.792	5.18	4	0.27
6	0.84	1.206	5.18	1	3.37
7	1.00	∞	5.92	8	0.73
TOTAL			37.00	37	15.32

Note: Column 2: $F_X(x) = 1 - \exp(-\lambda x)$; column 3: $x = -\ln[1 - F_X(x)]/\lambda$; and column 4: $E_i = n[F_X(x_i) - F_X(x_{i-1})]$.

Example 5-19: Chi-Square Test for an Exponential Distribution — Case 2

In this example, the sediment yield as discussed in Example 5-18 is considered. In this case, the data are not given, but a histogram is provided as shown in Figure 5-15, which has the shape of an exponential decay function. This example illustrates the effect on not having the original data (i.e., having only the histogram) on the results of the hypothesis testing.

The exponential density function is given in Equation 5-100, and cumulative function is given in Equation 5-101. The method-of-moments estimator of λ is the reciprocal of the mean (i.e., $\lambda = 1/\overline{X}$). Probabilities can be evaluated by integrating the density function $f_X(x)$ between the upper and lower bounds of each interval.

Using the sediment yield X data and the histogram of Figure 5-15, a test of hypothesis was made for the following hypothesis:

H_o: sediment yield has the following density function: $f_X(x) = 1.52 \exp(-1.52x)$.
H_A: $f_X(x)$ cannot be used to represent the sediment yield.

FIGURE 5-15 Histogram of sediment yield data.

In this case, the calculation of the computed chi-square value is shown in Table 5-10. Although the histogram included five cells, the last three cells were combined to ensure that all cells would have an expected frequency greater than 5. The computed chi-square statistic is 6.263. Two degrees of freedom are lost because n and $\overline{X}$ were used to compute the expected frequencies. Therefore, with only three cells, there is only 1 degree of freedom. For levels of significance of 5 and 1% and 1 degree of freedom, the critical values are 3.841 and 6.635, respectively. Thus the null hypothesis would be rejected for a 5% level of significance but accepted for 1%. This illustrates the importance of selecting the level of significance on the basis of a rational analysis of the importance of type I and II errors.

5.9.8. Kolmogorov-Smirnov One-Sample Test

A frequent problem in data analysis is verifying that the population can be represented by some specified probability density function. The chi-square goodness-of-fit test was introduced as one possible statistical test; but the chi-square test requires at least a moderate sample size of about 15, with much larger sample sizes preferred. The Kolmogorov-Smirnov one-sample test was also developed for verifying a population distribution and can be used with small samples.

The objective of the Kolmogorov-Smirnov one-sample test is to test the null hypothesis that the sample values of a random variable are likely to have been sampled from

TABLE 5-10 Goodness-of-Fit Test for Exponential Density Function — Case 2

Cell i	$\int_{x_{lower}}^{x_{upper}} f_X(x)\,dx$	Expected frequency, E_i	Observed frequency, O_i	$\dfrac{(O_i - E_i)^2}{E_i}$
Cell 1: $0 \le X \le 0.35125$	0.4137	15.30	22	$\dfrac{(22 - 15.30)^2}{15.30} = 2.93$
Cell 2: $0.35125 < X \le 1.01375$	0.3721	13.77	7	$\dfrac{(7 - 13.77)^2}{13.77} = 3.33$
Cell 3: $1.01375 < X \le 1.67625$	0.1359	5.03⎫	3⎫	
Cell 4: $1.67625 < X \le 2.33875$	0.0497	1.84⎬ 7.93	2⎬ 8	$\dfrac{(8 - 7.93)^2}{7.93} = 0.00$
Cell 5: $2.33875 < X$	0.0286	1.06⎭	3⎭	
Total	1.0000	37.00	37	6.263

a specified probability function; the null hypothesis must specify both the population distribution function and its parameters. The alternative hypothesis is accepted if the distribution function is unlikely to be the underlying function; this may be indicated if either the density function or the specified parameters are incorrect.

The test statistic, which is denoted as *KS*, is the maximum absolute difference between the values of the cumulative distributions of a random sample and the probability distribution function specified in the null hypothesis. Critical values of the test statistic are available only for selected values of the level of significance as shown in Appendix A, Table A-7.

The Kolmogorov-Smirnov one-sample test may be used for small samples and is generally more efficient than the chi-square goodness-of-fit test when the sample size is small. The test requires data on at least an ordinal scale, but is applicable for comparisons with continuous distributions. The Kolmogorov-Smirnov test uses the following six steps:

1. State the null and alternative hypotheses in terms of the proposed probability density function and its parameters.
2. The test statistic, *KS*, is the maximum absolute difference between the cumulative function of the sample (F_S) and the cumulative function of the probability function specified in the null hypothesis (F_X). The cumulative function for the sample (F_S) can be determined by rank-ordering the sample values from the smallest (x_1) to the largest (x_n) for a sample of size n. The sample cumulative function can be computed as

$$F_s(x) = \begin{cases} 0 & \text{for } x < x_1 \\ \dfrac{i}{n} & \text{for } x_i \le x < x_{i+1} \\ 1 & \text{for } x \ge x_n \end{cases} \qquad (5\text{-}102)$$

The resulting step function should be used to determine the maximum absolute difference (KS) with the computed $F_X(x)$ at any value of x. The maximum difference can be computed as [with $F_S(x_0) = 0$ and $F_S(x_n) = 1$]

$$KS = \max \left\{ \begin{array}{l} |F_X(x_1) - F_S(x_0)|, |F_X(x_1) - F_S(x_1)|, |F_X(x_2) - F_S(x_1)|, |F_X(x_2) - F_S(x_2)|, \dots, \\ |F_X(x_i) - F_S(x_{i-1})|, |F_X(x_i) - F_S(x_i)|, \dots, |F_X(x_n) - F_S(x_{n-1})|, |F_X(x_n) - F_S(x_n)| \end{array} \right\} \quad (5\text{-}103)$$

3. The level of significance should be set; values of 0.05 and 0.01 are usually used.
4. A random sample should be obtained and the sample cumulative probability function computed. After computing the cumulative probability function, the value of the test statistic should be computed.
5. The critical value, KS_α, of the test statistic is a function of α and the sample size, n. Table A-7 in Appendix A shows the critical values that can be used with this test.
6. If the computed value KS is greater than KS_α, the null hypothesis should be rejected.

Example 5-20: Water Quality

The following is a set of 13 measurements of a water quality parameter in ppm: {47, 53, 61, 57, 65, 44, 56, 52, 63, 58, 49, 51, 54}. If a level of significance of 5% is used, would it be safe to conclude that the sample is from a population that is normally distributed with a mean of 54.6 ppm and a standard deviation of 6.2 ppm? Table 5-11 shows the calculations for computing the maximum difference. The cumulative probabilities $F_S(x)$ and $F_X(x)$ are shown in Figure 5-16. The sample mean $(\overline{X})$ and the sample standard deviation (S) were computed to be 54.6 and 6.2, respectively. From Table 5-11 the value of the test statistic, which equals the largest absolute difference, is 0.078742. For a 5% level of significance, the critical value is $KS_\alpha = 0.368$. Since the test statistic KS is less than KS_α, the null hypothesis cannot be rejected, i.e., a normal probability distribution with a mean of 54.6 ppm and standard deviation of 6.2 ppm can be used as a model.

TABLE 5-11 Example Using the Kolmogorov-Smirnov One-Sample Test

Rank i	x_i	$F_s(x_i)$	Standard normal variate	$F_X(x_i)$	$|F_X(x_i) - F_s(x_{i-1})|$	$|F_X(x_i) - F_s(x_i)|$
1	44	0.076923	-1.7090	0.043741	0.043740884	0.03318219
2	47	0.153846	-1.2260	0.110118	0.033194626	0.04372845
3	49	0.230769	-0.9040	0.183011	0.029165271	0.04775781
4	51	0.307692	-0.5820	0.280285	0.049516107	0.02740697
5	52	0.384615	-0.4210	0.336871	0.029179031	0.04774405
6	53	0.461538	-0.2600	0.397416	0.012801112	0.06412196
7	54	0.538462	-0.0990	0.460544	0.000994266	0.07791734
8	56	0.615385	0.2229	0.588190	0.049728426	0.02719465
9	57	0.692308	0.3839	0.649462	0.034077401	0.04284568
10	58	0.769231	0.5449	0.707070	0.014762421	0.06216066
11	61	0.846154	1.0278	0.847973	0.078741806	0.00181873
12	63	0.923077	1.3497	0.911449	0.065295000	0.01162808
13	65	1	1.6717	0.952707	0.029629859	0.04729322

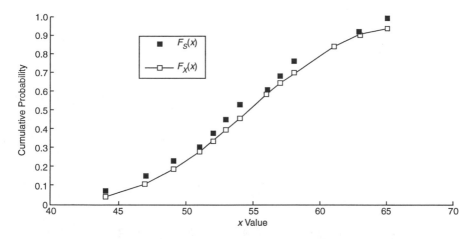

FIGURE 5-16 Example using the Kolmogorov-Smirnov One-Sample test.

5.10. APPLICATIONS

Nine examples are used to illustrate the six-step procedure for testing hypotheses. In each case, a statistical analysis of engineering data is used to make inferences about the populations. Statistical inferences about the mean of a random variable depend on knowledge of the variance of the population; a hypothesis test concerning a mean value is illustrated in the first example. The second and third examples deal with a comparison of two population means rather than a test of a single population statistic. The fourth and sixth emphasis illustrate a hypothesis test on the variance of a population. The fifth, seventh, eighth, and ninth examples illustrate the comparison of two sample variances.

5.10.1. Test of Mean Strength of Steel Prestressing Cables

Samples of steel prestressing cables to be used in a curved gravity dam are tested prior to construction to ensure that the shipment meets the design capacity. Design plans call for 100-wire cables with a mean capacity of 900 kips. Eight samples were randomly selected and tested, with the following capacities: 825, 900, 915, 850, 870, 930, 835, 885 kips.

If the cables selected for testing are representative of the cables supplied by the manufacturer, can we conclude that the design capacity is met?

The design capacity of 900 kips provides the null hypothesis, H_o: $\mu = 900$ kips. Since capacities greater than 900 kips would be acceptable, the problem demands a one-tailed test with the alternative hypothesis, H_A: $\mu < 900$ kips. Therefore, the two hypothesis are

$$H_o: \mu = 900 \text{ kips} \qquad \qquad (5\text{-}104a)$$

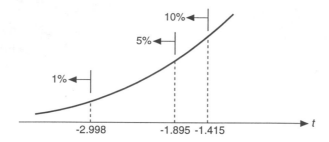

FIGURE 5-17 Distribution of test statistics.

$$H_A: \mu < 900 \text{ kips} \tag{5-104b}$$

If the null hypothesis is accepted, then the shipment of 100-wire cables is considered acceptable. If the null hypothesis is rejected, then the engineer should make adjustments, either in the form of a new shipment of cables or redesign to allow for lower-capacity cables.

Since the standard deviation of the capacity is not known from other studies, the t test should be applied. Analysis of the eight measurements yields a mean and standard deviation of 876.2 and 38.0 kips, respectively. Thus, the computed value of the t statistic is

$$t = \frac{\overline{X} - \mu_o}{S/\sqrt{n}} = \frac{876.2 - 900}{38/\sqrt{8}} = -1.771 \tag{5-105}$$

For a sample of eight, there are 7 degrees of freedom. Figure 5-17 shows the distribution of the test statistic and the region of rejection for levels of significance of 10, 5, and 1%. For a 5% level of significance, with a critical value of −1.895 according to Table A-2, the null hypothesis would be accepted. There is approximately a 6% chance of making a type I error. If a level of significance greater than about 6% were used, then the null hypothesis would have to be rejected.

5.10.2. Test of Mean Water Use of Hotels

Water use in hotels is measured on a basis of gallons per day per room (gpdr). Using annual data from 12 hotels in Baltimore, the mean and standard deviation are 225 and 35 gpdr, respectively. Data from seven hotels in Las Vegas give a mean and standard deviation of 420 and 90 gpdr, respectively. Based on these data, a test is made to access whether or not average water use in western cities is greater than that in eastern cities:

$$H_o: \mu_E = \mu_W \tag{5-106a}$$

$$H_A : \mu_E < \mu_W \qquad (5\text{-}106b)$$

The one-sided test on two means can be conducted using the test statistic of Equation 5-58 and the pooled variance of Equation 5-59 which is given by

$$S_p^2 = \frac{(12-1)(35)^2 + (7-1)(90)^2}{12+7-2} = 3651 \ (\text{gpdr})^2 \qquad (5\text{-}107)$$

Thus, the pooled standard deviation is 60.4 gpdr. The t statistic is given by

$$t = \frac{225 - 420}{60.43\left(\dfrac{1}{12} + \dfrac{1}{7}\right)^{0.5}} = -6.785 \qquad (5\text{-}108)$$

with 17 degrees of freedom. The critical t values for 5 and 1% levels of significance are −1.74 and −2.567, respectively, using Table A-2. Thus, even at a 1% level, the null hypothesis must be rejected, with the implication that the average water use in the hotels of western cities is greater than that in eastern cities. In assessing the validity of the implication, it is important to know whether or not water use in the two cities is representative of water use in their respective regions.

5.10.3. Test of Two Sample Means for Car Speed

The average speed of cars was measured at a particular location on an interstate highway before and after passage of a law allowing speeds up to 65 mph. One group argued that the average speed in excess of the 65-mph limit would be smaller than from the average speed in excess of the 55-mph limit before the law. Thus, the random variable is the difference between the average speed and the speed limit. For group 1, this would be $\overline{X}_1 - 55$; for group 2, this would be $\overline{X}_2 - 65$.

The comparison dictates the following null hypothesis:

$$H_o : \mu_1 = \mu_2 \qquad (5\text{-}109a)$$

where μ is the difference of the population mean from the speed limit. Given that the group believes that the difference will decrease for the 65-mph limit, the alternative hypothesis is

$$H_A : \mu_1 > \mu_2 \qquad (5\text{-}109b)$$

Thus, it is a one-tailed test with the direction that the difference is greater for the 55-mph limit.

The following 16 speeds were measured for both the 55- and the 65-mph limit:

$$X_1 = \{62, 59, 67, 66, 56, 63, 58, 67, 72, 54, 61, 63, 60, 57, 68, 64\} \quad \text{(5-110a)}$$

$$X_2 = \{73, 66, 68, 63, 76, 72, 68, 71, 75, 67, 74, 59, 70, 58, 75, 78\} \quad \text{(5-110b)}$$

For groups 1 and 2, the means are 62.31 and 69.56 mph, respectively, and the standard deviations are 4.91 and 5.91 mph, respectively. Using Equation 5-58 the computed value of the test statistic is

$$t = \frac{(62.31 - 55) - (69.56 - 65)}{S_p \left(\dfrac{1}{16} + \dfrac{1}{16} \right)^{0.5}} \quad \text{(5-111)}$$

where the pooled variance is computed from Equation 5-59 as

$$S_p^2 = \frac{15(4.91)^2 + 15(5.91)^2}{16 + 16 - 2} = 29.52 \quad \text{(5-112)}$$

Thus, the pooled standard deviation is 5.433 mph, and the computed value of the test statistic (t) is 1.432. The degrees of freedom is 30. For a level of significance of 5%, the critical t value for a one-sided test from Table A-2 is 1.697. Thus, the null hypothesis of equal excesses above the speed limit is accepted.

5.10.4. Variation of Effective Cohesion

The effective cohesion (C_e) is a decision element in problems of slope stability. Given the difficulty in determining the value of the effective cohesion at a site and the inherent variability of soil properties, including C_e, factors of safety are applied in slope stability problems. One engineer decides to estimate the effective cohesion from field sampling by taking five field samples and estimating C_e with each sample. However, as a practice, the engineer re-samples when the variance of the sample estimates of C_e is significantly greater than 10^4 (psf)^2.

On one project, the engineer uses five samples to estimate the effective cohesion, with the following values: 620, 410, 730, 570, and 380 lb/ft^2. Based on these measurements, which have a standard deviation of 146.5 psf, the engineer tests the following hypotheses:

$$H_o: \sigma^2 = 10^4 \text{ (psf)}^2 \quad \text{(5-113a)}$$

$$H_A: \sigma^2 > 10^4 \text{ (psf)}^2 \quad \text{(5-113b)}$$

The computed value of the test statistic according to Equation 5-67 is

$$\chi^2 = \frac{(n-1)S^2}{\sigma^2} = \frac{4(146.5)^5}{(100)^2} = 8.59 \tag{5-114}$$

which has $n - 1 = 4$ degrees of freedom. Using a 5% level of significance, the critical value from Table A-3 is 9.488, so the null hypothesis cannot be rejected. While the sample standard deviation is greater than the limiting value of 100 psf, it is not significantly greater. Thus, the engineer decides not to re-sample.

5.10.5. Variation of Infiltration Capacity

Two sites are being considered for an infiltration-based stormwater management system. In addition to the average infiltration capacity, it is desirable to have a relative uniform variation of capacity over the site where the facility is to be located. Because infiltration capacity is an important factor in siting, soil tests are made at feasible sites. For one project, two sites are being considered. In addition to the mean infiltration capacity, the variation is also computed. Because of other siting criteria, site A is currently favored over site B and will be selected if it meets the condition that the variability of the infiltration capacity at site A is not greater than that at site B. Thus, the following hypotheses are tested

$$H_o: \sigma_A^2 = \sigma_B^2 \tag{5-115a}$$

$$H_A: \sigma_A^2 > \sigma_B^2 \tag{5-115b}$$

If the null hypothesis is accepted, then site A will be used; if it is rejected, then site B will be used.

Measurements are made at the two sites with the following results: A = {0.9, 1.4, 1.0, 1.7} in./hr, and B = {1.4, 0.9, 1.2, 1.4} in./hr. This yields standard deviations of $S_A = 0.3697$ and $S_B = 0.2363$. The computed value of the test statistic is

$$F = \frac{S_A^2}{S_B^2} = \frac{(0.3697)^2}{(0.2363)^2} = 2.448 \tag{5-116}$$

with degrees of freedom of $k = v_A = n_A - 1 = 3$ and $u = v_B = n_B - 1 = 3$. The critical F value from Table A-4 is 9.28 for a 5% level of significance. Thus, we cannot reject the null hypothesis, so site A is selected as the better site for the planned stormwater management facility even though $S_A > S_B$.

5.10.6. Uniformity of Illumination

In addition to the mean illumination in an office building, the lighting engineer is concerned with uniformity of lighting. Except for specific-task locations, the light-source arrangement should provide uniform illumination. Existing lighting systems in similar offices provide a standard deviation of 40 lm/ft². For a new building, the engineer is considering a new lightning system and decides to test the uniformity prior to ordering the system. In order to make it worthwhile, the engineer believes that the uniformity provided by the new system will have to be significantly better; thus, the following hypotheses will govern the decision:

$$H_0: \sigma^2 = \sigma_o^2 \qquad (5\text{-}117a)$$

$$H_A: \sigma^2 < \sigma_o^2 \qquad (5\text{-}117b)$$

The null hypothesis suggests that the existing lighting system is as good as the new system. If the illumination provided by the new system has a smaller variance than that of the existing system, then the alternative hypothesis is appropriate. If the null hypothesis is rejected, then the new system should be provided.

To test the uniformity provided by the new system, the engineer decides to collect a random sample of eight measurements in an office building that has the new system and use a 1% level of significance to make a decision. The eight measurements are 62, 97, 83, 92, 54, 78, 86, and 66 lm/ft², which yield a standard deviation of 15.20 lm/ft². The computed value of the χ^2 test statistic according to Equation 5-67 is

$$\chi^2 = \frac{(n-1)S^2}{\sigma^2} = \frac{7(15.20)^5}{(40)^2} = 1.011 \qquad (5\text{-}118)$$

For $k = \nu = n - 1 = 7$ degrees of freedom and a 1% level of significance, the critical value according to Table A-3 is 1.239. Since the computed value is less than the critical value, the null hypothesis should be rejected. Thus, the new lighting system will be used.

5.10.7. Test of Two Variances for Car Speed

For the example of Section 5.10.3, the test statistic is based on the assumption that the variances of the two groups are equal. Furthermore, some argue that the higher speeds associated with the higher speed limit of 65 mph lead to a greater spread of speeds and that the accident rate increases as the spread of speeds increases. Thus, there is both a statistical and an engineering reason for testing for equality of variances for the data. The null hypothesis is

$$H_0: \sigma_1^2 = \sigma_2^2 \qquad\qquad (5\text{-}119a)$$

For the statistical assumption, the alternative hypothesis is

$$H_A: \sigma_1^2 \ne \sigma_2^2 \qquad\qquad (5\text{-}119b)$$

The computed F statistic (Equation 5-70) equals the ratio of the larger variance to the smaller variance, which is, i.e.,

$$F = \frac{S_2^2}{S_1^2} = \frac{(5.910)^2}{(4.909)^2} = 1.449 \qquad\qquad (5\text{-}120)$$

Both degrees of freedom equal 15, so from Table A-5 for a 5% level of significance, the critical F is 2.40; thus, the null hypothesis is accepted for a 10% level of significance. For the two-tailed alternative, the level of significance is twice the value shown on the table.

For the engineering assessment of the null hypothesis of equal variances, the alternative hypothesis is

$$H_A: \sigma_2^2 > \sigma_1^2 \qquad\qquad (5\text{-}121)$$

Therefore, the computed F statistic is the ratio S_2^2/S_1^2, which equals 1.449. The critical value from Table A-4 is 2.40, which means that the null hypothesis is accepted for a 5% level of significance. Thus, the test suggests that the spread of the speeds for the 65-mph limit is not significantly greater than the spread of the speeds for the 55-mph limit.

5.10.8. Test on Variances of Soil Friction Angle

Two methods of estimating the friction angle ϕ of soil are compared. The first method is more time-intensive to conduct but is considered to be more precise. The second method is easy to apply but it is believed to be less precise. A geotechnical engineer wishes to test whether or not the time-intensive method is more precise. From a homogeneous soil, five samples are obtained and analyzed with the first method, with the following estimates of ϕ: 0.34, 0.37, 0.35, 0.34, and 0.33. Five random samples obtained from the same soil and subjected to the second test yield the following estimates: 0.36, 0.32, 0.38, 0.31, and 0.34. The two-sample F test can be applied to evaluate the relative precision of the two methods of estimating ϕ. Since the second test is assumed to be less precise, the engineer expects its sample variance to be the greater, so it appears in the numerator of the F statistic of Equation 5-70 as follows:

$$F = \frac{S^2 \text{ of test 2}}{S^2 \text{ of test 1}} = \frac{0.00082}{0.00023} = 3.565 \qquad (5\text{-}122)$$

which would have degrees of freedom of $v_1 = v_2 = 4$. The critical F value from Table A-4 is 6.39 for a 5% level of significance. Since the computed F is less than the critical F, the null hypothesis cannot be rejected; the engineer concludes that the two tests are equally precise.

5.10.9. Test on Variances of Nitrogen Levels

In Example 5-9, a test on equality of two means was made for samples of nitrogen levels. The test was based on the assumption that the population variances are equal, so there is a valid reason for testing for equality of variances. The test on the two means was made because it was believed that the development would increase the mean nitrogen level. It would also seem reasonable to test whether or not development increases the variance; thus, there is a second reason for testing for equality of variances. Because of the second reason, a one-sided test on the variances can be used as follows:

$$H_o: \sigma_b^2 = \sigma_a^2 \qquad (5\text{-}123a)$$

$$H_A: \sigma_b^2 < \sigma_a^2 \qquad (5\text{-}123b)$$

where σ_a^2 and σ_b^2 are the variances for the after and before development nitrogen levels, respectively. Since the test is one sided (i.e., with a specified direction), the test value of Equation 5-70 becomes

$$F = \frac{S_a^2}{S_b^2} \qquad (5\text{-}124)$$

in which S_a^2 and S_b^2 are the sample variances for the after and before development nitrogen-level data, respectively. The degrees of freedom are $k = v_1 = n_a - 1 = 13$ and $u = v_2 = n_b - 1 = 10$. The sample standard deviations are $S_a = 0.87$ mg/l and $S_b = 0.36$ mg/l. Thus, the computed F is

$$F = \frac{(0.87)^2}{(0.36)^2} = 5.84 \qquad (5\text{-}125)$$

For levels of significance of 5 and 1%, the critical F values are 2.89 and 4.66, respectively. Thus the null hypothesis should be rejected at both levels.

The results of this test have two implications. First, it would suggest that development caused the variance of total nitrogen levels to increase. There may be an ecological consequence of this finding. Second, it would suggest that the assumption of equal variances that was used in the two-sample test on means of Section 5.5.3 was violated. Thus, a different test on two means may need to be investigated. There are parametric and nonparametric test alternatives that can be used instead, but these tests are beyond the scope of this book.

5.11. SIMULATION AND SAMPLING

In Section 4.5.2, the distribution of the correlation coefficient was approximated using simulation. The topics in this chapter stress the value of knowing the distribution of a statistic; specifically, the distribution function is used as the basis of testing hypotheses and in computing confidence limits. Knowledge of the distribution is also necessary to estimate sample size requirements.

For some statistics, such as the mean from a normal population, the distribution is known from theory. In other cases, theory cannot identify the underlying distribution. The correlation coefficient is an example of a statistic with a distribution function that cannot be identified from theory. In some cases, the distribution can be approximated. In other cases, a transformation can be made such that the resulting statistic has a known distribution or can be approximated with a known distribution. The chi-square goodness-of-fit test is another example where the test statistic is approximated by a known distribution function.

Some of the tests introduced in this chapter have a limitation or an assumption. For example, the t test on a single mean with an unknown population variance assumes that the mean was computed from a sample drawn from a normal population. Other tests are based on similar assumptions. The extent to which the assumption is important is rarely identified in textbooks, and conventional wisdom suggests that use of the test when the assumption is violated is wrong. Yet, if we only used statistical methods when all of the assumptions were met, statistical methods would be rarely used. Public opinion polls, with, for example, a stated accuracy of $\pm 3\%$, are one prime example where the underlying assumptions are likely to have been violated.

In summary, this brief discussion has identified two factors that are relevant to simulation. First, simulation can be a very useful tool for identifying the probability distribution of a test statistic. Second, simulation can be used to test the importance of assumptions that are assumed valid when using a statistical hypotheses test. The first use was shown in earlier chapters of this book (see Section 4.5.2). An example of the second factor follows.

Example 5-21: Assumption of Normality with the t Test

The t test for the sample mean of a single sample assumes that the sample is from a normal population. How important is this assumption? Would the t statistic provide reliable decisions if the data were samples from a chi-square distribution? Or an extreme-value distribution? Simulation can give some insight into this problem.

To test the assumption of normality, 20,000 samples, each of size 10, were sampled from normal, uniform, and exponential distributions. The data samples were generated using random number generators for each of the distributions. The sample mean and standard deviation of each set of ten values were computed and used to compute the t value with Equation 5-54. Based on the t values computed for the 20,000 samples from a distribution, three cumulative distributions were developed. Histograms of the 20,000 t values were formed and were used to obtain critical (i.e., significance) values for 10, 5, 2.5, and 1% of the area in the upper tails. The critical values for samples from the three populations are as follows, along with the true t statistic for 9 degrees of freedom (v):

Level of significance		Critical value for		
$\alpha(\%)$	$t_{\alpha,v}$	Normal	Uniform	Exponential
10	1.383	1.384	1.375	1.050
5	1.833	1.852	1.830	1.362
2.5	2.262	2.220	2.286	1.635
1	2.821	2.759	2.910	1.982

The accuracy of the simulated values can be expected to decrease as the level of significance becomes smaller. For a 10% level of significance, approximately 2000 sample values would be in the upper tail. Therefore, the sample size is large enough to get a reasonably, accurate estimate of the critical value. For the 1% level of significance, only 200 values would be in the tail of the distribution, such that one or two sample values could cause a noticeable change in the simulated critical value. The simulated values from the samples with a normal distribution produced absolute values of the errors of 0.001, 0.019, 0.042, and 0.062 for the 10, 5, 2.5, and 1% levels of significance, respectively. Greater accuracy could be achieved using more than 20,000 samples.

The errors for the statistics computed from the samples generated from a uniform distribution are 0.008, 0.003, 0.024, and 0.089. These errors are comparable to those from the normal-distribution samples. From these results, it appears that the t test would provide reasonable decisions when sampling is done from a uniform distribution. Both the normal and uniform distribution are symmetric. Could the t test be used with other symmetric distributions without severely violating the accuracy of the critical t values? Simulation could be used to answer this question.

The simulated t values computed from samples generated using an exponential population clearly indicate that the t test cannot be used when sampling from an exponential population. The errors are 0.333, 0.471, 0.627, and 0.839 for the four levels of significance, which are much larger than the errors shown for the normal and uniform critical values. The simulated critical values are in error when compared with the t values by amounts much greater than can be expected from sampling error. All of the critical values are much less than the true values. Thus, decisions made with the upper tail of the t test for samples from an exponential population can lead to rejection of the null hypothesis much less frequently than should be expected. From these results, would you expect that the result would generally be true for all nonsymmetric distributions?

5.12. PROBLEMS

5-1. Find the mean, variance, and standard deviation for the following values of a discrete random variable X and the corresponding probability mass function $(p_X(x))$:

x	0	1	2	3	4
$p_X(x)$	0.35	0.39	0.19	0.06	0.01

5-2. Given the values of the cumulative mass function $F_X(x)$ for the discrete random variable X, compute the mean, variance, and standard deviation:

x	0	1	2	3	4
$F_X(x)$	0.55	0.88	0.97	0.99	1.00

5-3. A continuous random variable has the following density function: $f_X(x) = 3x^2/26$ for the internal $1 \leq x \leq 3$. Find the mean, variance, and standard deviation of x.

5-4. Find the mean, variance, and standard deviation of the eight largest values of sediment yield given in Example 5-18.

5-5. Show the calculations of the mean and standard deviation for the data of Equation 5-110a.

5-6. Use the method of moments to derive an estimator of b for the following density function:

$$f_X(x) = \begin{cases} 2x/b^2 & \text{for } 0 \leq x \leq b \\ 0 & \text{otherwise} \end{cases}$$

5-7. Using the histogram for the daily evaporation data of Example 5-7, evaluate the parameters of a population having a uniform distribution using the method of moments ($\overline{X} = 0.1387$, and $S = 0.0935$).

5-8. Determine the maximum likelihood estimators of the parameters of (a) the normal distribution, and (b) the lognormal distribution using an approach similar to Example 5-5.

5-9. Write and execute a computer program to derive an approximation of the sampling distribution of the mean. Use a random number generator that generates standard normal deviates with a mean of zero and a variance of one. Generate 1000 samples of size n and compute the mean of each sample. Develop a histogram using the 1000 values and plot both the histogram and the normal distribution. Determine the number of sample means that lie in each tail for probabilities of 0.01, 0.025, 0.05, and 0.10. Execute the program for sample sizes of 5, 10, and 25.

5-10. A normal population is known to have mean of 10. (a) If the standard deviation of the population is 5, what is the probability that a random sample of size 25 will have a mean of 12 or higher? (b) What is the probability in (a) if $n = 100$? (c) If the standard deviation of the population is 10, what is the probability that a random sample of size 25 will have a mean of 12 or higher? (d) What is the probability in (c) for a mean of 6 or lower?

5-11. Write and execute a computer program to derive an approximation of the sampling distribution of the variance. Use a random number generator that generates standard normal deviate with a mean of zero and a variance of one. Generate 1000 samples of size n and compute the variance of each sample. Develop a histogram using the 1000 values and plot both the histogram and the chi-square distribution for $(n-1)$ degrees of freedom. Determine the number of sample variances that lie in each tail for probabilities of 0.01, 0.025, 0.05, and 0.10. Execute the program for sample sizes n of 5, 10, and 25.

5-12. A random variable X has a normal population with a mean μ and a variance of 3. (a) If samples of 10 are taken from the population, sketch the distribution of the variance. (b) Approximately what proportion of 10,000 samples, each with a sample size of 10, would have sample variances greater than 5.6?

5-13. A random variable X has a standard normal distribution. If many samples of 5 are drawn from the population, what sample variance will be exceeded 5% of the time? How does this compare with the 40 samples in Table 5-3c?

5-14. What are the characteristics of a null hypothesis? Alternative hypothesis?

5-15. What four factors influence the critical value of a test statistic? Show pictorially how each factor affects the critical value.

5-16. Define the region of rejection in terms of (a) values of the test statistic; (b) proportions of the area of the probability density function of the test statistic; (c) the region of acceptance; and (d) the critical value(s) of the test statistic.

5-17. What factors influence the selection of the level of significance? Discuss each factor.

5-18. A can of paint is advertised to cover an area of 400 ft². What are the implications of type I and type II errors?

5-19. Using the first sample of 5 in Table 5-3a, test the null hypotheses that $\mu = 0$. Use a two-sided alternative hypothesis and a 5% level of significance.

5-20. From Table 5-3a, find the sample that has a sample mean that deviates from 0 by the largest amount. Test the mean to determine if it is significantly different from 0. Use a two-sided alternative hypothesis and a 5% level of significance. Assuming the samples are from a N(0,1) population, is it reasonable that the null hypothesis would be rejected? Explain.

5-21. Two tests were provided for testing the hypothesis of a mean value against a standard, the Z test of Equation 5-47 and the t test of Equation 5-54. What are the differences of the two tests in terms of the purpose, data analysis requirements, underlying assumptions, and the critical value? Under what circumstances can the Z test be used as an alternative to the t test?

5-22. A sample of 20 yields a mean of 32.4. Test the two-sided hypothesis that the sample was drawn from a population with a mean of 35 for the following cases: (a) if the variance of the population is 33; and (b) if the variance of the population is unknown, but the sample variance is 33. Use a level of significance of 5%.

5-23. A random sample of 10 has a mean of 110. Test the null hypothesis that $\mu = 120$ against the alternative hypothesis that $\mu > 120$ at the 5% level of significance for the following cases: (a) assuming that the population standard deviation of 18 is known; and (b) assuming that the population standard deviation is unknown and the sample value is 18.

5-24. A random sample of 25 has a mean of 4.8 with a standard deviation of 0.32. Test the null hypothesis that $\mu = 4.95$ against the alternative hypothesis that $\mu < 4.95$ at the 1% level of significance.

5-25. A random sample of 10 yields a mean and standard deviation, respectively, of 73.6 and 7.9. Assuming a two-sided test, test the hypothesis that the sample is from a population with a mean of 80. Use a level of significance of (a) 5% and (b) 1%.

5-26. A public water supply official claims that the average household water use is greater than 350 gal/day. To test this assertion, a random sample of 200 households were questioned. If the random sample showed an average use of 359 gal/day and a standard deviation of 35 gal/day, would it be safe to conclude at the 1% level of significance that is greater than 350 gal/day?

5-27. A random sample of 12 has a mean and standard deviation of 240 and 30, respectively. Test the null hypothesis that $\mu = 215$ against the alternative hypothesis that $\mu > 215$ at a level of significance of (a) 5% and (b) 1%.

5-28. From the samples in Table 5-3, compare the sample means of the last two samples of 5 (1.412 vs. 0.094). Use a two-sided alternative hypothesis and a 5% level of significance.

5-29. The standard deviation of the 58 annual maximum discharges for the Piscataquis River (see Table 5-6) is 4095 ft^3/sec. Test the hypothesis that the population standard deviation is 3500 ft^3/sec. Make a two-tailed test and use a 5% level of significance.

5-30. The standard deviation of the 58 manual maximum discharges for the Piscataquis River (see Table 5-6) is 4095 ft^3/sec. Regional studies indicate that the standard deviation for that watershed should be at least 3500 ft^3/sec. Using a 1% level of significance, study whether or not the sample value of σ can reasonably be at least 3500 ft^3/sec.

5-31. Using the data from Table 5-3, test the significance of the two sample variances for the last two of the 40 samples (0.421 vs. 1.192). Use a 10% level of significance.

5-32. Compare the six steps of hypothesis testing with the six steps used to compute a confidence interval. Note the similarities and differences between the two procedures.

5-33. An engineering firm manufactures a missile component that will have a life of use that is approximately normally distributed with a population standard deviation of 3 hr. If a random sample of ten components has mean life of use of 13.4 hr, find a two-sided 99.5% confidence interval for the population mean of all components manufactured by this firm.

5-34. The mean life of a random sample of 20 tires of a certain brand is 27,500 mi. (a) If the standard deviation of the population is known to be 3500 mi, what is the two-sided 90% confidence on the mean? (b) If the variance of the population is not known and the standard deviation of the random sample is 3500 mi, what is the two-sided 90% confidence interval?

5-35. Twenty-five engineering students independently measured the melting point of lead. The mean and standard deviation of the 25 measurements were 329.6 and 5.1°C, respectively. Using a level of significance of 10%, what is the error that can be expected if the sample mean is used as an estimate of the true melting point of lead?

5-36. A random sample of test scores of the GRE exam by 25 students produces a mean of 575 and standard deviation of 60. (a) Determine the 99% confidence interval on the mean score for all students at the university. (b) What assertion can be made about the magnitude of the error, if we state that the mean score of all students at the university is 575, assuming that we wish to be 99% confident in our assertion?

5-37. A laboratory testing of five samples on the content of a certain water quality indicator results in the following values: 3.41, 3.06, 2.94, 3.27, and 3.32 mg/l. (a) Find the 95% confidence interval for the mean content of this indicator. (b) What assumptions are made?

5-38. Write and execute a computer program that generates 1000 samples from a standard normal distribution, with each sample of size n. For each sample compute the mean, standard deviation, and two-sided confidence intervals for confidence levels of 90 and 99%. For each confidence level, determine the number of samples for which the confidence intervals do not cover the true mean. Execute the program for sample sizes of 10, 25, and 50.

5-39. Using the first sample of 5 in Table 5-3, compute a two-sided, 95% confidence interval for the variance for the first sample ($S^2 = 1.764$). Does the interval cover the true variance in this case?

5-40. A lab technician wishes to determine the mean time to perform a lab experiment. What sample size will be needed to be 96% confident that the sample mean will be within 10 min of the true mean? Previous studies have indicated that $\sigma = 25$ min.

5-41. An efficiency expert is interested in knowing the mean number of days of sick leave taken by workers in the auto industry. How many workers' sick-leave records would have to be checked to be able to state that the sample mean will be within 3.0 days of the true mean? Assume a 5% level of significance and a value of the population standard deviation of 4.7 days.

5-42. A manufacturer produces a stereo component that will have a length of life that is normally distributed with a population standard deviation of 70 hr. (a) Show the change in the width of the 90% confidence interval for the population mean of all components produced by this manufacturer as a function of sample size. (b) What sample size is needed to ensure that the confidence interval is no greater than 40 hr?

5-43. For the conditions of Problem 5-41, compute the sample size that would be needed if the variance of the population is not known but the sample standard deviation would be approximately 4.7 days.

5-44. Make a normal frequency analysis for the January rainfall on Wildcat Creek (1954 to 1971). Use the Weibull plotting formula. Estimate: (a) the 100-year January rainfall (inches); and (b) the probability that the January rainfall will exceed 7 in. in any 1 year.

1954	6.43	1963	5.26
1955	5.01	1964	8.51
1956	2.13	1965	3.93
1957	4.49	1966	6.20
1958	3.59	1967	5.19
1959	3.99	1968	6.55
1960	8.62	1969	5.19
1961	2.55	1970	4.29
1962	4.56	1971	5.43

5-45. The following data are the annual minimum, 7-day low flows from Shoal Creek, TN (1926 to 1959). Perform a lognormal frequency analysis using logarithmic transformation to the base 10. Use the Weibull plotting formula. Estimate (a) the 25- and 100-year low flows, and (b) the probability that the 7-day low flow will be less than 60.

1926	99	1943	56
1927	90	1944	108
1928	116	1945	123
1929	142	1946	120
1930	99	1947	116
1931	63	1948	98
1932	128	1949	145
1933	126	1950	202
1934	93	1951	133
1935	83	1952	111
1936	83	1953	101
1937	111	1954	72
1938	112	1955	80
1939	127	1956	80
1940	97	1957	97
1941	71	1958	125
1942	84	1959	124

5-46. Perform a lognormal frequency analysis using logarithmic transformation to the base 10 of the total annual runoff (inches) from Wildcat Creek (1954 to 1970). Use the Weibull plotting formula. Estimate (a) the 50-year annual runoff; and (b) the probability that the annual runoff will be less than 5 in.

1954	8.56	1962	13.68
1955	6.86	1963	15.74
1956	14.13	1964	22.73
1957	11.76	1965	13.28
1958	10.74	1966	18.42
1959	7.72	1967	19.49
1960	11.48	1968	13.09
1961	16.76	1969	15.77
		1970	9.51

5-47. Use the data in Table 5-6 for the Piscataquis River with the sample mean and standard deviation (i.e., parameters) of 8620 and 4128 ft^3/sec, respectively, to test for the goodness of fit of the normal distribution. Use the chi-square test.

5-48. Use the data in Table 5-6 for the Piscataquis River with the sample log-mean (to the base 10) and log-standard deviation of 3.8894 and 0.2031 ft³/sec, respectively, to test for the goodness of fit of the lognormal distribution. Use the chi-square test. Note that sample log-mean is the mean of the transformed x sample values using $\log(x)$, and sample log-standard deviation is the standard deviation of the transformed x sample values using $\log(x)$.

5-49. The histogram of annual maximum discharges for the Piscataquis River is shown in Table 5-6 and Figure 5-8. Test whether or not it can be assumed that this random variable has a lognormal distribution with (i.e., parameters) $\log\text{-}\mu = 3.8$ and $\log\text{-}\sigma = 0.25$. Assume a 1% level of significance and use the chi-square test.

5-50. Using the sediment yield data presented in the histogram of Figure 5-15, test whether or not the sediment yield can be represented by a uniform distribution with $a = 0.0$ and $b = 2.5$. Use a level of significance of 5% and the chi-square test.

5-51. The histogram of annual maximum discharges for the Piscataquis River is shown in Table 5-6 and Figure 5-8. Test whether or not it can be assumed that this random variable has a normal distribution with $\mu = \overline{X}$ and $\sigma = S$. Assume a 10% level of significance and use the chi-square test.

5-52. Use the data in Table 5-6 for the Piscataquis River with the sample mean and standard deviation of 8620 and 4128 ft³/sec, respectively, to test for the goodness of fit of the normal distribution. Use the Kolmogorov-Smirnov test.

5-53. Use the data in Table 5-6 for the Piscataquis River with the sample log-mean and log-standard deviation (i.e., parameters) of 3.8894 and 0.2031, respectively, to test for the goodness of fit of the lognormal distribution. Use the Kolmogorov-Smirnov test.

5-54. Write a pseudocode that could be used to simulate the critical values for the normal, uniform, and exponential distributions presented in Example 5-21.

5-55. Simulate the critical values for the lower tail of the t statistic for the normal, uniform, and exponential distributions following the procedure of Example 5-21. Use a sample size of 10.

5-56. Gamma distributed random numbers g_i for scale parameter b and shape parameter c can be generated from uniform variates u_i as follows:

$$g_i = -b \ln\left(\prod_{i=1}^{c} u_i\right)$$

Using a sample size of 10, simulate the 10 and 5% critical values of the t statistic $t = (\overline{X} - \mu)/(S/\sqrt{n})$ where $\mu = bc$. With $b = 1$, perform the analyses for $c = 3, 6,$ and 10. Assess the applicability of the test for each of these populations. Would it be appropriate to use the t test for any of these populations? Explain.

Chapter 6
Regression Analysis

TABLE OF CONTENTS

6.1. INTRODUCTION

In analyzing single random variables, it is important to identify the underlying distribution of the random variable and make estimates of its parameters. Additionally, decision making requires an analysis of the uncertainty in the knowledge of the population. Hypothesis testing and interval estimation, i.e., confidence intervals, are means of assessing the significance of and quantifying this uncertainty.

In many cases, variation in the value of a random variable is associated with variation in one or more additional variables. By establishing the relationship between two or more variables, it may be possible to reduce the uncertainty in an estimate of the random variable of interest. Thus, before making a decision based on the analysis of observations on a random variable alone, it is wise to consider the possibility of systematic associations between the random variable of interest and other variables that have a causal relationship with it. These associations among variables can be understood using correlation and regression analyses.

To summarize these concepts, the variation in the value of a variable may be either random or systematic in its relationship with another variable. Random variation represents uncertainty. If the variation is systematically associated with one or more other variables, the uncertainty in estimating the value of a variable can be reduced by identifying the underlying relationship. Correlation and regression analyses are important statistical methods to achieve these objectives. They should be preceded by a graphical analysis to determine (1) if the relationship is linear or nonlinear, (2) if the relationship is direct or indirect, and (3) if any extreme events might control the relationship.

6.2. CORRELATION ANALYSIS

Correlation analysis provides a means of drawing inferences about the strength of the relationship between two or more variables. That is, it is a measure of the degree to which the values of these variables vary in a systematic manner. Thus it provides a quantitative index of the degree to which one or more variables can be used to predict the values of another variable.

It is just as important to understand what correlation analysis cannot do as it is to know what it can do. Correlation analysis does not provide an equation for predicting the value of a variable. Also, correlation analysis does not indicate whether a relationship is causal; that is, it is necessary for the investigator to determine whether there is a cause-and-effect relationship between the variables. Correlation analysis only indicates whether the degree of common variation is significant.

Correlation analyses are most often performed after an equation relating one variable to another has been developed. This happens because the correlation coefficient, which is a quantitative index of the degree of common linear variation, is used as a measure of the goodness of fit between the prediction equation and the data sample used to derive the prediction equation. But correlation analyses can be very useful in model formulation, and every model development should include a correlation analysis.

6.2.1. Graphical Analysis

The first step in examining the relationship between variables is to perform a graphical analysis. Visual inspection of the data can identify the following information:

1. Degree of common variation, which is an indication of the degree to which the two variables are related
2. Range and distribution of the sample data points
3. Presence of extreme events
4. Form of the relationship between the two variables
5. Type of relationship

Each of these factors is of importance in the statistical analysis of sample data and in decision making.

When variables show a high degree of association, we assume that a causal relationship exists. If there is a physical reason to suspect that a causal relationship exists, the association demonstrated by the sample data provides empirical support for the assumed relationship. Common variation implies that, when the value of one of the random variables is changed, the value of the other variable will change in a systematic manner. For example, an increase in the value of one variable occurs when the value of the other variable increases. If the change in the one variable is highly predictable from a given change in the other variable, there is a high degree of common variation. Figure 6-1 shows graphs of different samples of data for two variables having different degrees of common variation. The common variation is measured in the figure using the correlation coefficient (R). In Figure 6-1a and 6-1e the degree of common variation is very high; thus the variables are said to be correlated. In Figure 6-1c there is no correlation between the two variables because, as the value of X is increased, it is not certain whether Y will increase or decrease. In Figure 6-1b and 6-1d, the degree of correlation is moderate; in Figure 6-1b, evidently Y will increase as X is increased, but the exact change in Y for a change in X is difficult to estimate from just looking at the graph. A more quantitative description of the concept of common variation is needed.

It is important to use a graphical analysis to identify the range and distribution of the sample data points, so that the stability of the relationship can be assessed and so that we can assess the ability of the data sample to represent the distribution of the population. If the range of the data is limited, a computed relationship may not be stable; that is, it may not apply to the distribution of the population. The data set of Figure 6-2 shows a case where the range of the sample is much smaller than the expected range of the population. If an attempt is made to use the sample to project the relationship between the two random variables, a small change in the slope of the relationship causes a large change in the predicted estimate of Y for values of X at the extremes of the range of the population. A graph of two random variables might alert the investigator to a sample in which the range of the sample data may cause stability problems in a derived relationship between two random variables, especially when the relationship is projected beyond the range of the sample data.

It is important to identify extreme events in a sample of data for several reasons. First, extreme events can dominate a computed relationship between two variables. For example, in Figure 6-3a, the extreme point suggests a high correlation between X and Y; in this case, the cluster of points acts like a single observation. In Figure 6-3b, the extreme point causes a poor correlation between the two random variables; since the mean of the cluster of points is nearly the same as the value of Y of the extreme point, the data of Figure 6-3b suggest that a change in X is not associated with a change in Y. A correlation coefficient is more sensitive to an extreme point when the sample size is small. An extreme event may be due to (1) errors in recording or plotting the data or (2) a legitimate observation in the tail of the distribution. Therefore, an extreme event must be identified and the reason for the event determined. Otherwise, it is not possible to properly interpret the results of the correlation analysis.

Relationships can be linear or nonlinear. Since the statistical methods to be used for the two forms of a relationship differ, it is important to identify the form. Additionally, the most frequently used correlation coefficient depends on a linear relationship existing between the two random variables; thus low correlation may

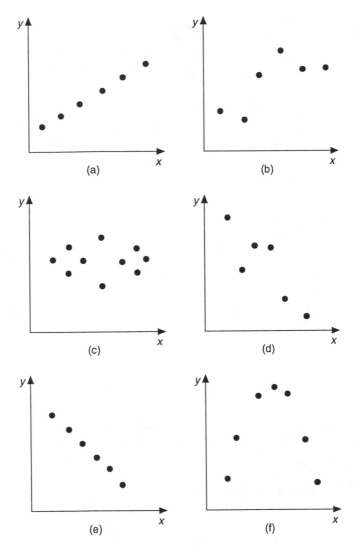

FIGURE 6-1 Different degrees of correlation between variables (X and Y):
(a) $R = 1.0$; (b) $R = 0.5$; (c) $R = 0.0$; (d) $R = -0.5$; (e) $R = -1.0$; (f) $R = 0.3$.

result for a nonlinear relationship even when the relationship is obvious. For example, the bivariate relationship of Figure 6-1f suggests a very predictable trend in the relationship between Y and X; but the correlation is poor and is certainly not as good as that in Figure 6-1a, even though the two plots suggest equal levels of predictability.

Graphs relating pairs of variables can be used to identify the type of the relationship. Linear trends can be either direct or indirect, with an indirect relationship indicating a decrease in Y as X increases. This information is useful for checking the rationality of the relationship, especially when dealing with data sets that include more than two variables. A variable that is not dominant in the physical relationship

FIGURE 6-2 Instability in the relationship between random variables.

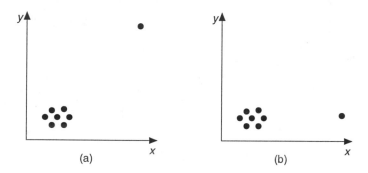

FIGURE 6-3 Effect of an extreme event in a data sample on the correlation:
(a) high correlation; (b) low correlation.

may demonstrate a physically irrational relationship with another variable because of the values of the other variable affecting the physical relationship.

6.2.2. Bivariate Correlation

Correlation is the degree of association between the elements of two samples of data, that is, between observations on two variables. Correlation coefficients provide a quantitative index of the degree of linear association. Examples of variables that are assumed to have a causal relationship and significant correlation are (1) the cost of living and wages and (2) the volumes of rainfall and flood runoff. However, examples that have no cause-and-effect relationship but may have significant correlation are (1) the crime rate and the sale of chewing gum over the past two decades and (2) annual population growth rates in 19th-century France and annual cancer death rates in the U.S. in the 20th century.

Many indexes of correlation exist. The method that is used most frequently is the *Pearson product-moment* correlation coefficient. Nonparametric correlation indexes include the contingency coefficient, the Spearman rank correlation coefficient, and the Kendall rank correlation coefficient. Only the Pearson correlation coefficient is considered in this chapter.

6.2.3. Separation of Variation

A set of observations on a random variable Y has a certain amount of variation, which may be characterized by the variance of the sample. The variance equals the sum of squares of the deviations of the observations from the mean of the observations divided by the degrees of freedom. Ignoring the degrees of freedom, the variation in the numerator can be separated into two parts: (1) variation associated with a second variable X and (2) variation not associated with X. That is, the total variation (TV), which equals the sum of the squares of the sample data points about the mean of the data points, is separated into the variation that is explained by variation in the second variable (EV) and the variation that is not explained, that is, the unexplained variation, UV. Thus TV can be expressed as

$$TV = EV + UV \tag{6-1}$$

Using the general form of the variation of a random variable, each of the three terms in this equation can be represented by a sum of squares as follows:

$$\sum_{i=1}^{n} \left(y_i - \overline{Y} \right)^2 = \sum_{i=1}^{n} \left(\hat{y}_i - \overline{Y} \right)^2 + \sum_{i=1}^{n} \left(y_i - \hat{y}_i \right)^2 \tag{6-2}$$

where y_i = an observation on the random variable, $\hat{y}_i$ = the value of Y estimated from the best linear relationship with the second variable X, and $\overline{Y}$ = the mean of the observations on Y. These variation components are shown in Figure 6-4; the dashed lines illustrate the deviations of Equation 6-2. Figure 6-4a shows the variation of the observations about the mean value; this represents the variation that exists in the sample and that may potentially be explained by the variable X; it reflects the left side of Equation 6-2. The variation in Figure 6-4b represents the variation of the line about the mean and thus corresponds to the first term of the right side of Equation 6-2. If all of the sample points fall on the line, the explained variation will equal the total variation. The variation in Figure 6-4c represents the variation of the points about the line; therefore, it corresponds to the second term on the right side of Equation 6-2. This is the variation that is not explained by the relationship between Y and X. If this unexplained variation equals the total variation, the line does not explain any of the variation in the sample, and the relationship between X and Y is not significant.

6.2.4. Correlation: Fraction of Explained Variation

The separation of variation concept is useful for quantifying the correlation coefficient by Pearson. Specifically, dividing both sides of Equation 6-1 by the total variation TV gives

$$1 = \frac{EV}{TV} + \frac{UV}{TV} \tag{6-3}$$

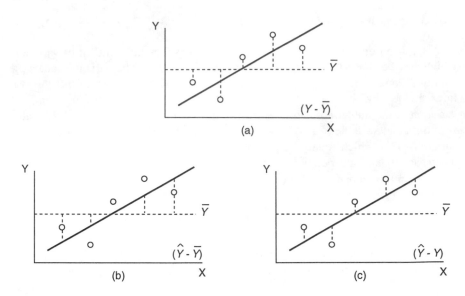

FIGURE 6-4 Separation of variation: (a) total variation; (b) explained variation; (c) unexplained variation.

The ratio EV/TV represents the fraction of the total variation that is explained by the linear relationship between Y and X; this is called the *coefficient of determination* and is given by

$$R^2 = \frac{\text{EV}}{\text{TV}} = \frac{\sum_{i=1}^{n}(\hat{y}_i - \overline{Y})^2}{\sum_{i=1}^{n}(y_i - \overline{Y})^2} \tag{6-4}$$

The square root of the ratio is the correlation coefficient, R. If the explained variation equals the total variation, the correlation coefficient will equal 1. If the relationship between X and Y is inverse, as in Figure 6-1e, and the explained variation equals the total variation in magnitude, R will equal −1. These represent the extremes, but both values indicate a perfect association, with the sign only indicating the direction of the relationship. If the explained variation equals zero, R equals zero. Thus a correlation coefficient of zero, which is sometimes called the null correlation, indicates no linear association between the two variables X and Y.

6.2.5. Computational Form for Correlation Coefficient

While Equation 6-4 provides the means to compute a value of the correlation coefficient, it can be shown that Equation 6-4 can be rearranged to the following form using a linear model for $\hat{Y}$ that minimizes the unexplained variation (UV):

$$R = \frac{\sum_{i=1}^{n} x_i y_i - \frac{1}{n}\left(\sum_{i=1}^{n} x_i\right)\left(\sum_{i=1}^{n} y_i\right)}{\sqrt{\sum_{i=1}^{n} x_i^2 - \frac{1}{n}\left(\sum_{i=1}^{n} x_i\right)^2} \sqrt{\sum_{i=1}^{n} y_i^2 - \frac{1}{n}\left(\sum_{i=1}^{n} y_i\right)^2}}$$ (6-5)

This equation is used most often because it does not require prior computation of the means, and thus the computational algorithm is easily programmed for a computer. The linear model $\hat{Y}$ that minimizes UV is based on the principle of least squares, as discussed in Section 6.4.

6.2.6. Distribution of the Correlation Coefficient

It should be recognized that the estimated correlation coefficient is a random variable, and each random variable must have a distribution function. The distribution of R is a function of the sample size n and the population value of the correlation coefficient ρ. Unfortunately, the correlation coefficient does not have a recognizable probability distribution.

6.2.6.1. Effect of ρ (n Constant)

Figure 6-5 shows the distribution of R for three values of ρ, with n held constant at 10. The most likely sample value is the population value. For $\rho = 0$, a sample value near zero (i.e., $\rho = 0$) is most likely. For $\rho = 0.5$, the probability of a sample value at or near zero is considerably smaller than where $\rho = 0$. For $\rho = 0.9$, there is very little chance of obtaining a sample estimate that is near zero, but a value near 0.9 is highly likely.

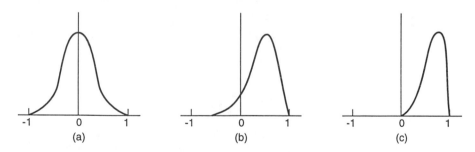

FIGURE 6-5 Effect of ρ ($n = 10$) on the distribution of R: (a) $\rho = 0$; (b) $\rho = 0.5$; (c) $\rho = 0.9$.

6.2.6.2. Effect of n (ρ Constant)

Figure 6-6 shows the distribution of R for sample sizes of 3, 4, and 25, with ρ held constant at zero. Figure 6-6a shows that for a sample size of 3, a large sample

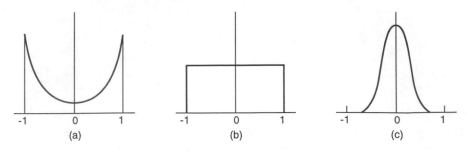

FIGURE 6-6 Effect of n ($\rho = 0$) on the distribution of R: (a) $n = 3$; (b) $n = 4$; (c) $n = 25$.

estimate of the correlation coefficient is more likely than a value near zero even when the population value equals zero. For a sample size of 4 (Figure 6-6b), all values of R are equally likely. As n increases to 10 (Figure 6-5a) and 25 (Figure 6-6c), sample estimates near zero become more likely. In summary, it is important to recognize that for small sample sizes, large sample estimates of R are very likely to occur and that even high correlation may not represent a significant relationship between the two random variables.

6.2.7. Hypothesis Testing

A correlation coefficient of zero indicates that there is no linear relationship between the two variables. Thus it would be of interest to make a two-sided test of the hypotheses:

$$H_o : \rho = 0 \tag{6-6}$$

$$H_A : \rho \neq 0 \tag{6-7}$$

It can be shown that for $n > 2$ these hypotheses can be tested using a t statistic that is given by

$$t = \frac{R}{\sqrt{\dfrac{1 - R^2}{n - 2}}} \tag{6-8}$$

where $t =$ the value of a random variable that has approximately a t distribution with $(n - 2)$ degrees of freedom. The correlation coefficient is considered statistically significant if the computed t value is greater in absolute value than a critical value having a t distribution and a level of significance of α; that is, the null hypothesis is rejected when the absolute value of the computed t statistic is greater than the critical value. One-sided tests of hypothesis are also possible using the test statistic above. Appendix A-5 provides the critical values of R for both one-sided and two-sided tests. A value of R greater than the critical value is considered significant (i.e., the null hypothesis of Equation 6-6 should be rejected). For negative values of R, the absolute value of the critical value in Table A-5 should be used.

In many cases, it is of interest to examine whether or not one can assume that the population has a value of the correlation coefficient ρ_o other than zero. The two-sided hypotheses for this test are

$$H_o: \rho = \rho_o \tag{6-9a}$$

$$H_A: \rho \neq \rho_o \tag{6-9b}$$

One-sided tests can also be performed using the same procedure as the two-sided alternative. The test statistic for the hypotheses of Equation 6-9 is

$$R' = 0.5 \ln\left(\frac{1+R}{1-R}\right) \tag{6-10}$$

in which R = the sample correlation coefficient. The random variable R' has a mean $\overline{R}'$ and standard deviation $S_{R'}$ given by

$$\overline{R}' = 0.5 \ln\left(\frac{1+\rho_o}{1-\rho_o}\right) \tag{6-11}$$

$$s_{R'} = \frac{1}{\sqrt{n-3}} \tag{6-12}$$

in which n = the sample size. A standard normal transformation yields

$$Z = \frac{R' - \overline{R}'}{s_{R'}} \tag{6-13}$$

in which Z = the value of a random variable having a standard normal distribution with a mean of zero and a standard deviation of 1. If either the sample (R) or population (ρ_o) correlation coefficient is negative, the absolute value should be used. The null hypothesis is rejected where the absolute value of the computed z coefficient of Equation 6-13 is greater than the critical value for the selected level of significance.

Example 6-1: Hypothesis Testing of Correlation Coefficients

The objective of this example is to illustrate both the hypothesis tests for correlation coefficients and the problem of using small sample sizes to estimate the correlation between two random variables. Assume that a sample of five yields a correlation coefficient of 0.87, i.e., $R = 0.87$. The null hypothesis of Equation 6-6 is tested using the t statistic of Equation 6-8 as follows:

$$t = \frac{0.87}{\sqrt{\dfrac{1 - 0.87^2}{5 - 2}}} = 3.056 \tag{6-14}$$

For a 5% level of significance, the critical t value for 3 degrees of freedom and $\alpha/2 = 0.025$ obtained from Table A-2 is 3.182; therefore, the null hypothesis should be accepted, and there is no reason to believe that the sample does not come from a population having a correlation coefficient of zero.

The same sample can be used to test the hypotheses of Equation 6-9 with $\rho_o = 0.99$. The computations for Equations 6-10 to 6-13 are

$$R' = 0.5 \ln\left(\frac{1 + 0.87}{1 - 0.87}\right) = 1.3331 \tag{6-15}$$

$$\overline{R}' = 0.5 \ln\left(\frac{1 + 0.99}{1 - 0.99}\right) = 2.6467 \tag{6-16}$$

$$s_{R'} = \frac{1}{\sqrt{5 - 3}} = 0.7071 \tag{6-17}$$

$$z = \frac{1.3331 - 2.6467}{0.7071} = -1.8577 \tag{6-18}$$

For a 5% level of significance, the critical z value obtained from Table A-1 is 1.96. Therefore, the null hypothesis that ρ equals 0.99 should be accepted.

In summary, the sample of five was used to illustrate the computation of hypothesis tests on the population correlation coefficient. However, it is more important to note important characteristics of hypothesis testing for correlation coefficients. First, estimates of R obtained from small samples may not be close to the population value; in this case, the sample value of 0.87 was not significantly different from either 0.0 or 0.99. This suggests that with a small sample, an engineer could prove almost anything, including an opinion based on personal bias. Second, the 5% level of significance may not be the appropriate level. Third, it is inadequate just to investigate the null hypothesis for null correlation; confidence intervals on ρ should be calculated.

Example 6-2: Computational Illustration

A sample of five observations, which is given as the values of X and Y in Table 6-1, is used to illustrate the computational methods outlined herein. The last row of Table 6-1 contains column summations. Using Equation 6-4, the correlation is

TABLE 6-1 Calculation of the Correlation Coefficient

x_i	y_i	$(x_i - \overline{X})$	$(y_i - \overline{Y})$	$(x_i - \overline{X})$ $(y_i - \overline{Y})$	$\hat{y}_i$	$\hat{y}_i - \overline{Y}$	$(\hat{y}_i - \overline{Y})^2$	$(\hat{y}_i - \overline{Y})^2$	$x_i y_i$	x_i^2	y_i^2
1	2	−2	−1	2	1.2	−1.8	3.24	1	2	1	4
2	1	−1	−2	2	2.1	−0.9	0.81	4	2	4	1
3	3	0	0	0	3.0	0.0	0.00	0	9	9	9
4	4	1	1	1	3.9	0.9	0.81	1	16	16	16
5	5	2	2	4	4.8	1.8	3.24	4	25	25	25
15	15			9			8.10	10	54	55	55

$$R = \sqrt{\frac{8.10}{10}} = 0.9 \tag{6-19}$$

The predicted values, $\hat{y}_i$, were obtained by fitting a regression equation to the data points; the regression method is discussed in the remaining sections of this chapter. Therefore, Equation 6-5 yields

$$R = \frac{54 - \dfrac{15(15)}{5}}{\sqrt{55 - \dfrac{15(15)}{5}} \sqrt{55 - \dfrac{15(15)}{5}}} = 0.9 \tag{6-20}$$

Thus, 81% of the total variation in Y is explained by variation in X.

6.3. INTRODUCTION TO REGRESSION

In many engineering projects, estimates of random variables are required. For example, in the design of irrigation projects, it is necessary to provide estimates of evaporation. If we are fortunate to have a past record of daily evaporation rates from a pan, the mean of these observations multiplied by some proportionality constant (a pan coefficient) may be our best estimate. However, the error in such an estimate can be significant. Because the standard deviation of the observations is a measure of the variation in the past, it is our best estimate of the future variation and thus the error in a single measurement. The error in a mean value is also a function of the standard deviation.

If the random variable can be related to other variables, it may be possible to reduce the error in a future estimate. For example, evaporation is a function of the temperature and humidity of the air mass that is over the water body. Thus, if measurements of the air temperature and the relative humidity are also available, a relationship, or model, can be developed. The relationship may provide a more accurate estimate of evaporation for the conditions that may exist in the future.

The process of deriving a relationship between a random variable and measured values of other variables is called *optimization*, or *model calibration*. It makes use of the fundamental concepts of calculus and numerical analysis. The objective of optimization is to find the values of a vector of unknowns that provides the minimum or maximum value of some function. Before examining regression analysis, it is necessary to discuss the fundamentals of statistical optimization.

6.3.1. Elements of Statistical Optimization

In statistical optimization, the function to be optimized is called the *objective function*, which is an explicit mathematical function that describes what is considered the optimal solution. A mathematical model relates a random variable, called the *criterion* or *dependent variable*, to a vector of unknowns, called regression coefficients, and a vector of *predictor variables*, which are sometimes called *independent variables*. The predictor variables are usually variables that have a causal relationship with the criterion variable. For example, if we are interested in using measurements of air temperature and humidity to predict evaporation rates, the evaporation would be the criterion variable and the temperature and relative humidity would be the predictor variables. While the objective function of statistical optimization corresponds to the explicit function, the vector of unknowns in statistical optimization corresponds to the unknowns in the explicit function. The unknowns are the values that are necessary to transform values of the predictor variable(s) into a predicted value of the criterion variable. It is important to note that the objective function and the mathematical model are two separate explicit functions. The third element of statistical optimization is a data set. The *data set* consists of measured values of the criterion variable and the predictor variable(s). In summary, statistical optimization requires (1) an objective function, which defines what is meant by best fit; (2) a mathematical model, which is an explicit function relating a criterion variable to vectors of unknowns and predictor variable(s); and (3) a matrix of measured data.

As an example, we may attempt to relate evaporation (E), the criterion variable, to temperature (T), which is the predictor variable, using the equation

$$\hat{E} = b_o + b_1 T \tag{6-21}$$

in which b_o and b_1 = the unknown coefficients, and $\hat{E}$ = the predicted value of E. To evaluate the unknowns b_o and b_1, a set of simultaneous measurements on E and T would be made. For example, if we were interested in daily evaporation rates, we may measure both the total evaporation for each day in a year and the corresponding mean daily temperature; this would give us 365 observations from which we could estimate the two regression coefficients. An objective function would have to be selected to evaluate the unknowns; for example, regression minimizes the sum of the squares of the differences between the predicted and measured values of the criterion variable. Other criterion functions, however, may be used.

6.3.2. Zero-Intercept Model

Assume that the criterion variable Y is related to a predictor variable X using the linear model

$$\hat{Y} = bX \tag{6-22}$$

in which b = the unknown and $\hat{Y}$ = the predicted value of Y. The objective function, F, is defined as minimizing the sum of the squares of the differences (e_i) between the predicted values $(\hat{y}_i)$ and the measured values (y_i) of Y as follows:

$$F = \min \sum_{i=1}^{n} e_i^2 = \min \sum_{i=1}^{n} (\hat{y}_i - y_i)^2 \tag{6-23}$$

where n = the sample size.

To derive a value for b, differentiate $\sum_{i=1}^{n} e_i^2$ with respect to the unknown, set the derivative equal to zero, and solve for the unknown. Substituting Equation 6-22 into Equation 6-23 yields

$$F = \min \sum_{i=1}^{n} (\hat{y}_i - y_i)^2 = \min \sum_{i=1}^{n} (bx_i - y_i)^2 \tag{6-24}$$

The derivative of $\sum_{i=1}^{n} e_i^2$ with respect to the unknown b is

$$\frac{\partial \sum_{i=1}^{n} e_i^2}{\partial b} = 2 \sum_{i=1}^{n} (bx_i - y_i)(x_i) \tag{6-25}$$

Setting Equation 6-25 to 0 and solving for b yields

$$b = \frac{\sum_{i=1}^{n} x_i y_i}{\sum_{i=1}^{n} x_i^2} \tag{6-26}$$

The value of b can be computed using the two summations, $\sum_{i=1}^{n} x_i y_i$ and $\sum_{i=1}^{n} x_i^2$.

Example 6-3: Fitting a Zero-Intercept Model

The following table gives the data that can be used for fitting the value of b:

i	x_i	y_i
1	3	2
2	5	3
3	6	4

The following table shows the necessary computations:

i	x_i	x_i^2	y_i	$x_i y_i$
1	3	9	2	6
2	5	25	3	15
3	6	36	4	24
Column summation		70		45

Therefore, the value of b equals $\dfrac{45}{70}$ or 0.64286. The following equation could then be used to derive a future estimate of Y for a future value of X:

$$\hat{Y} = 0.64286X \qquad (6\text{-}27)$$

It is important to note that Equation 6-26 is valid only for the linear model of Equation 6-22 that does not include an intercept; Equation 6-26 would not be valid for Equation 6-21, since Equation 6-21 includes an intercept coefficient (b_o).

6.3.3. Regression Definitions

The objective of regression is to evaluate the coefficients of an equation relating the criterion variable to one or more other variables, which are called the *predictor variables*. The predictor variables are variables in which their variation is believed to cause or agree with variation in the criterion variable. A predictor variable is often called an *independent variable*; this is a misnomer in that independent variables are usually neither independent of the criterion variable nor independent of other predictor variables.

The most frequently used linear model relates a criterion variable Y to a single predictor variable X by the equation

$$\hat{Y} = b_0 + b_1 X \qquad (6\text{-}28)$$

in which b_0 = the intercept coefficient and b_1 = the slope coefficient; b_0 and b_1 are called *regression coefficients* because they are obtained from a regression analysis. Because Equation 6-28 involves two variables, Y and X, it is sometimes referred to as the *bivariate model*. The intercept coefficient represents the value of Y when X equals zero. The slope coefficient represents the rate of change in Y with respect to change in X. Whereas b_0 has the same dimensions as Y, the dimensions of b_1 equal the ratio of the dimensions of Y to X.

The linear multivariate model relates a criterion variable to two or more predictor variables:

$$\hat{Y} = b_0 + b_1 X_1 + b_2 X_2 + \ldots + b_p X_p \qquad (6\text{-}29)$$

in which p = the number of predictor variables, X_i = the ith predictor variable, b_i = the ith slope coefficient, and b_0 is the intercept coefficient, where i =1, 2, ..., p. The coefficients b_i are often called *partial regression coefficients* and have dimensions equal to the ratio of the dimensions of Y to X_i. Equation 6-28 is a special case of Equation 6-29.

6.4. PRINCIPLE OF LEAST SQUARES

6.4.1. Definitions

The values of the slope and intercept coefficients of Equations 6-28 and 6-29 can be computed using the principle of least squares. The principle of least squares is a process of obtaining *best* estimates of the coefficients and is referred to as a regression method. Regression is the tendency for the expected value of one of two jointly correlated random variables to approach more closely the mean value of its set than any other. The principle of least squares is used to regress Y on either X or the X_i values of Equations 6-28 and 6-29, respectively. To express the principle of least squares, it is important to define the error, e, or residual, as the difference between the predicted and measured values of the criterion variable:

$$e_i = \hat{y}_i - y_i \qquad (6\text{-}30)$$

in which $\hat{y}$ = the ith predicted value of the criterion variable, y_i = the ith measured value of Y, and e_i = the ith error. It is important to note that the error is defined as the measured value of Y subtracted from the predicted value. Some computer programs use the measured value minus the predicted value. This definition is avoided because it indicates that a positive residual implies underprediction. With Equation 6-30 a positive residual indicates overprediction, while a negative residual indicates underprediction. The objective function for the principle of least squares is to minimize the sum of the squares of the errors:

$$F = \min \sum_{i=1}^{n} (\hat{y}_i - y_i)^2 \tag{6-31}$$

in which n = the number of observations on the criterion variable, i.e., the sample size.

6.4.2. Solution Procedure

After inserting the model of $\hat{Y}$ in Equation 6-31, the objective function of Equation 6-31 can be minimized by taking the derivatives with respect to each unknown, setting the derivatives equal to zero, and then solving for the unknowns. The solution requires the model for predicting y_i to be substituted into the objective function. It is important to note that the derivatives are taken with respect to the unknown b_i and not the predictor variables X_i.

To illustrate the solution procedure, the model of Equation 6-28 is substituted into the objective function of Equation 6-31, which yields

$$F = \min \sum_{i=1}^{n} (b_0 + b_1 x_i - y_i)^2 \tag{6-32}$$

The derivatives of the sum of the squares of the errors with respect to the unknowns b_0 and b_1 are, respectively,

$$\frac{\partial F}{\partial b_0} = 2 \sum_{i=1}^{n} (b_0 + b_1 x_i - y_i) = 0 \tag{6-33a}$$

$$\frac{\partial F}{\partial b_1} = 2 \sum_{i=1}^{n} (b_0 + b_1 x_i - y_i) x_i = 0 \tag{6-33b}$$

Dividing each equation by 2, separating the terms in the summations, and rearranging yields the set of normal equations:

$$n b_0 + b_1 \sum_{i=1}^{n} x_i = \sum_{i=1}^{n} y_i \tag{6-34a}$$

$$b_0 \sum_{i=1}^{n} x_i + b_1 \sum_{i=1}^{n} x_i^2 = \sum_{i=1}^{n} x_i y_i \tag{6-34b}$$

All the summations in Equation 6-34 are calculated over all values of the sample. Sometimes the index values of the summation are omitted when they refer to summation over all elements in a sample of size n. Also, the subscripts for X and Y can be omitted but are inferred. The two unknowns b_0 and b_1 can be evaluated by solving the following two simultaneous equations:

$$b_1 = \frac{\sum_{i=1}^{n} x_i y_i - \frac{1}{n} \sum_{i=1}^{n} x_i \sum_{i=1}^{n} y_i}{\sum_{i=1}^{n} x_i^2 - \frac{1}{n} \left(\sum_{i=1}^{n} x_i \right)^2} \tag{6-35a}$$

$$b_0 = \overline{Y} - b_1 \overline{X} = \frac{\sum_{i=1}^{n} y_i}{n} - \frac{b_1 \sum_{i=1}^{n} x_i}{n} \tag{6-35b}$$

Example 6-4: Least-Squares Analysis

The data of Table 6-2 can be used to illustrate the solution of the least-squares principle for a data set of five observations on two variables. Substituting the sums of Table 6-2 into the normal equations of Equation 6-34 gives

$$5 b_0 + 50 b_1 = 10 \tag{6-36a}$$

$$50 b_0 + 510 b_1 = 109 \tag{6-36b}$$

Solving Equations 6-36a and 6-36b for the unknowns yields the model

$$\hat{Y} = -7 + 0.9X \tag{6-37}$$

The errors, which are shown in Table 6-2, can be computed by substituting the measured values of X into Equation 6-37 to obtain predicted values that are usd in Equation 6-30 to obtain the errors. In solving a linear model using the least-squares principle, the sum of the errors always equals zero, which is shown in Table 6-2. The sum of the squares of the errors equals 1.9, which is a minimum.

Note that if a line is passed through the first three observations of Table 6-2 the error for each of these three observations is zero. Such a model would be

$$\hat{Y} = -8 + 1.0X \tag{6-38}$$

While three of the residuals are zero, the sum of the squares of the errors for Equation 6-38 equals 2.0, which is greater than the corresponding sum for Equation 6-37. Thus the least-squares solution always provides the straight line with the smallest sum of squares of the errors.

TABLE 6-2 Least-Squares Computations

i	x_i	y_i	x_i^2	y_i^2	x_iy_i	e_i	e_i^2
1	12	4	144	16	48	-0.2	0.04
2	11	3	121	9	33	-0.1	0.01
3	10	2	100	4	20	0.0	0.00
4	9	0	81	0	0	1.1	1.21
5	8	1	64	1	8	-0.8	0.64
Column summation	50	10	510	30	109	0.0	1.90

6.5. RELIABILITY OF THE REGRESSION EQUATION

Having evaluated the coefficients of the regression equation, it is of interest to evaluate the reliability of the regression equation. The following criteria should be assessed in evaluating the model: (1) the correlation coefficient, (2) the standard error of estimate, (3) the F statistic for the analysis of variance; (4) the rationality of the coefficients and the relative importance of the predictor variables, both of which can be assessed using the standardized partial regression coefficients; and (5) the degree to which the underlying assumptions of the regression model are met.

6.5.1. Correlation Coefficient

As suggested in Section 6.2, the correlation coefficient (R) is an index of the degree of linear association between two random variables. The magnitude of R indicates whether the regression provides accurate predictions of the criterion variable. Thus R is often computed before the regression analysis is performed in order to determine whether it is worth the effort to perform the regression. However, R is always computed after the regression analysis because it is an index of the goodness of fit. The correlation coefficient measures the degree to which the measured and predicted values agree and is used as a measure of the accuracy of future predictions. It must be recognized that if the measured data are not representative of the population, that is, data that will be observed in the future, the correlation coefficient cannot be indicative of the accuracy of future predictions.

The square of the correlation coefficient (R^2) equals the percentage of the variance in the criterion variable that is explained by the predictor variable. Because of this physical interpretation, R^2 is a meaningful indicator of the accuracy of predictions.

6.5.2. Standard Error of Estimate

In the absence of additional information, the mean is the best estimate of the criterion variable; the standard deviation S_Y of Y is an indication of the accuracy of prediction. If Y is related to one or more predictor variables, the error of prediction is reduced from S_Y to the standard error of estimate, S_e. Mathematically, the standard error of estimate equals the standard deviation of the errors, has the same units as Y, and is given by

$$S_e = \sqrt{\frac{1}{\nu} \sum_{i=1}^{n} (\hat{y}_i - y_i)^2} \qquad (6\text{-}39)$$

in which ν = the degree of freedom, which equals the sample size minus the number of unknowns estimated by the regression procedure. For the bivariate model of Equation 6-28, $p = 1$ and $\nu = n - 2$. For the general linear model with an intercept, Equation 6-29, there are $(p + 1)$ unknowns; thus $\nu = n - p - 1$. In terms of the separation of variation concept discussed previously, the standard error of estimate equals the square root of the ratio of the unexplained variation to the degrees of freedom. It is important to note that S_e is based on $(n - p - 1)$ degrees of freedom, while the value, S_Y, is based on $(n - 1)$ degrees of freedom. Thus in some cases S_e may be greater than S_Y. To assess the reliability of the regression equation, S_e should be compared with the bounds of zero and S_Y. If S_e is near S_Y, the regression has not been successful. If S_e is much smaller than S_Y and is near zero, the regression analysis has improved the reliability of prediction.

The standard error of estimate is sometimes computed using the following relationship:

$$S_e = S_Y \sqrt{1 - R^2} \qquad (6\text{-}40)$$

Equation 6-40 must be considered as only an approximation to Equation 6-39 because R is based on n degrees of freedom and S_Y is based on $(n - 1)$ degrees of freedom.

Using the separation-of-variation concept, the exact relationship between S_e, S_Y, and R can be computed as follows:

$$TV = EV + UV \qquad (6\text{-}41)$$

The total variation is related to the variance of Y by

$$S_Y^2 = \frac{TV}{n-1} \qquad (6\text{-}42)$$

The square of the correlation coefficient is the ratio of the explained variation (EV) to the total variation (TV):

$$R^2 = \frac{EV}{TV} \qquad (6\text{-}43)$$

The standard error of estimate is related to the unexplained variation (UV) by

$$S_e^2 = \frac{UV}{n-p-1} \tag{6-44}$$

Equation 6-42 can be solved for TV, which can then be substituted into both Equations 6-41 and 6-43. Equation 6-44 can be solved for UV, which is also substituted into Equation 6-41. Solving for S_e^2 yields

$$S_e^2 = \left(\frac{n-1}{n-p-1}\right) S_Y^2 (1-R^2) \tag{6-45}$$

Thus Equation 6-45 is a more exact relationship than Equation 6-40. However, for large sample sizes the difference between the two estimates is small.

Although S_e may actually be greater than S_Y, in general, S_e is within the range from zero to S_Y. When a regression equation fits the data points exactly, S_e equals zero; this corresponds to a correlation coefficient of 1. When the correlation coefficient equals zero, S_e equals S_Y. As was indicated previously, S_e may actually exceed S_Y when the degrees of freedom have a significant effect. The standard error of estimate is often preferred to the correlation coefficient because S_e has the same units as the criterion variable and its magnitude is a physical indicator of the error; the correlation coefficient is only a standardized index and does not properly account for degrees of freedom lost because of the regression coefficients.

6.5.3. Analysis of Variance

The regression coefficient (b_1) for the predictor variable X in Equation 6-28 is a measure of the change in Y that results from a change in X. Recognizing that the regression coefficient b_1 is dependent on the units of both Y and X, it is reasonable to ask whether or not a change in X causes a significant change in Y. This is easily placed in the form of the following statistical hypothesis test:

$$H_o: \beta_1 = 0 \tag{6-46a}$$

$$H_A: \beta_1 \neq 0 \tag{6-46b}$$

in which β_1 = the population value of the slope coefficient. The hypotheses are designed to test whether or not the relationship between Y and X is significant; this is exactly the same as testing the hypothesis that the population correlation coefficient equals zero (i.e., Equations 6-6 and 6-7).

The hypothesis test based on Equations 6-46a and 6-46b relies on the computations involved in the separation of variation concept. Thus it represents an analysis of variance (ANOVA). The total variation is represented by the total sum of squares SS_T:

$$SS_T = \sum_{i=1}^{n} y_i^2 - \frac{\left(\sum_{i=1}^{n} y_i\right)^2}{n} \tag{6-47}$$

The total sum of squares is separated into the regression (SS_R), or the explained variation, and the error (SS_E) sums of squares, or the unexplained variation, as follows:

$$SS_R = b_1 \left(\sum_{i=1}^{n} x_i y_i - \frac{\sum_{i=1}^{n} x_i \sum_{i=1}^{n} y_i}{n} \right) \tag{6-48}$$

$$SS_E = SS_T - SS_R \tag{6-49}$$

in which b_1 = the computed value of the slope coefficient. The mean squares, which correspond to variances, equal the sums of squares divided by the degrees of freedom, which are shown in Table 6-3.

The null hypothesis is tested for significance using the ratio of the mean square for regression to the mean square for error, as follows:

$$F = \frac{MS_R}{MS_E} \tag{6-50}$$

in which F = the computed value of a random variable having an F distribution with 1 and ($n - 2$) degrees of freedom. For a stated level of significance, α, the null hypothesis is accepted if F is less than the critical value F_α obtained from Table A-4 of Appendix A. The region of rejection consists of all values of F greater than F_α.

It is of interest to compare R, S_e, and the F from the ANOVA. The value of R is based on the ratio of EV to TV. The value of S_e is based on UV. The computed value of F is based on EV and UV. Thus the three indices that are used to examine the goodness of fit of the data to the linear model are based on the individual elements of the separation of variation concept (see Section 6.2.3).

6.5.4. Standardized Partial Regression Coefficients

Because the partial regression coefficient, b_1 of Equation 6-28, is dependent on the units of both Y and X, it is often difficult to use its magnitude to measure the rationality of the model. The partial regression coefficient can be standardized by

TABLE 6-3 ANOVA Table for a Bivariate Regression

Souce of variation	Sum of squares (SS)	Degrees of freedom	Mean square (MS)	F
Regression	SS_R	1	$MS_R = \dfrac{SS_R}{1}$	$\dfrac{MS_R}{MS_E}$
Error	SS_E	$n - 2$	$MS_E = \dfrac{SS_E}{n-2}$	
Total	SS_T	$n - 1$	—	

$$t = \frac{b_1 S_X}{S_Y} \qquad (6\text{-}51)$$

in which t is called the standardized partial regression coefficient, and S_X and S_Y are the standard deviations of the predictor and criterion variables, respectively. Because the units of b_1 are equal to the ratio of the units of Y to the units of X, t is dimensionless. For the bivariate regression model, t equals the correlation coefficient of the regression of the standardized value of Y on the standardized value of X. This suggests that t is a measure of the relative importance of the corresponding predictor variable. The value of t has the same sign as b_1. Therefore, if the sign of t is irrational, we can conclude that the model is not rational. Although it is more difficult to assess the rationality of the magnitude of a regression coefficient than it is to assess the rationality of the sign, the magnitude of the standardized partial regression coefficient can be used to assess the rationality of the model. For rational models, t must vary in the range $-1 \leq t \leq 1$. Because t equals R for the bivariate model of Equation 6-28, values of t outside this range should not occur; this is not true for multivariate models, such as Equation 6-29.

Example 6-5: Errors in Regression Models

The data of Table 6-2 were previously used to compute the standard error of estimate and the correlation coefficient; they can also be used to illustrate the use of the analysis of variance (ANOVA) for the analysis of the bivariate regression model. Table 6-4 shows the necessary ANOVA computations. For 1 and 3 degrees of freedom, the critical F values for levels of significance of 0.05 and 0.01 are from Table A-4, 10.13 and 34.12, respectively. Because the computed value of 12.8 is larger than the F value for $\alpha = 0.05$, the null hypothesis can be rejected at the 5% level. However, in comparing the computed F with the critical F for $\alpha = 1\%$, the null hypothesis must be accepted at the 1% level of significance. This example illustrates that it is important to evaluate the consequences of selecting a level of significance arbitrarily.

TABLE 6-4 Example of ANOVA Computations

Souce of variation	Sum of squares (SS)	Degrees of freedom	Mean square	F
Regression	$0.9\left[109 - \dfrac{(50)(10)}{5}\right] = 8.1$	1	8.1	12.8
Error	$10 - 8.1 = 1.9$	3	0.633	
Total	$30 - \dfrac{(10)^2}{5} = 10$	4	—	

The standardized partial regression coefficient, which is computed using Equation 6-51, equals

$$t = 0.9 \sqrt{\frac{510 - \dfrac{(50)^2}{5}}{30 - \dfrac{(10)^2}{5}}} = 0.9 \qquad (6\text{-}52)$$

Using the equations given previously for computing the correlation coefficient, it is easy to show that t equals R.

6.5.5. Assumptions Underlying the Regression Model

The principle of least squares assumes that the errors, that is, the differences between the predicted and measured values of the criterion variable, (1) are independent of each other, (2) have zero mean, (3) have a constant variance across all values of the predictor variables, and (4) are normally distributed. If any of these assumptions are violated, we must assume that the model structure is not correct. Violations of these assumptions are easily identified using statistical analyses of the residuals.

When the sum of the residuals does not equal zero, it reflects a bias in the model. The regression approach, when applied analytically, requires the sum of the residuals for a linear model to equal 0. For a nonlinear model, the sum of the errors may not equal 0, which suggests bias. When an inadequate number of significant digits is computed for the regression coefficients, the sum of the residuals may not equal 0, even for a linear model. However, a model may be biased even when the sum of the residuals equals zero. For example, Figure 6-7 shows an X-Y plot in which the trend of the data is noticeably nonlinear, with the linear regression line also shown. If the errors e_i are computed and plotted vs. the corresponding values of the predictor variable, a noticeable trend appears (Figure 6-8). The errors are positive for both low and high values of X, while the errors show a negative bias of prediction for interme-

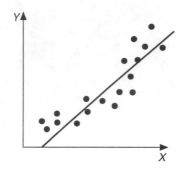

FIGURE 6-7 Biased linear regression model.

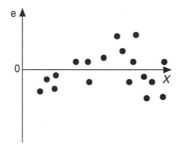

FIGURE 6-8 Residual plot for a biased linear regression model.

diate values of X. While the sum of the residuals equals 0, the trends in the residuals suggest a biased model that should be replaced by a model that has a different structure.

The population model for linear regression has the form

$$Y = \beta_0 + \beta_1 X + \varepsilon \tag{6-53}$$

in which β_0 and β_1 are the population statistics for the intercept and slope coefficients, respectively; and ε is a random variable having a normal distribution with zero mean and a constant variance. The variance of ε equals the square of the standard error of estimate. Therefore, the population of the residuals is defined as

$$\varepsilon \sim N\left(0, \sigma_e^2 = \text{constant}\right) \tag{6-54}$$

If the error variance (σ_e^2) is not constant across all values of the predictor and criterion variable, the underlying model assumptions are not valid and a more accurate model structure should be identified. The data shown in Figure 6-9 reveal a trend in the error variance; specifically, as X increases, the error variance increases. Thus the following relationship is apparent:

$$\varepsilon \sim N\left[0, \sigma_e^2 = f(X)\right] \tag{6-55}$$

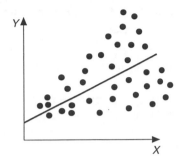

FIGURE 6-9 Model with a nonconstant error variance.

FIGURE 6-10 Model in which $\varepsilon \sim N(0,\ \text{constant } \sigma_e^2)$.

Although an exact test does not exist for testing a data set for a nonconstant error variance, it is not unreasonable to separate the errors into two or more sets based on the value of X and use a hypothesis test for comparing variances to detect a significant difference. A systematic method for selecting either the number of groups or the points at which the separation will be made does not exist. Therefore, one should attempt to have large subsamples and a rational explanation for choosing the points of separation.

The linear regression procedure also assumes that the errors are normally distributed across the range of the values of the predictor variables. The normality assumption is illustrated in Figure 6-10. If the errors are not normally distributed, the model structure should be examined. An exact procedure for testing for normality does not exist. However, in a manner similar to the test for constant variance, the residuals could be separated into groups that have similar values of X, and tests, such as a rank-order frequency analysis, the chi-square goodness-of-fit test, or the Kolmogorov-Smirnov test, could be performed to study the normality question.

The fourth assumption deals with the independence of the observations. This is not usually a problem except when time or space is the predictor variable. In such cases, the measurement of Y for one time period may not be completely independent of the measurement for an adjacent time period. The runs test and/or the serial correlation coefficient may be used to check for statistical independence. Statistical independence should not be used in place of an examination of the physical independence of the data.

6.6. RELIABILITY OF POINT ESTIMATES OF THE REGRESSION COEFFICIENTS

The regression coefficients are important because of their physical significance; the slope coefficient represents the effect of change in the predictor variable on the criterion variable, and the intercept coefficient equals the value of the criterion variable when the predictor variable equals zero. Because of their physical significance, it is important to examine the reliability of the coefficients. Confidence intervals and tests of hypotheses are useful for indicating the quality of the regression statistics.

6.6.1. Sampling Distributions of the Regression Coefficients

The computed coefficients b_0 and b_1 are sample estimates of the population parameters b_0 and b_1, respectively. Recognizing that b_0 and b_1 are random variables, it is important to examine their probability density functions. Assuming that the true error variance σ_e^2 is not known, it could be shown that the coefficients are random variables having t distributions with $(n-2)$ degrees of freedom, mean estimates equal to the estimated values of b_0 and b_1, and the following error variances:

$$S_{e,b_0}^2 = \frac{S_e^2 \sum_{i=1}^{n} x_i^2}{n \sum_{i=1}^{n} (x_i - \overline{X})^2} \tag{6-56}$$

and

$$S_{e,b_1}^2 = \frac{S_e^2}{\sum_{i=1}^{n} (x_i - \overline{X})^2} \tag{6-57}$$

Therefore, two-sided confidence intervals for the coefficients are

$$b_0 \pm t_{\alpha/2,n-2} S_{e,b_0} \tag{6-58}$$

and

$$b_1 \pm t_{\alpha/2,n-2} S_{e,b_1} \tag{6-59}$$

in which α is the level of significance.

Example 6-6: Confidence Intervals for Regression Coefficients

The data of Table 6-2 resulted in the regression equation of Equation 6-37, with slope and intercept coefficients of 0.9 and –7.0, respectively. Since the sum of squares equals 1.9, the standard error of estimate is given by

$$S_e = \sqrt{\frac{1.9}{5-2}} = 0.7958 \tag{6-60}$$

Therefore, the error variances of the regression coefficients are

$$S_{e,b_0}^2 = \frac{(0.7958)^2 (510)}{5(10)} = 6.4596 \tag{6-61}$$

and

$$S_{e,b_1}^2 = \frac{(0.7958)^2}{10} = 0.06333 \tag{6-62}$$

The width of the confidence intervals is a function of the level of significance; for a level of significance of 5%, the two-sided confidence intervals on b_0 and b_1 are

$$-7.0 - 3.182(6.4596)0.5 \le b_0 \le -7.0 + 3.182(6.4596)0.5 \tag{6-63a}$$

$$-15.087 \le b_0 \le 1.0873 \tag{6-63b}$$

and

$$0.9 - 3.182(0.06333)0.5 \le b_1 \le 0.9 + 3.182(0.06333)0.5 \tag{6-64a}$$

$$0.09924 \le b_1 \le 1.7008 \tag{6-64b}$$

One can conclude with 95% confidence that the true value is within the bounds set by Equations 6-63 and 6-64. However, because of the small sample size, the confidence intervals are extremely wide.

6.6.2. Hypothesis Test on the Slope Coefficient

In addition to interval estimates of a statistic, tests of hypotheses are also of value in decision making. If one is interested in a particular value of β_1, say, $\tilde{\beta}_1$, the following null hypothesis would be of interest:

$$H_o: \beta_1 = \tilde{\beta}_1 \qquad (6\text{-}65)$$

This can be tested against either one- or two-sided alternative hypotheses. The null hypothesis of Equation 6-65 can be tested with the following test statistic:

$$t = \frac{b_1 - \tilde{\beta}_1}{S_{e,b_1}} \qquad (6\text{-}66)$$

in which t = the value of a random variable having a t distribution with $(n - 2)$ degrees of freedom. For a given level of significance, the following summarizes the decision process for the three possible alternative hypotheses:

If H_A is	Reject H_o if :
$\beta_1 < \tilde{\beta}_1$	$t < -t_\alpha$
$\beta_1 > \tilde{\beta}_1$	$t > t_\alpha$
$\beta_1 \neq \tilde{\beta}_1$	$t < -t_{\alpha/2}$ or $t > t_{\alpha/2}$

The selection of a one- or two-sided test depends on the problem. The selection of the level of significance should also be selected by considering the physical problem and the implications of making errors in the decision. The computed value can be compared with the critical value according to the decision criteria given in the tabulation above.

If $\tilde{\beta}_1$ equals zero, one may conclude that variation in X does not cause variation in Y; that is, X and Y are independent. Thus the following hypothesis may be of special interest:

$$H_o: \beta_1 = 0 \qquad (6\text{-}67)$$

This can be tested against either one- or two-sided alternative hypotheses by setting $\tilde{\beta}_1 = 0$. This null hypothesis corresponds exactly to the null hypothesis that $\rho = 0$. This is equivalent to the test of Section 6.5.3.

Example 6-7: Hypothesis Testing of Slope

While the sample estimate of β_1 in Example 6-6 was 0.9, three of the five points in the sample discussed previously fell on a straight line defined by Equation 6-38. Thus it may be of interest to test the null hypothesis that $\beta_1 = 1$ against the two-sided alternative hypothesis that $\beta_1 \neq 1$. The computed value of the test statistic would be

$$t = \frac{0.9 - 1.0}{\sqrt{0.06333}} = -0.397 \qquad (6\text{-}68)$$

For a level of significance of 10% and 3 degrees of freedom, the critical t value from Table A-2 is 2.353. For the two-sided alternative, the computed t value lies in the region of acceptance. Therefore, one could not reject the null hypothesis that the true value of β_1 equals 1.0. However, the value of 0.9 is still the best estimate.

6.6.3. Hypothesis Test on the Intercept Coefficient

Because of the physical significance of the intercept, it may be of value to test the hypothesis that the population intercept β_0 equals some specific value, $\tilde{\beta}_0$:

$$H_o: \beta_0 = \tilde{\beta}_0 \tag{6-69}$$

The null hypothesis can be tested against either a one- or two-sided alternative hypothesis using the following test statistic:

$$t = \frac{b_0 - \tilde{\beta}_0}{S_{e,b_0}} \tag{6-70}$$

in which t is the value of a random variable having a t distribution with $(n - 2)$ degrees of freedom. For a given level of significance, the following summarizes the decision process for the three possible alternative hypotheses:

If H_A is	Reject H_o if :
$\beta_0 < \tilde{\beta}_0$	$t < -t_\alpha$
$\beta_0 > \tilde{\beta}_0$	$t > t_\alpha$
$\beta_0 \neq \tilde{\beta}_0$	$t < -t_{\alpha/2}$ or $t > t_{\alpha/2}$

6.7. CONFIDENCE INTERVALS OF THE REGRESSION EQUATION

Confidence intervals provide a useful way for assessing the quality of a point estimate. Whereas a hypothesis test provides a "yes-no" response to a hypothesis, a confidence interval is an interval within which we believe a given population parameter is located. In regression analysis there are several confidence intervals of interest that include (1) a confidence interval for the line as a whole, (2) a confidence interval for a single point on the line, and (3) a confidence interval for a single future value of Y corresponding to a chosen value of X.

6.7.1. **Confidence Interval for a Line as the Whole**

When the entire line is of interest, such as when an experiment is repeated many times and a regression equation is computed for each sample, it may be of interest to compute a confidence interval on the line as a whole; such a confidence interval permits one to simultaneously make confidence statements about estimates of Y for a number of values of the predictor variable. For a specified level of confidence, m points x_{ai} ($i = 1, 2, ..., m$) are selected such that the points adequately cover the range of interest of the predictor variable; usually, seven to ten values are adequate for delineating the confidence interval. The F statistic is obtained for $(2, n - 2)$ degrees of freedom and a level of significance of $\alpha = 1 - \gamma$. For each value of x_{ai} selected, the estimated values of the criterion variable $\hat{y}_{ai}$ and the confidence limits $\hat{y}_{ci}$ are computed as follows:

$$\hat{y}_{ai} = \overline{Y} + b_1\left(x_{ai} - \overline{X}\right) \tag{6-71}$$

$$\hat{y}_{ci} = \hat{y}_{ai} \pm S_e \sqrt{2F\left[\frac{1}{n} + \frac{\left(x_{ai} - \overline{X}\right)^2}{\displaystyle\sum_{i=1}^{n}\left(x_i - \overline{X}\right)^2}\right]} \tag{6-72}$$

The regression line and the confidence interval can then be plotted for each value of x_{ai}.

6.7.2. **Confidence Interval Estimate for a Single Point on the Line**

If a particular value of the predictor variable, say x_o, is of special importance, a confidence interval on the value of the criterion variable corresponding to x_o may be of interest. In such a case the two-sided, γ-percent confidence interval on the estimated value of Y corresponding to x_o can be computed as

$$\hat{y} \pm S_e\, t_{\alpha/2} \sqrt{\frac{1}{n} + \frac{\left(x_o - \overline{X}\right)^2}{\displaystyle\sum_{i=1}^{n}\left(x_i - \overline{X}\right)^2}} \tag{6-73}$$

in which $t_{\alpha/2}$ is the value of a random variable having a t distribution with $(n - 2)$ degrees of freedom with a level of significance α, which equals $1 - \gamma$. The estimated value of Y, which is denoted by $\hat{y}$, is computed by

$$\hat{y} = \bar{Y} + b_1\left(x_o - \bar{X}\right) \tag{6-74}$$

If the experiment was repeated many times, γ percent of the computed intervals would cover the true value of Y corresponding to x_o. It is important to note that this confidence interval is concerned with a mean value; that is, the line represents the mean value of the distribution of Y values for a given x_o and the confidence interval of Equation 6-73 is for the line at the point x_o only. The interval is computed only for the point x_o.

6.7.3. Confidence Interval Estimate for a Future Value

A third confidence interval of interest can be used to evaluate the accuracy of a single (future) value of Y corresponding to a chosen value of X, say, x_f. The two-sided, γ-percent confidence interval for an estimated value $\hat{y}$, which can be computed from Equation 6-74 with $x_o = x_f$, is given by

$$\hat{y} \pm S_e\, t_{\alpha/2}\sqrt{1 + \frac{1}{n} + \frac{\left(x_f - \bar{X}\right)^2}{\displaystyle\sum_{i=1}^{n}\left(x_i - \bar{X}\right)^2}} \tag{6-75}$$

It is of interest to compare the form of Equations 6-73 and 6-75. They differ by a factor of 1.0 in the square root radical. The larger confidence interval that results from the 1.0 factor reflects the less certainty resulting from the estimation of a single future value of Y rather than the mean value computed for the confidence interval of Equation 6-73. Equation 6-75 provides a confidence interval on $\hat{y}$ only for the single point x_f.

6.7.4. Summary of Confidence Intervals

Confidence intervals provide a measure of reliability. Three interval estimates have been described: (1) the confidence intervals for the line as a whole, (2) the confidence intervals for the mean value of Y corresponding to a specific value x_o, and (3) the confidence interval for a value of Y corresponding to a future estimate of the predictor variable x_f. It should be of interest to distinguish between these three using a very simple example. Assume that the problem involves finding the relationship between the height and weight of graduate students at a university, with weight being the criterion variable. If the problem was to determine the true relationship, a confidence interval for the line as a whole would be appropriate for assessing the overall accuracy. If the mean value of those students who were 6 ft tall was of special interest, it might be of interest to compute the confidence interval for the point on the line corresponding to $x_o = 6$ ft; this confidence interval would indicate the accuracy of the estimated value of the weight (Y) for a height (x_o) of 6 ft. If the problem of interest is "the weight of the very next graduate student who is 6 feet tall and passes by the door," the estimated weight would be the same as that computed for the previous problem. However, because it is a single value for the weight of a specific person, the confidence interval would be wider than that for the average of all graduate students who are 6 ft tall. This confidence interval would be computed using Equation 6-75.

Example 6-8: Confidence Intervals in Regression

Three types of confidence intervals were described in the previous paragraphs; the selected type of confidence interval depends on the type of decision that must be made. For the data of Table 6-2, the confidence interval on the line as a whole can be computed using Equation 6-72. If it can be assumed that values of the predictor variable X are likely to be in the range from 6 to 16, it is reasonable to compute the confidence interval within these bounds. Values of x_{ai} are selected in increments of two because six values should be sufficient to draw the confidence interval. For a 5% level of significance and degrees of freedom of 2 and 3, the F statistic from Table A-4 is 9.55. For the standard error of 0.7958, the 95% confidence interval for any value of x_{ai} is

$$\hat{y}_{ai} \pm 0.7958 \sqrt{2(9.55)} \left[\frac{1}{5} + \frac{(x_{ai} - 10)^2}{10} \right]^{0.5} \qquad (6\text{-}76a)$$

$$\hat{y}_{ai} \pm 3.478 \left[\frac{1}{5} + \frac{(x_{ai} - 10)^2}{10} \right]^{0.5} \qquad (6\text{-}76b)$$

The computed values of $\hat{y}_{ai}$ are determined using Equation 6-37 with $X = x_{ai}$. The computations are given in Table 6-5, and the confidence interval is shown in Figure 6-11. The confidence interval is smallest at the mean value of X (i.e., $\overline{X} = 10$) and increases in width as X differs from the mean. This results because of the greater confidence in estimated values near the mean. For values beyond the range of the data, the confidence interval is especially wide. The confidence interval is highly nonlinear because of the small sample size. For large samples, the width of the confidence interval is not characterized by such noticeable nonlinearity. For large samples it is not necessary to use as many values of x_{ai} to get a reasonable delineation of the confidence interval.

If a particular value of X and the corresponding mean value of Y are of special interest, the confidence interval of Equation 6-73 can be used to show the accuracy of the value of Y on the regression line that corresponds to the X value. If a value

TABLE 6-5 Computation of Confidence Interval for the Line as a Whole

x_{ai}	y_{ai}	$\dfrac{(x_{ai}-10)^2}{10}$	Half-width	Lower limit	Upper limit
6	−1.6	1.6	4.67	−6.3	3.1
8	0.2	0.4	2.69	−2.5	2.9
10	2.0	0.0	1.56	0.4	3.6
12	3.8	0.4	2.69	1.1	6.5
14	5.6	1.6	4.67	0.9	10.3
16	7.4	3.6	6.78	0.6	14.2

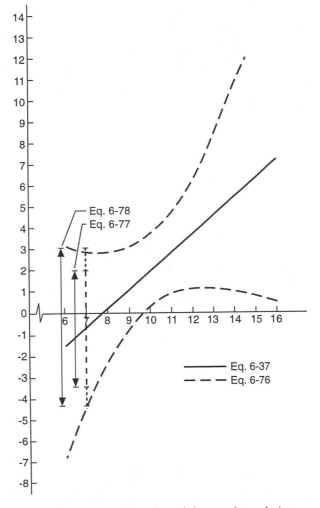

FIGURE 6-11 Confidence intervals in regression analysis.

of X of 7 (i.e., $x_o = 7$) is of special interest with the data of Table 6-2, the 95% confidence interval at that point is

$$\hat{y} \pm 3.812(0.7958)\left[\frac{1}{5} + \frac{(7-10)^2}{10}\right]^{0.5} \tag{6-77a}$$

$$\hat{y} \pm 2.66 \tag{6-77b}$$

The predicted value of Y at X equal to 7 is –0.7. Therefore, the confidence limits are –3.4 and 2.0 as shown in Figure 6-11. The width of the confidence interval for the mean value is smaller than the width of the confidence interval of the line as a whole at $X = 7$, because we are only interested in a single value of X and not for all values of X.

If a particular occurrence of the variable X and the corresponding value of Y is of special interest, the confidence interval of Equation 6-75 can be used to assess the expected accuracy of the predicted value. If the specific value of X is 7 (i.e., $x_f = 7$), the 95% confidence interval is

$$\hat{y} \pm 3.812(0.7958)\left[1 + \frac{1}{5} + \frac{(7-10)^2}{10}\right]^{0.5} \tag{6-78a}$$

$$\hat{y} \pm 3.67 \tag{6-78b}$$

Thus, for the predicted value of –0.7, the confidence limits are –4.4 and 3.0, which are shown in Figure 6-11. The confidence interval of Equation 6-78 is wider than that of Equation 6-77 because Equation 6-78 is intended as a measure of the accuracy of the predicted value of Y for a specific occurrence of X, while Equation 6-78 measures the accuracy of the mean value of Y at $X = 7$.

6.8. CORRELATION VS. REGRESSION

Before discussing other topics in regression, it is important to emphasize the differences between correlation and regression. Regression is a means of calibrating the unknown coefficients of a prediction equation, whereas correlation provides a measure of goodness of fit. Thus regression is a method for model calibration, while correlation would have usefulness in model formulation and model verification. When using regression, it is necessary to specify which variable is the criterion and which is the predictor; when using correlation, it is not necessary to make such a distinction. The distinction is necessary with regression because a regression equation is not transformable unless the correlation coefficient equals 1.0. That is, if Y is the criterion variable when the equation is calibrated, the equation cannot be rearranged algebraically to get an equation for predicting X. Specifically, Equation 6-29 can be rearranged to predict X as

$$Y = b_0 + b_1 X \tag{6-79a}$$

$$Y - b_0 = b_1 X \tag{6-79b}$$

Therefore,

$$X = \frac{Y - b_0}{b_1} \tag{6-79c}$$

or

$$X = -\frac{b_0}{b_1} + \frac{1}{b_1} Y \tag{6-79d}$$

$$X = a_0 + a_1 Y \tag{6-79e}$$

If X is regressed on Y, the resulting coefficients will not be the same as the coefficients obtained by regressing Y on X (Equation 6-79a) and then setting $a_o = -\frac{b_o}{b_1}$ and $a_1 = \frac{1}{b_1}$. The correlation for the regression of Y on X is the same as the correlation for the regression of X on Y.

Computationally, the correlation coefficient is a function of the explained and total variation; the slope coefficient of a regression equation is related to the correlation coefficient by

$$b_1 = \frac{RS_Y}{S_X} \tag{6-80}$$

in which S_X and S_Y are the standard deviations of X and Y, respectively.

After a regression equation is calibrated, it is very important to examine the rationality of the regression coefficients; specifically, is the predicted value of Y rational for all reasonable values of X? Since the slope coefficient of the regression equation represents the rate of change of Y with respect to X, is the effect of X on Y rational? A regression equation that is not considered rational should be used with caution, if at all. In addition to checking for rationality, the goodness-of-fit statistics, R and S_e, should be computed. Also, a hypothesis test, either ANOVA or a test on ρ, should be made. Confidence intervals can be computed on both the regression coefficients and the line and used to assess the accuracy of predictions made with the regression equation. If the expected accuracy is not acceptable, we may elect to collect more data or to develop a model that uses other predictor variables.

6.9. APPLICATIONS OF BIVARIATE REGRESSION ANALYSIS

6.9.1. Estimating Trip Rate

A traffic planner needs estimates of trip rates for residential areas. The planner conducts studies in ten residential areas of different densities of development, with the following measured rates:

X	3	4	8	8	13	15	18	22	24	27
Y	4.5	3.3	3.5	2.3	3.8	2.6	2.7	1.6	1.9	1.7

where X = residential density (households per acre) and Y = trip rate (daily trips per household). The regression of trip rate on density yielded the following linear equation:

$$\hat{Y} = 4.100 - 0.09226X \qquad (6\text{-}81)$$

The negative slope coefficient indicates that the trip rate decreases by 0.092 daily trips per household for each unit increase in the number of households per acre. The standard error of estimate is 0.603, which is an improvement over the standard deviation of 0.965 ($S_e/S_y = 0.625$). The correlation coefficient is -0.808, which yields a normalized explained variance (R^2) of 0.653. The standard error ratio of the slope coefficient ($S_{e,b_1}/b_1$) equals 0.258, which suggests that the coefficient is accurate. Figure 6-12 shows the measured data and the regression line; the moderate variation of the measured points about the line confirms the moderate correlation between residential density and trip rate.

6.9.2. Breakwater Cost

A construction engineering firm has recently had a number of contracts for the construction of breakwaters in coastal environments. To reduce the effort in developing cost estimates for future clients, the company compiled a data record of 14 breakwater construction projects. The data consist of the length of the breakwater and the cost for rock, rock spurs, groins, marsh plants, and beach fill (see Table 6-6). The cost (Y) was regressed on the length of the breakwater (X):

$$\hat{Y} = 253.4 + 0.08295X \qquad (6\text{-}82)$$

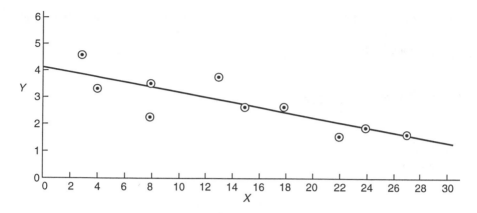

FIGURE 6-12 Bivariate regression of trip rate (Y, daily trip per household) on residential density (X, households per acre).

TABLE 6-6 Bivariate Regression Analysis of Breakwater Cost

X	Y	$\hat{Y}$	e
450	230	291	61
625	490	305	−185
670	150	309	159
730	540	314	−226
810	280	321	41
880	120	326	206
1020	380	338	−42
1050	170	340	170
1100	530	345	−185
1175	290	351	61
1230	460	355	−105
1300	230	361	131
1350	470	365	−105
1510	360	379	19

The standard error of estimate is $\$149.1 \times 10^3$, which is slightly greater than the standard deviation of the measured costs ($S_y = \$145.5 \times 10^3$), that is, $S_e/S_y = 1.024$. Since S_e is greater than S_y, it would be more accurate to use the mean cost ($\$335.7 \times 10^3$) than the value estimated with the regression equation. If we use Equation 6-5 to compute the correlation coefficient, we get a value of 0.176; however, Equation 6-45 shows that we have essentially no reliability in the estimated values. Thus, for practical purposes, we have a correlation of zero. The adjusted correlation computed with Equation 6-45 is a more realistic measure of reliability than is the value of R computed with the unadjusted, but more commonly used, equation (Equation 6-5), because Equation 6-45 accounts for differences in the degrees of freedom. The standard error ratio of the slope coefficient, $S_{e,b_1}/b_1$, is 1.61; this exceptionally large value is another indication of the poor reliability of the linear, bivariate equation.

Why does the computed equation give such poor results? Certainly, breakwater length is a determinant of cost. However, it appears that other factors are more important and necessary to get accurate estimates of the cost.

6.9.3. Stress-Strain Analysis

Table 6-7 includes nine measurements of the observed stress and the resulting axial strain for an undisturbed sample of a Boston Blue clay. A linear equation was computed by regressing the strain (Y) on the stress (X):

$$\hat{Y} = -8.29 + 0.6257X \tag{6-83}$$

TABLE 6-7 Bivariate Regression Analysis of Stress-Strain Data

X	Y	$\hat{Y}$	e
10.8	0.1	−1.5	−1.6
14.1	0.4	0.5	0.1
16.3	1.1	1.9	0.8
17.5	1.4	2.7	1.3
18.6	2.7	3.4	0.7
19.8	3.6	4.1	0.5
22.1	5.7	5.5	−0.2
23.8	6.8	6.6	−0.2
24.3	8.3	6.9	−1.4

The equation has a standard error of estimate of 1.04%, which is much smaller than the standard deviation of the measured strains, that is, $S_e/S_y = 0.35$. The correlation coefficient of 0.945 also suggests that the linear equation provides a reasonable fit to the measured data. An R^2 of 0.89 indicates that almost 90% of the total variation is explained by the linear regression; the standard error ratio of 0.35 suggests that the fit is not as good as suggested by the R^2. The predicted strains and residuals are given in Table 6-7 and shown in Figure 6-13. While most of the errors are small, there is a noticeable bias in the residuals, with negative residuals for small and large stresses. While the mean bias ($\bar{e}$) is zero, the local biases indicate that a nonlinear structure may be more appropriate. Another problem with the linear model is the negative estimates of strain that occur for stresses of less than 13.244 psi. The negative strain is physically not meaningful, so the model is not rational for small strains.

6.9.4. Project Cost vs. Time

For large projects that require years to complete, the construction engineer needs to schedule project finances just as it is necessary to schedule construction activities.

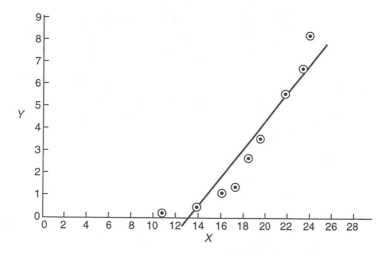

FIGURE 6-13 Axial strain (Y) of a Boston Blue clay due to stress (X).

While costs vary with time, a construction engineer believes that a linear model can be used to represent the relationship between the value of work completed (Y) and the time since the start of the project (X). Since no work would be completed at the start of the project, a zero-intercept model would seem appropriate. The engineer compiled semiannual cost for a recent 6-year project; the data are given in Table 6-8. The linear, zero-intercept, bivariate model was fit to the data as follows:

$$\hat{Y} = 0.50421X \tag{6-84}$$

where X = time in months from start of project, and $\hat{Y}$ = predicted value of work in $\$ \times 10^6$. The model is shown in Figure 6-14 and the residuals are given in Table 6-8. The model underpredicts for the middle part of the period while it overpredicts for the early and latter parts of the time period. The standard error of estimate is $\$7.73 \times 10^6$, which gives $S_e/S_y = 0.757$; this indicates that the model does not provide highly accurate estimates of Y. Additionally, the sum of the errors is not zero, so the model is biased. Whereas a linear model with an intercept will provide an unbiased model, a zero-intercept model can be biased, as it is for this example with a mean error of $-\$0.52 \times 10^6$. This means that, on the average, Equation 6-84 tends to underpredict by $520,000. A correlation coefficient should not be computed for a biased model. The correlation coefficient assumes a linear model with an intercept, so it would be incorrect to use either Equation 6-5 or 6-45 to compute a correlation coefficient for Equation 6-84.

Because of the poor accuracy of the model of Equation 6-84, the following linear model with an intercept was fit to the data of Table 6-8:

$$\hat{Y} = 2.36 + 0.4570X \tag{6-85}$$

TABLE 6-8 Bivariate Regression Analyses of Project Cost (Y) Vs. Time (X)

		Equation 6-84		Equation 6-85	
X	Y	$\hat{Y}$	e	$\hat{Y}$	e
6	1.5	3.03	1.53	5.10	3.60
12	4.3	6.05	1.75	7.84	3.54
18	11.2	9.08	-2.12	10.59	-0.61
24	15.5	12.10	-3.40	13.33	-2.17
30	19.1	15.13	-3.97	16.07	-3.03
36	21.8	18.15	-3.65	18.81	-2.99
42	24.0	21.18	-2.82	21.55	-2.45
48	26.2	24.20	-2.00	24.30	-1.90
54	26.9	27.23	0.33	27.04	0.14
60	27.8	30.25	2.45	29.78	1.98
66	30.1	33.28	3.18	32.52	2.42
72	33.8	36.30	2.50	35.26	1.46

Note: X = time (months) from start of project,
Y = value of work completed ($ × 10⁶),
$\hat{Y}$ = predicted value, and e = error in predicted value = $\hat{Y} - Y$.

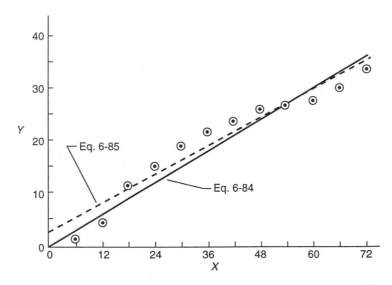

FIGURE 6-14 Regression of project cost (Y, $10⁶) on time from start of project (X, months) for zero-intercept (Equation 6-84) and linear (Equation 6-85) models.

This model is also shown in Figure 6-14 and the residuals are given in Table 6-8. Equation 6-85 is unbiased and has a standard error of $2.651 × 10⁶, which gives $S_e/S_y = 0.26$. The model explains (R^2) 94% of the variation in Y and has a correlation coefficient of 0.969. In spite of the reasonably accurate estimates suggested by the goodness-of-fit statistics, the model still has local biases in the residuals (see Table 6-8), with underpredictions for the range $18 \leq X \leq 48$ months. However, even with the local biases, the model of Equation 6-85 may be sufficiently accurate for making project-planning estimates. The nonzero intercept of Equation 6-84 suggests a cost of 2.36×10^6 at the time the project is initiated. While a value of $0 is expected, the nonzero intercept may suggest that the model may not be realistic for the initial part of a project.

In comparing the two models, the standard error ratio would suggest that the zero-intercept model is considerably less accurate than the nonzero-intercept model. However, in comparing the residuals of the two models, there appears to be very little difference. The averages of the absolute values of the residuals are 2.5 and 2.2 for the Equations 6-84 and 6-85, respectively. The nonzero-intercept model is slightly better, but greater accuracy may be achieved with another model structure.

TABLE 6-9 Regression Analysis

		Equation 6-86		Equation 6-87	
X	Y	$\hat{Y}$	e	$\hat{Y}$	e
640	0.063	0.066	0.003	—	—
1930	0.061	0.050	-0.011	0.058	-0.003
1980	0.057	0.049	-0.008	0.055	-0.002
2010	0.049	0.049	0.000	0.053	0.004
2080	0.043	0.048	0.005	0.049	0.006
2120	0.048	0.048	0.000	0.046	-0.002
2200	0.039	0.047	0.008	0.041	0.002
2250	0.042	0.046	0.004	0.038	-0.004

Note: X = distance from nuclear power facility, Y = concentration of ^{241}Am in soil samples, $\hat{Y}$ = predicted value of Y, and e = residual = $\hat{Y} - Y$.

6.9.5. Effect of Extreme Event

Small samples very often contain an extreme event, which can greatly influence the resulting equation. Table 6-9 contains eight measurements of the concentration of ^{241}Am in soil samples and the distance of the soil sample from the nuclear facility. The eight measurements were used to regress the ^{241}Am concentration (Y) on the distance (X):

$$\hat{Y} = 0.07366 - 0.0000123X \tag{6-86}$$

The negative slope coefficient indicates that the concentration decreases with distance. The standard error of estimate is 0.00694, which gives $S_e/S_y = 0.764$. This suggests poor prediction ability. The correlation coefficient of -0.706 means that

50% of the total variation in Y is explained by the regression equation. The standard error ratio of the slope coefficient, $S_{e,b_1}/b_1$, equals 0.41, which is moderate and suggests a less than adequate model. The predicted values of Y and the errors for Equation 6-86 are shown in Table 6-9. Except for the value for a distance of 640, the predicted values range from 0.046 to 0.050 picocuries per gram (pCi/g); however, the measured values for these distances ranged from 0.039 to 0.061 pCi/g. These predicted values are not much different from the mean of 0.050 pCi/g. The batch of seven points acts as a single point.

A second model was developed using seven of the eight values. By eliminating the extreme event, the following model was fitted:

$$\hat{Y} = 0.1755 - 0.0000611X \tag{6-87}$$

This model has a standard error of 0.00412 pCi/g ($S_e/S_y = 0.51$). The correlation coefficient in this case is -0.885, which means that 78.3% of the total variation in the seven measurements was explained by the regression equation of Equation 6-87. The standard error ratio for the slope coefficient, $S_{e,b_1}/b_1$, was 0.235, which is considerably better than the corresponding value for the model of Equation 6-86. The residuals for Equation 6-87 are small, and the predicted values range from 0.038 to 0.058 pCi/g, which is much greater than the range for the model of Equation 6-86. But for a distance of 640, which is the measured value not included in fitting Equation 6-87, the value predicted with Equation 6-87 is 0.136 pCi/g; this is considerably higher than the measured value of 0.063 pCi/g.

In summary, in spite of the goodness-of-fit statistics for both models, neither model is an accurate representation of the system. This is a result of both the small sample size ($n = 8$) and the distribution of the data. It is undesirable to have a cluster of points and then one extreme event. Such a distribution of values makes it difficult to attain a model that is representative over a wide range of values of the predictor variable.

6.9.6. Variable Transformation

Studies have shown that fuel consumption (Y, gallons per mile) is related to the vehicle velocity (X, miles per hour) by

$$\hat{Y} = b_o + \frac{b_1}{X} \tag{6-88}$$

By creating a new variable $W = 1/X$, a relationship between Y and X can be fitted. The data of Table 6-10 can be used to illustrate the fitting of a model that requires a transformation of the predictor variable.

A linear regression analysis of Y on W produces the following equation:

$$\hat{Y} = 0.03939 + \frac{0.9527}{X} \tag{6-89}$$

TABLE 6-10 Bivariate Regression of Fuel Consumption (Y) on Speed (X)

X	W	Y	$\hat{Y}$	e
3.4	0.2940	0.320	0.3190	−0.0010
8.1	0.1230	0.140	0.1570	0.0170
16.4	0.0610	0.130	0.0975	−0.0325
23.7	0.0422	0.094	0.0796	−0.0144
31.3	0.0319	0.062	0.0698	0.0078
38.8	0.0258	0.066	0.0640	−0.0020
42.3	0.0236	0.051	0.0619	0.0109
49.0	0.0204	0.056	0.0588	0.0028
54.6	0.0183	0.055	0.0568	0.0018
58.2	0.0172	0.049	0.0558	0.0068
66.1	0.0151	0.051	0.0538	0.0028

Note: X = fuel consumption (gallons/mile), $W = 1/X$, and Y = mean speed (miles/hour).

The standard error of estimate for Equation 6-89 is 0.0141 gal/mi, which is 17.5% of the standard deviation of Y (i.e., $S_e/S_y = 0.175$). Accurate estimates are also suggested by the correlation coefficient of 0.986 ($R^2 = 0.972$). The standard error ratio for the slope coefficient, $S_{e,b_1}/b_1$, is 0.056, which indicates that the computed value of b_1 is highly accurate. The residuals (see Table 6-10) are small and a trend is not apparent.

In this case, only the predictor variable was transformed, which does not create problems. Problems can arise when it is necessary to transform the criterion variable in order to put the relationship into a linear form.

6.10. MULTIPLE REGRESSION ANALYSIS

What options are available to improve the accuracy of predictions if the accuracy from a bivariate regression is still not sufficient for the design problem? Multivariate systems are one possible option to this problem. Where one predictor variable may not be adequate, several predictor variables may provide sufficient prediction accuracy. Thus one reason for using multivariate models rather than a bivariate model is to reduce the standard error of estimate.

Multivariate data analyses are also of value when a goal of the analysis is to examine the relative importance of the predictor variables. If a predictor variable is not included in the model, a regression analysis cannot determine the sensitivity of the criterion variable to variation in that predictor variable. Also, if a predictor variable that has not been included in the model is correlated with another predictor variable that is included in the model, the regression coefficient for the former variable may reflect the effects of both the former and latter predictor variables. Thus predictor variables that have only slight correlation with the criterion variable are often included in the model so that a complete analysis may be made.

How does a multivariate analysis compare with a bivariate analysis? Actually, the two are very similar. First, the same least-squares objective function is used to calibrate the regression coefficients. Second, bivariate correlations are still computed. Third, the data should be properly screened using graphical analyses prior to selecting a model structure. The major difference between multivariate and bivariate analyses is the necessity to account for the interdependence (i.e., correlation) of the predictor variables.

6.10.1. Correlation Matrix

The first step in a multivariate analysis is to perform a graphical analysis. This includes developing plots of the criterion variable vs. each of the predictor variables in the data set. Also, plots of each pair of predictor variables should be made; such plots will help to identify the characteristics of the interdependence between predictor variables. It will become evident that the interdependence is extremely important in the analysis of multivariate systems. All the plots should be examined for the characteristics identified previously for bivariate analyses, for example, extreme events, nonlinearity, and random scatter.

After a graphical analysis, the bivariate correlation coefficients should be computed for each pair of variables; this includes the correlation between the criterion variable and each predictor variable, as well as the correlation between each pair of predictor variables. The correlations are best presented in matrix form. The correlation matrix is a means of presenting in an organized manner the correlations between pairs of variables in a data set; it appears as

$$
\begin{array}{c|ccccccc}
 & X_1 & X_2 & X_3 & \cdots & X_p & Y \\
\hline
X_1 & 1.0 & r_{12} & r_{13} & \cdots & r_{1p} & r_{1Y} \\
X_2 & & 1.0 & r_{23} & \cdots & r_{2p} & r_{2Y} \\
X_3 & & & 1.0 & \cdots & r_{3p} & r_{3Y} \\
\vdots & & & & \ddots & \vdots & \vdots \\
X_p & & & & & 1.0 & r_{pY} \\
Y & & & & & & 1.0
\end{array}
\tag{6-90}
$$

Note that the correlation matrix, which includes p predictor variables (X_i, $i = 1, 2, ..., p$) and the criterion variable (Y), is shown in upper-triangular form because the matrix is symmetric (i.e., $r_{ij} = r_{ji}$); also, the elements on the principal diagonal equal 1.0 because the correlation between a variable and itself is unity. The matrix is $(p + 1) \times (p + 1)$. Whereas the correlation matrix shown above uses the sample correlation coefficients (r_{ij}), the correlation matrix for the population can be indicated using ρ_{ij} as the elements.

Example 6-9: Sediment Yield Data

The sediment data base includes four predictors and a criterion variable, the sediment yield. The ultimate objective would be to develop a model for estimating the sediment yield from small watersheds. The predictor variables are (1) the precipitation/temperature ratio (X_1), which reflects the vegetation potential of the area; (2) the average watershed slope (X_2), which reflects the erosion potential due to the momentum of surface runoff; (3) the soil particle size index (X_3), which reflects the coarse-particle composition of the soil; and (4) the soil aggregation index (X_4), which reflects the dispersion characteristics of the small particles. Since vegetation retards erosion, we would expect r_{1y} to be negative. Similarly, coarse particles are more difficult to transport, so r_{3Y} should also be negative. As the slope increases, the water will move with a greater velocity and more soil can be transported, so r_{2Y} should be positive. Erosion should increase as the dispersion characteristics of small particles increases, so r_{4y} should be positive.

The correlation matrix for 37 measurements of these five variables is given in Table 6-11. In general, the intercorrelations are low, with the largest in absolute value being for r_{14} (−0.445); this suggests that more than one predictor variable will probably be used to obtain accurate estimates of the criterion variable. Many of the intercorrelations are very low, less than 0.2; this is usually a desirable characteristic. The predictor-criterion correlations are low to moderate; the largest is 0.570. Using the square of the predictor-criterion correlations, the fraction of variance explained ranges from 0.062 to 0.325. These low values would also suggest that more than one predictor variable will be necessary to accurately estimate the sediment yield.

TABLE 6-11 Correlation Matrix of the Sediment Data Base

	X_1	X_2	X_3	X_4	Y
X_1: precipitation/temperature ratio	1.000	0.340	−0.167	−0.445	−0.297
X_2: average slope		1.000	−0.051	−0.185	0.443
X_3: soil particle size			1.000	0.069	−0.253
X_4: soil aggregation index				1.000	0.570
Y: sediment yield					1.000

Example 6-10: Evaporation Data

The evaporation data set includes four predictor variables and a criterion variable, which is the pan evaporation from a small pond in southeastern Georgia. The predictor variables are (1) the mean daily temperature in degrees Fahrenheit (°F) X_1; (2) the wind speed in miles per day X_2; (3) the radiation in equivalent inches X_3; and (4) the vapor pressure deficit X_4. The wind speed attempts to define the rate at which the drier air is moved over the pond to replace the more moist air. The temperature and the radiation measure the potential energy of the water molecules and their potential for escaping from the water surface. The vapor pressure deficit represents the potential for the overlying air mass to accept water molecules.

The correlation matrix, which was based on 354 daily measurements, is shown in Table 6-12. The correlation matrix provides a sharp contrast to that from the sediment data because the matrix of Table 6-12 is characterized by high intercorrelation and moderate predictor-criterion correlations. The high intercorrelations suggest that only one or two predictor variables will be necessary to estimate evaporation rates. The moderate predictor-criterion correlations suggest that the potential accuracy of predictions is only moderate. The fraction of the variance explained by X_4 is 0.403. Because most of the intercorrelations are high, it is unlikely that the variation explained by the four predictor variables will be much greater than the 40% explained by just one of the predictor variables.

TABLE 6-12 Correlation Matrix for the Evaporation Data Base

	X_1	X_2	X_3	X_4	Y
X_1: temperature (°F)	1.000	−0.219	0.578	0.821	0.581
X_2: wind speed (miles/day)		1.000	−0.261	−0.304	−0.140
X_3: radiation (equivalent inches)			1.000	0.754	0.578
X_4: vapor pressure deficit				1.000	0.635
Y: pan evaporation (inches)					1.000

6.10.2. Calibration of the Multiple Linear Model

The three components of regression analysis are the model, the objective function, and the data set. The data set consists of a set of n observations on p predictor variables and one criterion variable, where n should be, if possible, at least four times greater than p. The data set can be viewed as a matrix having dimensions of n by $(p + 1)$. The principle of least squares is used as the objective function. The model, in raw score form, is

$$\hat{Y} = b_0 + b_1 X_1 + b_2 X_2 + \ldots + b_p X_p \tag{6-91}$$

in which X_j $(j = 1, 2, \ldots, p)$ are the predictor variables, b_j $(j = 1, 2, \ldots, p)$ are the partial regression coefficients, b_0 is the intercept coefficient, and Y is the criterion variable. Using the least-squares principle and the model of Equation 6-91, the objective function becomes

$$F = \min \sum_{i=1}^{n} e_i^2 = \min \sum_{i=1}^{n} \left(b_0 + \sum_{j=1}^{p} b_j x_{ij} - y_i \right)^2 \tag{6-92}$$

in which F is the value of the objective function. It should be noted that the predictor variables include two subscripts, with i indicating the observation and j the specific predictor variable.

The method of solution is to take the $(p + 1)$ derivatives of the objective function, Equation 6-92, with respect to the unknowns, b_j $(j = 0, 1, ..., p)$, setting the derivatives equal to zero and solving for the unknowns. A set of $(p + 1)$ normal equations is an intermediate result of this process.

As an example, consider the case where $p = 2$; thus Equation 6-91 reduces to

$$\hat{Y} = b_0 + b_1 X_1 + b_2 X_2 \tag{6-93}$$

Also, the objective function, Equation 6-92, is given by

$$F = \min \sum_{i=1}^{n} e_i^2 = \min \sum_{i=1}^{n} \left(b_0 + b_1 x_{i1} + b_2 x_{i2} - y_i\right)^2 \tag{6-94}$$

The resulting derivatives are

$$\frac{\partial \sum_{i=1}^{n} e_i^2}{\partial b_0} = 2 \sum_{i=1}^{n} \left(b_0 + b_1 x_{i1} + b_2 x_{i2} - y_i\right)(1) = 0 \tag{6-95a}$$

$$\frac{\partial \sum_{i=1}^{n} e_i^2}{\partial b_1} = 2 \sum_{i=1}^{n} \left(b_0 + b_1 x_{i1} + b_2 x_{i2} - y_i\right)\left(x_{i1}\right) = 0 \tag{6-95b}$$

$$\frac{\partial \sum_{i=1}^{n} e_i^2}{\partial b_2} = 2 \sum_{i=1}^{n} \left(b_0 + b_1 x_{i1} + b_2 x_{i2} - y_i\right)\left(x_{i2}\right) = 0 \tag{6-95c}$$

Rearranging Equations 6-95 yields a set of normal equations:

$$nb_0 + b_1 \sum_{i=1}^{n} x_{i1} + b_2 \sum_{i=1}^{n} x_{i2} = \sum_{i=1}^{n} y_i \tag{6-96a}$$

$$b_0 \sum_{i=1}^{n} x_{i1} + b_1 \sum_{i=1}^{n} x_{i1}^2 + b_2 \sum_{i=1}^{n} x_{i1} x_{i2} = \sum_{i=1}^{n} x_{i1} y_i \tag{6-96b}$$

$$b_0 \sum_{i=1}^{n} x_{i2} + b_1 \sum_{i=1}^{n} x_{i1} x_{i2} + b_2 \sum_{i=1}^{n} x_{i2}^2 = \sum_{i=1}^{n} x_{i2} y_i \tag{6-96c}$$

The solution of the three simultaneous equations yields values of b_0, b_1, and b_2.

6.10.3. Standardized Model

When the approach of the previous section is used and the means and standard deviations of the predictor and criterion variables are significantly different, round-off error, which results from the inability to maintain a sufficient number of significant digits in the computations, may cause the partial regression coefficient to be erroneous. Thus most multiple regression analyses are made using a standardized model:

$$Z_Y = t_1 Z_1 + t_2 Z_2 + \ldots + t_p Z_p \qquad (6\text{-}97)$$

in which t_j $(j = 1, 2, \ldots, p)$ are called standardized partial regression coefficients, and Z_Y and the Z_j $(j = 1, 2, \ldots, p)$ are the criterion variable and the predictor variables, respectively, expressed in standardized form; specifically, for $i = 1, 2, \ldots, n$, they are computed by

$$Z_Y = \frac{Y_i - \overline{Y}}{S_Y} \qquad (6\text{-}98)$$

and

$$Z_j = \frac{X_j - \overline{X}_j}{S_j} \qquad (6\text{-}99)$$

in which S_Y is the standard deviation of the criterion variable and S_j $(j = 1, 2, \ldots, p)$ are the standard deviations of the predictor variables. It can be shown that the standardized partial regression coefficients (i.e., the t_j's) and the partial regression coefficients (i.e., the b_j's) are related by

$$b_j = \frac{t_j S_Y}{S_j} \qquad (6\text{-}100)$$

The intercept coefficient can be computed by

$$b_o = \overline{Y} - \sum_{j=1}^{p} b_j \overline{X}_j \qquad (6\text{-}101)$$

Thus the raw score model of Equation 6-91 can be computed directly from the standardized model, Equation 6-97, and Equations 6-100 and 6-101.

Although the differences between regression and correlation have been emphasized, the correlation matrix can be used in solving for the standardized partial regression coefficients. The solution is represented by

$$R_{11} \times t = R_{12} \tag{6-102}$$

in which R_{11} is the $p \times p$ matrix of intercorrelations between the predictor variables, t is a $p \times 1$ vector of standardized partial regression coefficients, and R_{12} is the $p \times 1$ vector of predictor-criterion correlation coefficients. Since R_{11} and R_{12} are known while t is unknown, it is necessary to solve the matrix equation, Equation 6-102, for the t vector. This involves premultiplying both sides of Equation 6-102 by R_{11}^{-1} (i.e., the inverse of R_{11}) and using the matrix identities, $R_{11}^{-1} \times R_{11} = I$ and $I \times t = t$, where I is the unit matrix, as follows:

$$R_{11}^{-1} \times R_{11} \times t = R_{11}^{-1} \times R_{12} \tag{6-103a}$$

$$I \times t = R_{11}^{-1} \times R_{12} \tag{6-103b}$$

$$t = R_{11}^{-1} \times R_{12} \tag{6-103c}$$

It is important to recognize from Equation 6-103c that the elements of the t vector are a function of both the intercorrelations and the predictor-criterion correlation coefficients. If $R_{11} = I$, then $R_{11}^{-1} = I$, and $t = R_{21}$; this suggests that since the t_j values serve as weights on the standardized predictor variables, the predictor-criterion correlations also reflect the importance, or "weight", that should be given to a predictor variable. However, when R_{11} is very different from I (i.e., when the intercorrelations are significantly different from zero), the t_j values will provide considerably different estimates of the importance of the predictors than would be indicated by the elements of R_{12}.

6.10.4. Intercorrelation

It is evident from Equation 6-103c that intercorrelation can have a significant effect on the t_j values and thus on the b_j values. In fact, if the intercorrelations are significant, the t_j values can be irrational. It should be evident that irrational regression coefficients can lead to irrational predictions. Thus it is important to assess the rationality of the coefficients.

The irrationality results from the difficulty in taking the inverse of the R_{11} matrix that is necessary for Equation 6-103c; this corresponds to the round-off-error problem associated with the solution of the normal equations of Equations 6-96a to 6-96c. A matrix in which the inverse cannot be evaluated is called a singular matrix. A near-singular matrix is one in which one or more pairs of the standardized normal equations are nearly identical (i.e., linearly dependent).

The determinant of a square matrix, such as a correlation matrix, can be used as an indication of the degree of intercorrelation. The determinant is a unique scalar value that characterizes the intercorrelation of the R_{11} matrix; that is, the determinant is a good single-valued representation of the degree of linear association between the normal equations.

Consider the following four matrices which differ in their level of intercorrelation:

$$A_1 = \begin{bmatrix} 1 & 0 & 0 \\ 0 & 1 & 0 \\ 0 & 0 & 1 \end{bmatrix} \tag{6-104}$$

$$A_2 = \begin{bmatrix} 1.0 & 0.5 & 0.5 \\ 0.5 & 1.0 & 0.5 \\ 0.5 & 0.5 & 1.0 \end{bmatrix} \tag{6-105}$$

$$A_3 = \begin{bmatrix} 1.0 & 0.9 & 0.9 \\ 0.9 & 1.0 & 0.9 \\ 0.9 & 0.9 & 1.0 \end{bmatrix} \tag{6-106}$$

$$A_4 = \begin{bmatrix} 1.0 & 0.9 & 0.0 \\ 0.9 & 1.0 & 0.0 \\ 0.0 & 0.0 & 1.0 \end{bmatrix} \tag{6-107}$$

The determinants of these matrices are

$$|A_1| = 1 \tag{6-108a}$$

$$|A_2| = 0.5 \tag{6-108b}$$

$$|A_3| = 0.028 \tag{6-108c}$$

$$|A_4| = 0.19 \tag{6-108d}$$

These values indicate that (1) the determinant of a correlation matrix will lie between 0.0 and 1.0; (2) if the intercorrelations are zero, the determinant of a correlation matrix equals 1.0; (3) as the intercorrelations become more significant, the determinant approaches zero; and (4) when any two rows of a correlation matrix are nearly identical, the determinant approaches zero.

6.10.5. Criteria for Evaluating a Multiple Regression Model

After a multiple regression model has been calibrated, we may ask how well the linear model represents the observed data. The following criteria should be used in answering this question: (1) the rationality of the coefficients; (2) the coefficient of multiple determination (R^2); (3) the standard error of estimate, which is usually compared with S_y; (4) the relative importance of the predictor variables; and (5) the characteristics of the residuals. The relative importance of each of these five criteria may vary with the problem as well as with the analyst.

The coefficient of multiple determination equals the fraction of the variation in the criterion variable that is explained by the regression equation. Thus it is the square of the correlation coefficient. Mathematically, it is defined as

$$R^2 = \frac{\sum_{i=1}^{n}(\hat{y}_i - \bar{Y})^2}{\sum_{i=1}^{n}(y_i - \bar{Y})^2}$$

(6-109)

It may be computed by

$$R^2 = \sum_{j=1}^{p} t_j r_{jY}$$

(6-110)

in which r_{jY} is the predictor-criterion correlation for predictor j. The value of R^2 is always in the range from zero to 1.0, with a value of zero indicating that Y is not related to any of the predictor variables. The coefficient of multiple determination must be at least as large as the square of the largest predictor-criterion correlation; if the intercorrelations are high, there will be little difference between the two, which would indicate that including more than one predictor variable in the regression equation does little to improve the accuracy of predictions. If the intercorrelations are low, including more than one predictor variable will probably provide greater accuracy than a bivariate model.

The standard error of estimate (S_e) is defined as the square root of the sum of the squares of the errors divided by the degrees of freedom (ν):

$$S_e = \left[\frac{1}{\nu} \sum_{i=1}^{n} (\hat{y}_i - y_i)^2 \right]^{0.5}$$

(6-111)

Previously, the degrees of freedom used to compute S_e were ($n - 2$); however, in general, the degrees of freedom are defined as

$$\nu = n - q$$

(6-112)

in which q is the number of unknowns. For the case where there are p partial regression coefficients and one intercept coefficient, $q = p + 1$ and $v = n - p - 1$. The S_e has the same units as the criterion variable and should be compared with S_Y in assessing the accuracy of prediction.

If we use Equation 6-100 to compute t_i for the bivariate case (i.e., one predictor variable), it will be evident that t, the standardized partial regression coefficient, equals the correlation coefficient. Even in the case where $p > 1$, t_i is still a measure of the common variation. However, Equation 6-103c indicates that the t_i values are a function of both the intercorrelation and the predictor-criterion correlations. When the intercorrelations are significant (i.e., R_{11} of Equation 6-103c is significantly different from I, the identity matrix), the t_i values may be irrational; thus they would not be valid measures of the relative importance of the corresponding predictor variable. Since the t value corresponds to a correlation coefficient in the bivariate case, it is not unreasonable to provide a similar interpretation for a multiple regression; thus t_i values should be considered irrational in magnitude if they are greater than 1.0 in absolute value. Even values of t near 1.0 may indicate irrationality. If a t value is irrational, the corresponding b value must be considered irrational, since b is a scaled value of t.

6.10.6. Analysis of Residuals

The residuals (i.e., $\hat{y} - y_i$) were shown to be an important criterion in assessing the validity of a bivariate regression equation; the same concepts apply in the analysis of the residuals for a multiple regression equation. Plots of the residuals may indicate an incorrect model structure, such as a nonlinear relationship between the criterion variable and one of the predictor variables. Additionally, the residuals may suggest a need to consider another predictor variable that was not included in the model. Plots of the residuals should be made to assess the assumptions of zero mean, constant error variation, independence, and normality. A frequency analysis using probability papers (see Section 5.9) of the residuals can be used to check for normality. A plot of the residuals vs. the predicted value of the criterion variable (i.e., e_i vs. $\hat{y}_i$) may identify either a violation of the assumption of constant variance or a lack of independence of the observations. The Durbin-Watson test can be used to test whether or not significant correlation exists among the residuals.

6.10.7. Computational Example

Although we rarely perform a multiple regression without the aid of a computer, it is instructive to see the manual computations at least one time. Table 6-13 provides values for the criterion variable (Y) and two predictor variables (X_1 and X_2); the sample consists of six observations (i.e., $n = 6$). Figure 6-15 shows the graphical relationship between each pair of variables; in general, linear associations are evident, although the scatter about the apparent linear relationships is significant. The correlation between the two predictor variables appears to be high relative to the degree of linearity between the criterion variable and the predictor variables.

TABLE 6-13 Data Base for Example of Two-Predictor Model

	Y	X_1	X_2	X_1^2	X_1X_2	X_1Y	X_2^2	X_2Y
	2	1	2	1	2	2	4	4
	2	2	3	4	6	4	9	6
	3	2	1	4	2	6	1	3
	3	5	5	25	25	15	25	15
	5	4	6	16	24	20	36	30
	6	5	4	25	20	30	16	24
Column summation	21	19	21	75	79	77	91	82

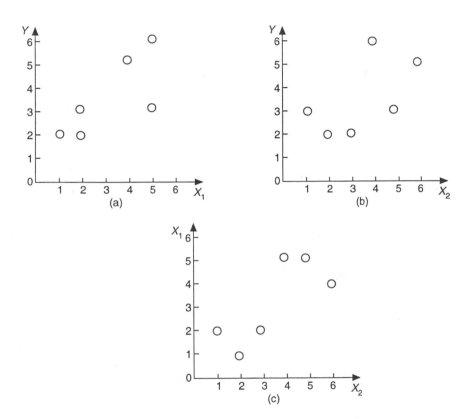

FIGURE 6-15 Graphs of variables for computational example: (a) Y vs. X_1; (b) Y vs. X_2; (c) X_1 vs. X_2.

After examining the plots, the correlation matrix (R) should be computed:

$$R = \begin{array}{c|ccc} & X_1 & X_2 & Y \\ \hline X_1 & 1.000 & 0.776 & 0.742 \\ X_2 & & 1.000 & 0.553 \\ Y & & & 1.000 \end{array} \qquad (6\text{-}113)$$

The correlation coefficient of 0.776 between X_1 and X_2 indicates that the intercorrelation is high. The determinant of the intercorrelation matrix equals 0.398, which should be considered as moderate; that is, it suggests that we should carefully examine the resulting regression equation for rationality.

The summations shown in Table 6-13 can be used to develop the normal equations (Equations 6-96a to 6-96c), as follows:

$$19b_0 + 75b_1 + 79b_2 = 77 \tag{6-114a}$$

$$21b_0 + 79b_1 + 91b_2 = 82 \tag{6-114b}$$

$$6b_0 + 19b_1 + 21b_2 = 21 \tag{6-114c}$$

The normal equations can be solved to obtain values for the coefficients of the following equation:

$$\hat{Y} = 1.30 + 0.75X_1 - 0.05X_2 \tag{6-115}$$

Similarly, the correlation matrix could be used to develop the normal equations for the standardized model:

$$\begin{bmatrix} 1.000 & 0.776 \\ 0.776 & 1.000 \end{bmatrix} \begin{bmatrix} t_1 \\ t_2 \end{bmatrix} = \begin{bmatrix} 0.742 \\ 0.553 \end{bmatrix} \tag{6-116}$$

which yields

$$t_1 + 0.776t_2 = 0.742 \tag{6-117a}$$

$$0.776t_1 + t_2 = 0.553 \tag{6-117b}$$

Solving for t_1 and t_2 yields the following standardized equation:

$$Z_Y = 0.786 Z_1 - 0.057 Z_2 \tag{6-118}$$

Both the raw score equation (Equation 6-115) and the standardized equation (Equation 6-118) indicate that as X_2 increases, Y decreases; this is in conflict with both the correlation coefficient and the graphical relationship shown in Figure 6-15. This irrationality is the result of the high intercorrelation between the predictor variables.

The correlation coefficient can be computed from either Equation 6-109 or 6-110. Equation 6-110 yields

$$R^2 = t_1 r_{1Y} + t_2 r_{2Y} \qquad (6\text{-}119a)$$

$$= 0.786(0.742) + (-0.057)(0.553) = 0.552 \qquad (6\text{-}119b)$$

Therefore, R equals 0.743. Predictor X_1 explained 55.1% (i.e., 0.742^2) of the variation in Y by itself. Thus the use of X_2 is not justified because it does not make a significant contribution to the explained variation and, even more important, it results in an irrational model. The standard error of estimate is

$$S_e = S_Y \sqrt{\left(\frac{n-1}{n-p-1}\right)(1-R)^2} = 1.643\sqrt{\left(\frac{5}{3}\right)(1-0.552)^2} \qquad (6\text{-}120a)$$

$$S_e = 1.643(0.864) = 1.420 \qquad (6\text{-}120b)$$

Thus the two-predictor model reduced the error variance by 25.3%; the standard error of estimate is about 86% of the standard deviation of the criterion variable. For the one-predictor model with X_1 as the predictor variable, the standard error of estimate would be 1.231, which is less than the standard error of estimate for the two-predictor model; this occurs because the ratio $(n-1)/(n-p-1)$ is much less for the one-predictor model. This example illustrates the importance of computing the standard error of estimate when the sample size is small. The standard error provides a more realistic measure of the goodness of fit than the correlation coefficient does.

6.11. REGRESSION ANALYSIS OF NONLINEAR MODELS

In most empirical analyses, linear models are attempted first because of the relative simplicity of linear analysis. Also, linear models are easily applied, and the statistical reliability is easily assessed.

Linear models may be rejected because of either theoretical considerations or empirical evidence. Specifically, theory may suggest a nonlinear relationship between a criterion variable and one or more predictor variables; for example, biological growth curves used by environmental engineers are characterized by nonlinear forms. When a model structure cannot be identified by theoretical considerations, empirical evidence can be used in model formulation and may suggest a nonlinear form. For example, the hydrologic relationship between peak discharge and the drainage area of a watershed has been found to be best represented by a log-log equation, which is frequently referred to as a power model. A nonlinear form may also be suggested by the residuals that result from a linear analysis; that is, if a linear model produces nonrandomly distributed residuals (i.e., a bias is apparent), the underlying assumptions are violated; a nonlinear functional form may produce residuals that satisfy the assumptions that underlie the principle of least squares.

One very common use of nonlinear models is the fitting of existing design curves that have significant nonlinearity. The equations are then used as part of computer software that performs a design procedure that, in the past, was solved manually.

6.11.1. Common Nonlinear Alternatives

Linear models were separated into bivariate and multivariate; the same separation is applicable to nonlinear models. It is also necessary to separate nonlinear models on the basis of the functional form. Although polynomial and power models are the most frequently used nonlinear forms, it is important to recognize that other model structures are available and may actually provide the correct structure. In addition to the power and polynomial structures, forms such as a square root, exponential, and logarithmic may provide the best fit to a set of data. Since the polynomial and power forms are so widely used, it is of importance to identify their structure:

1. Bivariate
   a. Polynomial

$$\hat{Y} = b_0 + b_1 X + b_2 X^2 + \ldots + b_p X^p \tag{6-121}$$

   b. Power

$$\hat{Y} = b_0 X^{b_1} \tag{6-122}$$

2. Multivariate
   a. Polynomial

$$\hat{Y} = b_0 + b_1 X_1 + b_2 X_2 + b_3 X_1^2 + b_4 X_2^2 + b_5 X_1 X_2 \tag{6-123}$$

   b. Power

$$\hat{Y} = b_0 X_1^{b_1} X_2^{b_2} \ldots X_p^{b_p} \tag{6-124}$$

In these relationships, Y is the criterion variable, X is the predictor variable in the bivariate case, X_i is the ith predictor variable in the multivariate case, b_j ($j = 0$, 1, ..., p) is the jth regression coefficient, and p is either the number of predictor variables or the order of the polynomial. The bivariate forms are just special cases of the multivariate form.

6.11.2. Calibration of Polynomial Models

The power and polynomial models are widely used nonlinear forms because they can be transformed in a way that makes it possible to use the principle of least squares. Although the transformation to a linear structure is desirable from the standpoint of calibration, it has important consequences in terms of assessing the goodness-of-fit statistics.

The bivariate polynomial of Equation 6-121 can be calibrated by forming a new set of predictor variables:

$$W_i = X^i \quad \text{for } i = 1, 2, \ldots, p \qquad (6\text{-}125)$$

This results in the model

$$\hat{Y} = b_0 + b_1 W_1 + b_2 W_2 + \ldots + b_p W_p \qquad (6\text{-}126)$$

The coefficients b_j can be estimated using a standard linear multiple regression analysis.

The multivariate polynomial of Equation 6-123 can also be solved using a multiple regression analysis for a set of transformed variables. For the model given by Equation 6-123, the predictor variables are transformed as follows:

$$W_1 = X_1 \qquad (6\text{-}127a)$$

$$W_2 = X_2 \qquad (6\text{-}127b)$$

$$W_3 = X_1^2 \qquad (6\text{-}127c)$$

$$W_4 = X_2^2 \qquad (6\text{-}127d)$$

$$W_5 = X_1 X_2 \qquad (6\text{-}127e)$$

The revised model has the form

$$\hat{Y} = b_0 + \sum_{j=1}^{5} b_j W_j \qquad (6\text{-}128)$$

It is important to note that the polynomial models do not require a transformation of the criterion variable.

The model given by Equation 6-123 has only two predictor variables (i.e., X_1 and X_2) and is a second-order equation. In practice, a model may have more predictor variable and may be of higher order. In some cases the interaction terms (i.e., X_iX_j) may be omitted to decrease the number of coefficients that must be calibrated. However, if the interaction terms are omitted when they are actually significant, the goodness-of-fit statistics may suffer, unless the variation is explained by the other terms in the model; when this occurs, the coefficients may lack physical significance.

6.11.3. Fitting a Power Model

The bivariate power model of Equation 6-122 can be calibrated by forming the following set of variables:

$$\hat{Z} = \ln \hat{Y} \tag{6-129a}$$

$$c = \ln b_0 \tag{6-129b}$$

$$W = \ln X \tag{6-129c}$$

These transformed variables form the following linear equation, which can be calibrated using bivariate linear regression analysis:

$$\hat{Z} = c + b_1 W \tag{6-130}$$

After values of c and b_1 are obtained, the coefficient b_0 can be determined using

$$b_0 = e^c \tag{6-131}$$

Base 10 logarithms can also be used; in such a case, Equations 6-129a and 6-129c will use the base 10 logarithm and Equation 6-131 will use a base 10 rather than e.

The multivariate power model of Equation 6-124 can be evaluated by making a logarithmic transformation of both the criterion and the predictor variables:

$$\hat{Z} = \ln \hat{Y} \tag{6-132}$$

$$c = \ln b_0 \tag{6-133}$$

$$W_i = \ln X_i \quad \text{for } i = 1, 2, \ldots, p \tag{6-134}$$

The resulting model has the form

$$\hat{Z} = c + \sum_{j=1}^{p} b_j W_j \tag{6-135}$$

The coefficients of Equation 6-135 can be evaluated using a multiple regression analysis. The value of b_0 can be determined by making the transformation of Equation 6-131. Again, a base 10 transformation can be used rather than a natural log transformation.

6.11.4. Goodness of Fit

Since most computer programs that perform multiple regression analyses include goodness-of-fit statistics as part of the output, it is important to recognize the meaning of these statistics. For nonlinear models in which the criterion variable $\hat{Y}$ is not transformed, the goodness-of-fit statistics are valid indicators of the reliability of the model; however, when the criterion variable is transformed, such as is necessary for the power model form, the principle of least squares is applied in the log-log space. As a result, the residuals that are used to compute the standard error of estimate, and therefore the correlation coefficient, are measured in the domain of the logarithm of $\hat{Y}$ and not the $\hat{Y}$ domain. Therefore, the goodness-of-fit statistics are not necessarily a reliable indicator of model reliability, especially since engineering decisions are made to predict Y and not $\log(Y)$. Therefore, when a model requires a transformation of the criterion variable $\hat{Y}$ in order to calibrate the coefficients, the goodness-of-fit statistics that are included with the multiple regression output should not be used as measures of reliability. Instead, values for the goodness-of-fit statistics should be recomputed using values of $\hat{Y}$ rather than $\log(\hat{Y})$. The correlation coefficient and the standard error of estimate should be computed using Equations 6-109 and 6-111, respectively. In summary, when an equation is calibrated with the criterion variable transformed, such as in the log-log space, the least-squares concepts apply only to the transformed space, not to the measurement nontransformed space. The sum of the residuals based on the logarithms of Y will equal zero; however, the sum of the residuals of the transformed values of Y will not equal zero, and the sum of the squares of the errors may not be a minimum in the Y domain even though it is in the $\log(Y)$ domain. Furthermore, the assumption of a constant variance may also not be valid. Because these basic assumptions of regression are not valid in the X-Y space, many practitioners object to data transformations. However, in spite of these theoretical considerations, the transformations may provide reasonable estimates in the X-Y space.

6.11.5. Analysis of Variance for Polynomial Models

Just as the analysis of variance (ANOVA) is used to examine the statistical adequacy of linear models, it can be used to test whether or not the polynomial models are adequate. Both the total and partial F statistics can be used. Table 6-14 provides a summary of the ANOVA test for first-, second-, and third-order polynomials that include a single predictor variable. However, the formulas are easily adapted for polynomial models involving more than one predictor variable, including analyses involving interaction terms. Although the partial F test can be used sequen-

TABLE 6-14 ANOVA Table for Polynomial Analysis

Model	Correlation coefficient	F test	H_o	Sum of squares	Degrees of freedom	Test statistic: F
First-order (linear) $Y = a + b_1 X$	R_1	Total and partial	$\beta_1 = 0$	Regression: R_1^2 Error: $1 - R_1^2$	1 $n - 2$	$F = \dfrac{R_1^2}{\left(\dfrac{1 - R_1^2}{n - 2}\right)}$
Second-order (quadratic) $Y = a + b_1 X + b_2 X^2$	R_2	Total	$\beta_1 = \beta_2 = 0$	Regression: R_2^2 Error: $1 - R_2^2$	2 $n - 3$	$F = \dfrac{\dfrac{1}{2} R_2^2}{\left(\dfrac{1 - R_2^2}{n - 3}\right)}$
		Partial	$\beta_2 = 0$	Regression: $R_2^2 - R_1^2$ Error: $1 - R_2^2$	1 $n - 3$	$F = \dfrac{R_2^2 - R_1^2}{\left(\dfrac{1 - R_2^2}{n - 3}\right)}$
Third-order (cubic) $Y = a + b_1 X + b_2 X^2 + b_3 X^3$	R_3	Total	$\beta_1 = \beta_2 = \beta_3 = 0$	Regression: R_3^2 Error: $1 - R_3^2$	3 $n - 4$	$F = \dfrac{\dfrac{1}{3} R_3^2}{\left(\dfrac{1 - R_3^2}{n - 4}\right)}$
		Partial	$\beta_3 = 0$	Regression: $R_3^2 - R_2^2$ Error: $1 - R_3^2$	1 $n - 4$	$F = \dfrac{R_3^2 - R_2^2}{\left(\dfrac{1 - R_3^2}{n - 4}\right)}$

tially to select terms for the final polynomial model, it is always important to remember that the selection of a level of significance is a somewhat arbitrary decision that can affect the resulting model.

6.11.6. Additional Model Forms

In addition to the polynomial and power models, other forms can provide good approximations to the underlying population relationship between two or more variables. An exponential model has the form

$$\hat{Y} = b_0 e^{b_1 X} \tag{6-136}$$

in which b_0 and b_1 are coefficients requiring values by fitting to data. Values for b_0 and b_1 can be obtained by linear bivariate regression after taking the logarithms of both sides of Equation 6-136 as follows:

$$\ln \hat{Y} = \ln b_0 + b_1 X \tag{6-137}$$

In this case, $\hat{Y}$ is transformed; thus the correlation coefficient and standard error of estimate apply to the $\ln \hat{Y}$ and not to $\hat{Y}$. Also, the intercept coefficient obtained from the regression analysis for Equation 6-137 must be transformed for use with Equation 6-136.

A logarithmic curve can also be used to fit values of Y and X:

$$\hat{Y} = b_0 + b_1 \ln X \tag{6-138}$$

In this case, $\hat{Y}$ is not transformed, so the values of the correlation coefficient and the standard error of estimate are valid indicators of the goodness of fit in the $\hat{Y}$ space.

6.12. APPLICATIONS

6.12.1. One-Predictor Polynomial of Sediment Yield vs. Slope

A computer program was used to derive first-, second-, and third-order polynomials that regress sediment yield on watershed slope. The correlation matrix is given in Table 6-15. It is evident that the intercorrelation coefficients are very significant and should lead to irrational regression coefficients; furthermore, because of the high intercorrelation it is unlikely that the explained variance will improve significantly as the order of the equation increases. Table 6-16 provides a

TABLE 6-15 Correlation Matrix:
Sediment Yield (Y) Vs. Slope (S)

	S	S^2	S^3	Y
S	1.000	0.908	0.791	0.443
S^2		1.000	0.972	0.443
S^3			1.000	0.428
Y				1.000

TABLE 6-16 Summary of Polynomial Analyses for Sediment Yield Vs. Slope Models

Model	R	Intercept b_o	Partial regression coefficients			Standardized partial regression coefficients		
			b_1	b_2	b_3	t_1	t_2	t_3
First-order	0.443	0.220	0.0263	—	—	0.443	—	—
Second-order	0.454	0.323	0.0136	2.31×10^{-4}	—	0.229	0.235	—
Third-order	0.499	0.042	0.0865	-0.0031	3.43×10^{-5}	1.456	-3.164	2.35

TABLE 6-17 ANOVA: Total F Test

Model	R^2	$1 - R^2$	First degrees of freedom	Second degrees of freedom	Total F	$F_{0.05}$	Decision
First-order	0.196	0.804	1	35	$\dfrac{0.196/1}{0.804/35} = 8.53$	4.12	Reject H_o
Second-order	0.206	0.794	2	34	$\dfrac{0.206/2}{0.794/34} = 4.41$	3.28	Reject H_o
Third-order	0.249	0.751	3	33	$\dfrac{0.249/3}{0.751/33} = 3.65$	2.90	Reject H_o

TABLE 6-18 ANOVA: Partial F Tests

Model	ΔR^2	$1 - R^2$	First degrees of freedom	Second degrees of freedom	Partial F	$F_{0.05}$	Decision
First-order	0.196	0.804	1	35	$\dfrac{0.196/1}{0.804/35} = 8.53$	4.12	Reject H_o
Second-order	0.010	0.794	1	34	$\dfrac{0.01/1}{0.794/34} = 0.43$	4.13	Accept H_o
Third-order	0.043	0.751	1	33	$\dfrac{0.043/1}{0.751/33} = 1.89$	4.14	Accept H_o

summary of the correlation coefficients, the standardized partial regression coefficients, and the partial regression coefficients. The second- and third-order models do not provide a significant increase in the explained variation when compared with the linear model.

The ANOVA results are given in Tables 6-17 and 6-18. The total F tests indicate that all three models are statistically significant at the 5% level of significance. However, the partial F tests indicate that including the quadratic term in the equation does not improve the statistical goodness of fit, since the computed F value is less than the critical value.

TABLE 6-19 Correlation Matrix:
Evaporation (E) Vs. Temperature (T)

	T	T^2	T^3	E
T	1.000	0.995	0.981	0.581
T^2		1.000	0.996	0.602
T^3			1.000	0.627
E				1.000

TABLE 6-20 Summary of Regression Analysis

Model	R	S_e^a	t_1	t_2	t_3	b_o	b_1	b_2	b_3
First-order	0.581	0.0763	0.581	—	—	-0.114	0.00383	—	—
Second-order	0.648	0.0715	2.160	2.76	—	0.425	-0.01420	0.000143	—
Third-order	0.654	0.0711	3.110	-8.44	5.98	-0.239	0.02050	-0.000439	3.14×10^{-6}

[a] $S_y = 0.0936$ in./day.

6.12.2. One-Predictor Polynomial of Evaporation vs. Temperature

The evaporation data were used to derive polynomials for the regression of evaporation on temperature. The correlation matrix, which is shown in Table 6-19, shows very high intercorrelation. Table 6-20 provides a summary of the regression analyses. Whereas the linear model explains approximately 34% of the variation in evaporation, the higher-order terms provide only a marginal increase in the explained variation. The quadratic model increases the explained variance by about 8%. The marginal improvement is also evident from the small decrease in the standard error of estimate. In judging the rationality of a model, it is of value to examine the predicted values of evaporation for values of the predictor variable that are likely to be observed. The three models were used to estimate the daily evaporation depth for selected temperatures (Table 6-21). Because water freezes at 32°F, evaporation rates near zero are expected; only the linear model gives a rational estimate at 32°F. Estimates made using the second-order model decrease with increases in temperature for temperatures up to about 50°F. Both the second- and third-order models provide especially high rates at 100°F. In summary, the high intercorrelations lead to irrational estimates when the higher-order forms are used.

Tables 6-22 and 6-23 summarize the ANOVA. The total F test (Table 6-22) indicates that all models are statistically significant. Similarly, the partial F tests (Table 6-23) indicate that each term is significant, and thus the third-order equation should be selected.

TABLE 6-21 Predicted Evaporation Rates (in./day)

Model	Temperature (°F)			
	32	50	75	100
First-order	0.009	0.078	0.173	0.269
Second-order	0.117	0.073	0.164	0.435
Third-order	0.070	0.081	0.154	0.561

TABLE 6-22 ANOVA: Total F Tests

Model	R^2	$1 - R^2$	First degrees of freedom	Second degrees of freedom	Total F	$F_{0.05}$	Decision
First-order	0.338	0.662	1	352	$\frac{0.338/1}{0.662/352} = 179.5$	3.86	Reject H_0
Second-order	0.420	0.580	2	351	$\frac{0.420/2}{0.580/351} = 127.1$	3.02	Reject H_0
Third-order	0.428	0.572	3	350	$\frac{0.428/3}{0.572/350} = 87.1$	2.62	Reject H_0

TABLE 6-23 ANOVA: Partial F Tests

Model	ΔR^2	$1 - R^2$	First degrees of freedom	Second degrees of freedom	Partial F	$F_{0.05}$	Decision
First-order	0.338	0.662	1	352	$\frac{0.338/1}{0.662/352} = 179.7$	3.86	Reject H_0
Second-order	0.082	0.580	1	351	$\frac{0.082/1}{0.580/351} = 49.6$	3.86	Reject H_0
Third-order	0.008	0.572	1	350	$\frac{0.008/1}{0.572/350} = 4.9$	3.86	Reject H_0

This example illustrates that different criteria for model selection may conflict, for example, the partial F tests and the standardized partial regression coefficients. For this example, it appears that the ANOVA is an inadequate criterion for model selection. Quite possibly, the 5% level of significance is not an adequate criterion; after all, use of the 5% level is based solely on convention and not on a rational analysis of the relationship between the level of significance, the sample size, and the physical separation criterion.

6.12.3. Single-Predictor Power Model

A data set that consists of five observations will be used to illustrate the fitting of a power model that includes a single predictor variable. The data set is given in

TABLE 6-24 Data Base for Single-Predictor Power Model

X	Y	lnX	lnY	Using Equation 6-141		Using Equation 6-140	
				$\hat{Y}$	e	ln $\hat{Y}$	e_e
2.718	7.389	1	2	−214.54	−221.929	6.048	−1.341
7.389	20.086	2	3	53.18	33.095	40.445	20.359
20.086	403.429	3	6	780.93	377.499	270.427	−133.002
54.598	2980.960	4	8	2759.03	−221.927	1807.960	−1173.000
148.413	8103.080	5	9	8136.00	32.923	12087.400	3984.310
					$\sum e = 0$		$\sum e_e \neq 0$

Table 6-24. The values were converted to natural logarithms and the coefficients of Equation 6-130 were calibrated using least squares:

$$\ln \hat{Y} = -0.1 + 1.9 \ln X \qquad (6\text{-}139)$$

A correlation coefficient of 0.9851 was computed for Equation 6-139. Equation 6-139 can be retransformed to the X-Y coordinate space:

$$\hat{Y} = 0.9048 X^{1.9} \qquad (6\text{-}140)$$

A linear analysis of the values of X and Y provides a correlation coefficient of 0.9975 and the following regression equation:

$$\hat{Y} = 370.3301 + 57.3164 X \qquad (6\text{-}141)$$

The residuals for Equations 6-140 and 6-141 are given in Table 6-24. Whereas the residuals for Equation 6-141 have a sum of zero, the sum of the residuals for Equation 6-140 is not equal to zero; however, the sum of the residuals for Equation 6-139 will equal zero in the natural-log space. The nonzero sum of the residuals indicates that the model is biased.

The correlation coefficient and standard error of estimate that are computed for the model in the natural-log space are not valid indicators of the accuracy of the estimates in the domain of the nontransformed criterion variable. For example, the correlation coefficient of 0.9851 computed with Equation 6-139 is not a measure of the accuracy of Equation 6-140. Values of Y that were estimated with Equation 6-140 are given in Table 6-24, and the resulting residuals are denoted as e_e. If the separation of variation concept is applied to Equation 6-140, the components are

$$\sum_{i=1}^{n} \left(y_i - \overline{Y} \right)^2 = \sum_{i=1}^{n} \left(y_i - \hat{y}_i \right)^2 + \sum_{i=1}^{n} \left(\hat{y}_i - \overline{Y} \right)^2 \qquad (6\text{-}142)$$

$$48,190,499.4 \neq 17,276,427.6 + 100,525,199.1 \qquad (6\text{-}143)$$

The separation of variation is not valid for Equation 6-140 because the model was calibrated by taking logarithms. The separation of variation does, however, apply to the logarithms. It is evident that if the correlation coefficient were computed as the ratio of the explained variation to the total variation (Equation 6-109), then from the values of Equation 6-143 the correlation coefficient would be greater than 1.0, which is not rational. Thus the standard error of estimate, rather than the correlation coefficient, should be used as a measure of accuracy:

$$S_e = \left(\frac{17,276,427.6}{5-2}\right)^{0.5} = 2399.75 \tag{6-144}$$

The standard deviation of the criterion variable is 3470.97; therefore, the power model is not highly accurate, and the computed correlation for the log model (i.e., 0.9851) certainly does not reflect the accuracy of the model for predicting Y.

6.12.4. Multivariate Power Model

Sediment yield data will be used to illustrate a power model that includes more than one predictor variable. The small particle size index cannot be included as a predictor variable because some of the observations equal zero; thus the model regresses sediment yield (Y) on three predictor variables: the precipitation/temperature ratio (P), the slope (S), and the soil index (I). The data are given in Table 6-25. The resulting regression equation in power form is

$$\ln \hat{Y} = -4.563 - 1.123 \ln P + 0.8638 \ln S - 0.1089 \ln I \tag{6-145}$$

Transforming Equation 6-145 to the X-Y space yields

$$\hat{Y} = 0.01043 P^{-1.123} S^{0.8638} I^{-0.1089} \tag{6-146}$$

The computer program provided a correlation coefficient of 0.725 and a standard error estimate of 1.019; however, these are for Equation 6-145 and may not be accurate indications of the accuracy of Equation 6-146. The standard error of estimate is expressed in logarithmic units. The standard error of estimate as measured in the X-Y space was computed as 0.595, which can be compared with a standard deviation of 0.817 for the 37 observations of sediment yield.

For a comparison, the following linear model was calibrated using the same three predictor variables:

$$\hat{Y}_{\text{Linear}} = 0.8920 - 1.428P + 0.03664S - 0.01308I \tag{6-147}$$

This resulted in a multiple correlation coefficient of 0.720 and a standard error of estimate of 0.592. Thus, for this case, there is no significant difference in the expected accuracy of the linear and power models.

TABLE 6-25 Sediment Yield Data[a]

P	S	I	Y
0.135	4.000	40.000	0.140
0.135	1.600	40.000	0.120
0.101	1.900	22.000	0.210
0.353	17.600	2.000	0.150
0.353	20.600	1.000	0.076
0.492	31.000	0.000	0.660
0.466	70.000	23.000	2.670
0.466	32.200	57.000	0.610
0.833	41.400	4.000	0.690
0.085	14.900	22.000	2.310
0.085	17.200	2.000	2.650
0.085	30.500	15.000	2.370
0.193	6.300	1.000	0.510
0.167	14.000	3.000	1.420
0.235	2.400	1.000	1.650
0.448	19.700	3.000	0.140
0.329	5.600	27.000	0.070
0.428	4.600	12.000	0.220
0.133	9.100	58.000	0.200
0.149	1.600	28.000	0.180
0.266	1.200	15.000	0.020
0.324	4.000	48.000	0.040
0.133	19.100	44.000	0.990
0.133	9.800	32.000	0.250
0.133	18.300	19.000	2.200
0.356	13.500	58.000	0.020
0.536	24.700	62.000	0.030
0.155	2.200	27.000	0.036
0.168	3.500	24.000	0.160
0.673	27.900	22.000	0.370
0.725	31.200	10.000	0.210
0.275	21.000	64.000	0.350
0.150	14.700	37.000	0.640
0.150	24.400	29.000	1.550
1.140	15.200	11.000	0.090
1.428	14.300	1.000	0.170
1.126	24.700	44.000	0.170

[a] Y = sediment yield, X_1 = climate variable, X_2 = watershed slope variable, and X_3 = soil aggregation variable.

From Flakman, E. M., J. Hydraulic Div. ASCE, 98(HY12), 2073, 1972.

6.12.5. Estimating Breakwater Costs

In Section 6.9.2, a bivariate model between the construction cost of a breakwater and length of the breakwater was developed. It was found to be inaccurate, with the

TABLE 6-26 Data for Breakwater Cost Estimation

Length, X_1 (ft)	Depth, X_2 (ft)	Cost, Y ($ × 10³)
450	6.3	230
625	9.6	490
670	3.1	150
730	10.1	540
810	5.4	280
880	3.5	120
1020	6.5	380
1050	2.9	170
1100	8.8	530
1175	3.9	290
1230	7.4	460
1300	3.3	230
1350	7.2	470
1510	4.6	360

suggestion given that other variables associated with cost were probably better determinants of the cost.

Table 6-26 includes the original measurements of length and cost along with a second predictor variable, the average depth of the breakwater. The correlation matrix is

	X_1	X_2	Y
X_1: length	1.000	−0.229	0.176
X_2: depth		1.000	0.905
Y: cost			1.000

While length and depth are poorly correlated, depth and cost have a high correlation, with R^2 equal to 0.819.

A multiple regression analysis yields the following equation:

$$\hat{Y} = -199.8 + 0.1904X_1 + 58.72X_2 \tag{6-148}$$

The multiple correlation coefficient equals 0.987, thus, $R^2 = 0.974$, which indicates that the two-predictor model provides accurate estimates of the cost. The standard error of estimate is $25.3 × 10³, with $S_e/S_y = 0.174$. The F statistic of 210 is exceptionally large and well beyond any F values for commonly used level of significance. The standard error ratios for the regression coefficients are small: $S_{e,b_1}/b_1 = 0.122$ and $S_{e,b_2}/b_2 = 0.050$. Overall, the goodness-of-fit statistics suggest an accurate model.

The standardized partial regression coefficients are $t_1 = 0.405$ and $t_2 = 0.998$. While t_2 is below 1, it is still very large.

The intercept is quite large in magnitude. While none of the predicted costs are negative, it is still possible because the intercept is negative. For example, at a length of 450 ft and depth of 2 ft, the estimated cost would be negative. Equation 6-148 should be used cautiously beyond the range of the data: $450 < X_1 < 1510$ ft and $2.9 < X_2 < 10.1$ ft.

It is interesting to compare the two-predictor multiple regression model with both one-predictor models. The length vs. cost model was discussed in Section 6.9.2. It provided very poor accuracy, $R^2 = 0.031$. The depth vs. cost model would explain 81.9% of the variation in cost (0.905^2). Since the multiple regression model explains 97.4%, adding the length to the model increases the explained variance by 15.5% ($97.4 - 81.9$); this is a significant increase, so the two-predictor model is preferable to either of the one-predictor models. Also, when the depth is added to the equation, the importance of the length increases. The t for the bivariate model is 0.176, and $t_1 = 0.405$ for the multiple regression model.

TABLE 6-27 Regression Analysis

X_1	X_2	Y	$\hat{Y}$	e
1	0	0.9	0.42	−0.48
2	0	1.8	1.87	0.07
3	0	2.4	3.32	0.92
4	0	4.3	4.76	0.46
5+	0	6.6	6.21	−0.39
1	1	1.9	1.13	−0.77
2	1	2.7	2.58	−0.12
3	1	3.6	4.03	0.43
4	1	5.4	5.47	0.07
5+	1	7.7	6.92	−0.78
1	2+	2.1	1.84	−0.26
2	2+	3.1	3.29	0.19
3	2+	3.7	4.74	1.04
4	2+	5.9	6.18	0.28
5+	2+	8.3	7.63	−0.67

6.12.6. Trip Generation Model

For a regional-planning study, a transportation engineer collects data on traffic. The data of Table 6-27 gives measurements of nonwork, home-based trip-production rates for suburban, medium-density households (Y) as a function of the number of persons per household (X_1) and the number of vehicles per household (X_2). In this case, the predictor variables are integers.

The correlation matrix for the sample of 15 is given by

	X_1	X_2	Y
X_1	1.000	0.000	0.931
X_2		1.000	0.264
Y			1.000

The intercorrelation between the two predictor variables is zero because the data were collected on a grid. One advantage of having no intercorrelation is that the potential for irrational regression coefficients is minimized. The predictor-criterion correlation coefficient for the number of persons per household is large ($R^2 = 0.868$), while that for the number of vehicles per household is small ($R^2 = 0.070$). This may suggest that the bivariate equation $\hat{Y} = f(X_1)$ may be as accurate as the multiple regression model $\hat{Y} = f(X_1, X_2)$.

A multiple regression analysis produced the following equation:

$$\hat{Y} = -1.023 + 1.4467X_1 + 0.7100X_2 \tag{6-149}$$

The correlation coefficient is 0.968, which gives $R^2 = 0.937$. Since the intercorrelation between X_1 and X_2 is zero, the sum of the squares of the two predictor-criterion correlation coefficients equals the coefficient of multiple determination:

$$R^2 = R_{1Y}^2 + R_{2Y}^2 \tag{6-150a}$$

$$0.937 = (0.931)^2 + (0.264)^2 \tag{6-150b}$$

This equality holds only when there is no intercorrelation. It also shows that the multiple regression model explains 7% more variation than the bivariate model based solely on X_1.

The standard error of estimate is 0.6153, which gives $S_e/S_y = 0.271$. The total F statistic is 89.6. For $k = v_1 = 2$ and $u = v_2 = 12$, this is statistically significant at even the smallest level of significance. Both of these statistics support the conclusion suggested by the correlation coefficient that the model is highly accurate.

The standardized partial regression coefficients are $t_1 = 0.931$ and $t_2 = 0.264$. Since the intercorrelation coefficient is 0, the t values equal the predictor-criterion correlation coefficients. Again, the number of persons per household is shown to be the more important predictor.

The intercept of Equation 6-149 is −1.023, which is irrational, since a value of 0 would be expected. If both X_1 and X_2 are 0, then Y should be 0. Furthermore, the magnitude of the intercept is large, that is, approximately 25% of the mean of Y and larger then the value of Y for $X_1 = 1$ and $X_2 = 0$. This suggests that another model structure may be needed to realistically represent the data.

The residuals are given in Table 6-27, and they also suggest that the linear model of Equation 6-149 is an incorrect structure. There are very noticeable local biases. For each value of X_2 the residuals for $X_1 = 1$ and $X_1 = 5$ are negative; additionally, for each value of X_2 the residuals for $X_1 = 3$ and $X_1 = 4$ are positive. The trends for each value of X_2 suggest that a nonlinear structure would be more appropriate.

TABLE 6-28 Multiple Regression Analysis of the Stream
Reaeration Coefficient (Y) on Stream Velocity (X_1, m/sec),
Water Depth (X_2, m), and Water Temperature (X_3, °C)

X_1	X_2	X_3	Y	$\hat{Y}$	e
1.4	1.24	10	2.16	2.69	0.80
0.6	0.87	9	2.50	2.57	0.07
2.2	1.87	13	2.54	3.55	1.01
1.5	0.99	11	2.60	3.06	0.46
0.5	0.85	14	2.71	2.97	0.26
1.6	0.97	8	3.00	2.83	−0.17
1.3	1.12	19	3.02	3.70	0.68
0.8	1.21	11	3.03	2.85	−0.18
1.6	1.64	14	3.27	3.41	0.14
2.3	1.91	6	3.42	2.98	−0.44
1.0	0.70	15	3.42	3.21	−0.21
1.9	1.62	12	3.45	3.34	−0.11
1.3	1.01	17	3.49	3.51	0.02
1.8	1.43	15	3.58	3.54	−0.04
1.7	1.05	15	3.61	3.47	−0.14
0.7	0.62	18	3.69	3.36	−0.33
2.2	1.58	12	3.85	3.43	−0.42
1.4	1.21	15	3.88	3.39	−0.49
1.8	2.07	20	3.99	4.04	0.05
1.5	1.59	16	4.51	3.55	−0.96
2.8	2.38	24	4.74	4.75	0.01

6.12.7. Estimation of the Reaeration Coefficient

To estimate the biochemical oxygen demand of a stream the environmental engineer needs an estimate of the reaeration coefficient. The reaeration coefficient varies with the stream velocity (X_1, in meters per second), the water depth (X_2, in meters), and the water temperature (X_3, in °C). Table 6-28 gives a set of 21 measurements of the three predictor variables and estimated reaeration coefficients (Y). The correlation matrix is

	X_1	X_2	X_3	Y
X_1	1.000	0.839	0.139	0.436
X_2		1.000	0.240	0.456
X_3			1.000	0.621
Y				1.000

The intercorrelation is high ($|R_{11}| = 0.276$) because of the high correlation between X_1 and X_2. The water temperature is not highly correlated with either velocity or depth. The regression analysis of the 21 sample values yields

$$\hat{Y} = 1.508 + 0.3315X_1 + 0.09731X_2 + 0.08672X_3 \qquad (6\text{-}151)$$

All the regression coefficients are rational in sign. Equation 6-151 results in a correlation coefficient of 0.715, which means that the equation explains 51.2% of the total variation. The standard error of estimate is 0.497 ($S_e/S_y = 0.758$), which suggests that the equation provides estimates of Y that are not highly accurate. For the ANOVA test on the null hypothesis H_o: $\beta_1 = \beta_2 = \beta_3 = 0$, the computed F is 5.935 for $k = v_1 = 3$ and $u = v_2 = 17$. Critical F values for selected levels of significance (α) and the decisions are as follows:

α	F_α	Decision
0.001	8.73	*Accept* H$_o$
0.005	6.16	*Accept* H$_o$
0.010	5.18	*Reject* H$_o$
0.050	3.20	*Reject* H$_o$

Thus, for the commonly used levels of 1 and 5%, the null hypothesis would be rejected; however, for more restrictive levels of less than 1%, the null hypothesis would be accepted. The limiting probability for acceptance is about 0.6%.

Of the three predictor variables, the water temperature (X_3) would provide the highest R^2 for a bivariate model ($R^2 = 0.386$). The multiple regression model of Equation 6-151 explains 51.2%. Thus the addition of X_1 and X_2 increases the explained variance by 12.6%. This means that more than one predictor variable should be used, but it does not necessarily mean that all three predictor variables are needed. It may be worthwhile developing two-predictor models: $\hat{Y} = f(X_1, X_3)$ and $\hat{Y} = f(X_2, X_3)$. Because X_1 and X_2 are highly correlated, both are probably not needed, so a two-predictor model may provide estimates that are just as accurate as the three-predictor model.

The standardized partial regression coefficients are $t_1 = 0.299$, $t_2 = 0.070$, and $t_3 = 0.563$. These suggest that X_3 is most important, followed by X_1, and finally X_2, which appears to be unimportant. This differs somewhat from the predictor-criterion correlation coefficients, which suggest that X_3 is most important, but that X_1 and X_2 are of equal importance. Since velocity is probably physically more important than depth, the t values probably give a more realistic assessment of the relative importance of the predictor variables than are given by the predictor-criterion correlation coefficients.

6.12.8. Estimating Slope Stability

Slope stability is very important to engineers who design with unrestrained slopes. The stability number is an index used by geotechnical engineers to assess the safety of a slope. Among others, the slope stability is a function of the angle of friction of the soil (X_1, in degrees) and the slope angle (X_2, in degrees). Twelve laboratory measurements were made of the stability number for selected values of the angle of friction and the slope angle (see Table 6-29). The correlation matrix for the twelve tests is

	X_1	X_2	Y
X_1	1.000	0.520	0.415
X_2		1.000	-0.543
Y			1.000

The intercorrelation is moderate; the predictor-criterion correlation coefficients are not high, with the two values only representing 17.2 and 29.5% of the total variation in the stability number.

The least-squares regression equation is

$$\hat{Y} = 0.04234 + 0.002842 X_1 - 0.005093 X_2 \tag{6-152}$$

The multiple correlation coefficient for Equation 6-152 is 0.981, which means that the equation explains 96.2% of the variation in the measured values of the stability number. The standard error of estimate is 0.0072, which gives $S_e/S_y = 0.214$. The goodness-of-fit statistics suggest that Equation 6-152 provides accurate estimates of the stability number. The total F statistic is 115.1, with $k = v_1 = 2$ and $u = v_2 = 9$; since the critical value for a level of significance of 0.1% is 16.4, the ANOVA would suggest that the two slope coefficients of Equation 6-152 are significantly different from 0.

Note that in spite of the relatively low predictor-criterion correlation coefficients (i.e., 0.415 and -0.543), the total R^2 is very high. The intercorrelation actually enhances the reliability of the equation.

The goodness-of-fit statistics are revealing, but they do not give an entirely accurate assessment of the model. The standardized partial regression coefficients for X_1 and X_2 are $t_1 = 0.96$ and $t_2 = -1.04$. Because the latter is greater than 1, it is considered irrational and suggests that b_2 gives an unrealistic indication of the rate

TABLE 6-29 Multiple Regression Analysis of Slope Stability Values

X_1	X_2	Y	$\hat{Y}$	e
10	5	0.053	0.045	-0.008
10	10	0.005	0.020	0.015
20	5	0.081	0.074	-0.007
20	12	0.036	0.038	0.002
20	18	0.012	0.008	-0.004
30	6	0.102	0.097	-0.005
30	13	0.058	0.061	0.003
30	21	0.021	0.021	0.000
40	9	0.105	0.110	0.005
40	14	0.079	0.085	0.006
40	21	0.052	0.049	-0.003
40	26	0.027	0.024	-0.003

Note: X_1 = friction angle (degrees), X_2 = slope angle (degrees), and Y = stability number.

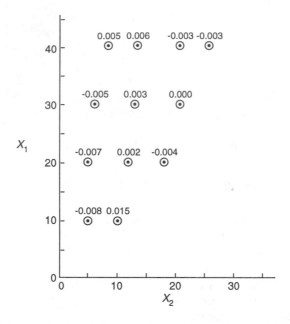

FIGURE 6-16 Residuals of a multiple regression model vs. predictor variables X_1 and X_2.

of change of Y with respect to X_2. The high accuracy indicated by the goodness-of-fit statistics suggests that the model is acceptable for estimation; however, the t_2 value suggests that the model may not be reliable upon extrapolation or for assessing model sensitivity.

The residuals (see Table 6-29) should be evaluated. Figure 6-16 shows a plot of the residuals as a function of the angle of friction (ordinate) and the slope angle (abscissa). While the sample size is probably too small to be very conclusive, there appears to be a region of positive residuals between friction angles of 9° and 15° and two regions of negative residuals, friction angles less than 6° and greater than 18°. This would suggest that the structure of the linear model may be incorrect.

The intercept also appears to be a problem. If both X_1 and X_2 are zero, then the stability number equals 0.042, which does not seem reasonable.

6.13. SIMULATION AND PREDICTION MODELS

Prediction models based on statistical analysis are important in many engineering design methods. Traditionally, the linear models with an intercept have been used. As computers have enabled more complex models to be developed and fit to data, greater accuracy has been achieved because of the greater flexibility of the complex models. However, one advantage of the linear model is that theory provides the basis for computing confidence intervals and performing tests of hypotheses. Theoretical developments are not available for more complex models. Thus, while the more complex models may provide greater prediction accuracy, the lack of theory that would enable hypothesis tests to be made and confidence intervals to be computed is a drawback.

For complex models where theory does not provide a description of the underlying probability distributions, simulation can be used to derive the distribution and then develop methods for constructing confidence intervals and making hypothesis tests. While the full procedure is beyond the scope of this discussion, the use of simulation for the purpose of understanding the distribution of a regression-model statistic is presented using a simple example.

Example 6-11: Distribution of the Slope Coefficient of the Zero-Intercept Model

Equation 6-57 provides the standard error of the slope coefficient for the bivariate model with an intercept (Equation 6-28). If the zero-intercept model of Equation 6-22 were being used, could the standard error of Equation 6-57 be used to make hypothesis tests or construct confidence intervals on the slope coefficient? The answer is no. Equation 6-57 applies only to the bivariate model of Equation 6-28.

To learn more about the distribution of the slope coefficient (b) of the zero-intercept model, the following population model was assumed:

$$\hat{Y} = \beta X + Z\sigma_e \qquad (6\text{-}153)$$

where β is the slope coefficient, X is the predictor variable, Z is a standard normal deviate, and σ_e is the standard error of the population. For this example, a value of 1.4 was assumed for β. The variable X was assumed to be normally distributed with $\mu = 10$ and $\sigma_x = 2$. The standard error of the population model was assumed to equal 2.857. Sample values of X and Z were generated for sample sizes of 10. The model of Equation 6-153 was used to compute values of Y, and then the simulated values of X and Y were used to fit the slope coefficient b. A total of 100,000 samples (NS) of a size 10 for each sample (N) were generated using the program shown in Figure 6-17. The mean and standard deviation of the 100,000 values of b were 1.4006 and 0.08912, respectively. The standard deviation is the standard error of the slope coefficient. The computed mean of 1.4006 agrees with the assumed population value. In addition to the two moments, a histogram of the values was computed, and the critical values of the slope coefficient were obtained from the cumulative histogram for selected probabilities, these are given in Table 6-30 (column 2). The standard error of b was multiplied by the standard normal deviates (column 3) for the probabilities (column 1) to obtain the simulated values that would be used to compute confidence intervals or make hypothesis tests. A comparison of columns 2 and 4 of Table 6-30 indicates that the normal distribution can be applied to the analysis of the zero-intercept model.

The above analysis is limited in that it considered only one sample size and one standard error. To develop a general expression relating $S_e(b)$ and S_e, it would be

```
PROGRAM SIMDIST
C          Simulate Upper Tails of Distribution for Slope Coefficient
C          for Zero-intercept Model
C          N = Sample Size
C          NS = Number of Samples
C          ISEED = Seed Value for Simulation
           DIMENSION B(100000), Y(10), X(10), F(20)
           CHARACTER*75 FILE2
           IAD = 10
           IADS = 100000
           FILE2 = 'OUTPUT'
           OPEN (6,FILE=FILE2,STATUS='NEW')
C          INPUT ISEED, N, NS
           EN = N
C          delete first 10 random numbers
           RN = RNN(ISEED)
           DO 10 I = 1, 10
              J = INT(RN)
              RN = RNN(J)
    10     CONTINUE
C          generate samples
           DO 80 K = 1, NS
              S1 = 0.0
              S2 = 0.0
              DO 20 I = 1, N
                 J = INT(RN)
                 RN = RNN(J)
                 X(I) = 10. + RN * 2.
                 J = INT(RN)
                 RN = RNN(J)
                 Y(I) = 1.4 * X(I) + 2.857 * RN
                 S1 = S1 + X(I) * Y(I)
                 S2 = S2 + X(I) ** 2
C                 WRITE(6,19) K, I, X(I), Y(I), RN
    19           FORMAT(2I5, 3F10.5)
    20        CONTINUE
              B(K) = S1 / S2
C              WRITE(6,21) S1, S2, B(K)
    21        FORMAT(40X, 3F12.5)
    80     CONTINUE
C          Compute Statistics of Coefficient B and Print
C          STAT Subroutine not Provided
           CALL STAT
C          Construct Histogram for Coefficient B and Print
C          FREQ Subroutine not Provided
           CALL FREQ
           CLOSE (6,STATUS='KEEP')
           END

           FUNCTION RNN(K)
C          Generates random normal numbers (not provided)
           FUNCTION RAND(K)
C          Generates random numbers (not provided)
           SUBROUTINE STAT
C          Computes Mean and Standard Deviation for AN ARRAY
           SUBROUTINE FREQ
C          Determines the number of elements in each histogram cell
```

FIGURE 6-17 Outline of a computer program used in Example 6-11.

necessary to make many analyses, with S_e, n, and the characteristics of X allowed to vary. A general expression could then be identified. It is important to recognize that simulation provides a means for obtaining statistical results that cannot be obtained from theory because of the complexity of the problem.

TABLE 6-30 Analysis of Critical Values for the
Distribution of the Slope Coefficient of the Zero-
Intercept Model

(1)	(2)	(3)	(4)
			Simulated
Probability	Critical		values
α	value	z_α or $z_{1-\alpha}$	$-b + z\, S_e(b)$
0.005	1.171	-2.575	1.171
0.010	1.195	-2.327	1.193
0.050	1.254	-1.645	1.254
0.100	1.286	-1.282	1.286
0.900	1.515	1.282	1.515
0.950	1.547	1.645	1.547
0.990	1.604	2.327	1.608
0.995	1.631	2.575	1.630

6.14. PROBLEMS

6-1. Plot the following data, assess the degree of systematic variation, and assess
the appropriateness of using a linear model:

X	0.7	1.3	2.8	5.1	7.3	8.9	11.5
Y	1.1	3.8	7.2	7.9	9.7	8.8	9.6

6-2. Plot the following data, and assess the quality of the data for prediction:

X	0.7	0.8	1.3	1.2	1.7	1.9	4.1
Y	1.0	1.8	1.2	2.1	1.0	2.8	5.7

6-3. Using the following values of X and Y, show that the total variation (TV)
equals the sum of the explained variation (EV) and the unexplained variation
(UV):

X	2	5	7	6	2
Y	1	3	5	7	9

Assume that the regression equation relating Y and X is given by

$$\hat{Y} = 4.5849 + 0.09434X$$

6-4. Using the following values of X and Y, show that the total variation (TV)
equals the sum of the explained variation (EV) and the unexplained variation
(UV):

X	1	2	3	4	5	6
Y	2	2	3	5	4	5

Assume that the regression equation relating Y and X is given by

$$\hat{Y} = 1.2 + 0.657X$$

6-5. For the data in Problem 6-4, compute the correlation coefficient.

6-6. For the following observations on X and Y, compute the correlation coefficient.

X	-3	-2	-1	0	1	2
Y	2	2	3	0	-2	-1

6-7. For the data given in Problem 6-3, compute the correlation coefficient using the standard normal transformation definition of Equation 6-5.

6-8. Derive Equation 6-5 from Equation 6-4.

6-9. Given the following observations on X and Y, (a) construct a graph of X vs. Y, and (b) compute the Pearson correlation coefficient.

X	1	2	3	4	5
Y	1	1	2	4	4

6-10. Compute the correlation coefficient for the data of Problem 6-2.

6-11. Given the following paired observations on variables X and Y, and the values of the standardized variables Z_X and Z_Y, compute the Pearson correlation coefficient between each of the following pairs: (a) X and Y; (b) Z_X and Z_Y; (c) Z_X and X; (d) Z_X and Y; (e) X and $Y - 3$; and (f) X and $2Y$. Briefly discuss your results.

X	Y	Z_X	Z_Y
3	4	-1.5	-1.5
5	8	-0.5	0.5
6	7	0.0	0.0
7	6	0.5	-0.5
9	10	1.5	1.5

6-12. Given the following observations on random variable X and Y, (a) calculate the correlation coefficient; (b) using a level of significance of 5%, test the null hypothesis that the two variables are uncorrelated; (c) using a level of significance of 1%, test the null hypothesis that the population correlation is 0.8.

X	4	3	9	6	8
Y	3	6	8	8	10

6-13. The correlation coefficient computed from a sample of 15 observations equals 0.431. Test the null hypothesis using a level of significance of 5% that the correlation of the population is (a) 0.0; and (b) 0.65.

6-14. A sample of 12 observations on X and Y yields a correlation coefficient of 0.582. Using a level of significance of 5%, test the null hypothesis that the correlation coefficient of the population is (a) 0.0; and (b) –0.1.

6-15. Seven students in an introduction to statistics course received the following grades on the first two tests. Does the evidence suggest that knowledge of the material covered on the first test helped the students on the second test? Use a 5% level of significance.

Test 1	98	94	93	90	87	85	84
Test 2	96	94	91	85	93	86	90

6-16. The design engineer believes that a correlation of 0.8 is needed in the population to have the necessary prediction accuracy. A sample of 22 measurements on Y and X yields a sample correlation coefficient of 0.49. Is it safe to assume that the population correlation coefficient is at least 0.8? Use a 1% level of significance.

6-17. Using the data of Problem 6-2, fit the zero-intercept model of Equation 6-22.

6-18. Using the data of Problem 6-15, fit the zero-intercept model of Equation 6-22. Predict values of test 2 using values of test 1.

6-19. Given the following four pairs of observations

X	1	1	2	2	?
Y	1	2	1	2	?

Compute the correlation coefficient and the regression coefficients if (a) the fifth pair is $(Y = 8, X = 8)$; (b) the fifth pair is $(Y = 2, X = 8)$. (c) Plot both sets of points and draw the regression lines.

6-20. Show that Equation 6-38 results in a larger sum of squares of the errors for the data of Table 6-2 than the value obtained with Equation 6-37.

6-21. Given the following paired observations on Y and X, (a) graph Y vs. X; (b) calculate the correlation coefficient; (c) determine the slope and intercept coefficients of the linear regression line; (d) show the regression line on the graph of part (a); (e) compute the predicted value of Y for each observed value of X; (f) calculate $\sum e$ and S_e. (g) What is the predicted value of Y for $X = 5$? $X = 10$?

X	3	5	6	7	9
Y	4	8	7	6	10

6-22. Using the data set of Problem 6-12, compute the regression coefficients for the linear bivariate model with X as the predictor variable. Also, determine the standard error of estimate and perform an analysis of variance (use a 5% level of significance).

6-23. Compute the standard errors of the regression coefficients and the two-sided 95% confidence intervals for the regression coefficients of Problem 6-21.

6-24. Using the data of Problem 6-12 and the regression coefficients of Problem 6-22, compute standard errors of the regression coefficients and the two-sided, 90% confidence intervals for the coefficients.

6-25. Perform hypothesis tests for the significance of the regression coefficients for Problem 6-21. Use a 5% level of significance and alternative hypothesis that the coefficients are different from zero.

6-26. Perform hypothesis tests for the significance of the regression coefficients for Problem 6-21. Use a 10% level of significance and alternative hypothesis that the coefficients are different from zero.

6-27. Compute a two-sided, 95% confidence interval for the line as a whole using the data for Problem 6-21. Provide values of $\overline{X} - 2S_X$, $\overline{X} - S_X$, $\overline{X} + S_X$, and $\overline{X} + 2S_X$.

6-28. Compute a two-sided, 95% confidence interval for the line as a whole using the data for Problem 6-12. Provide values of $\overline{X} - 2S_X$, $\overline{X} - S_X$, $\overline{X} + S_X$, and $\overline{X} + 2S_X$.

6-29. Using the data from Problem 6-21, compute the two-sided, 95% confidence interval for the mean value of Y at $X = 5$. Also, compute a similar interval at $X = 11$. Compare the two intervals and explain the difference in their widths.

6-30. Using the data from Problem 6-12, compute the two-sided, 95% confidence interval for the mean value of Y at $X = 6$. Also, compute a similar interval at $X = 0$. Compare the two intervals and explain the difference in their widths.

6-31. Compare correlation analysis and regression analysis as tools in decision making.

6-32. Using the data set given below, regress (a) Y on X; (b) X on Y; (c) Compute the correlation coefficient for each regression; (d) Transform the regression of Y on X into an equation for computing X, and compare these coefficients with the coefficients for the regression of X on Y; (e) Why do the coefficients of parts (b) and (d) differ?

X	2	1	3	2
Y	1	2	4	5

6-33. The data shown are the results of a tensile test of a steel specimen, where Y is elongation in thousands of an inch that resulted when the tensile force was X thousands of pounds. Fit the data with a linear model and make a complete assessment of the results (use $\alpha = 5\%$ where necessary).

X	0.8	1.6	3.1	4.4	6.3	7.9	9.2
Y	2.8	4.9	6.5	8.1	8.8	9.1	8.9

6-34. The stream reaeration coefficient (Y) is related to the water temperature (T) in °C. Fit the following with a linear model and make a complete assessment of the results (use $\alpha = 10\%$ where necessary):

X	14	23	17	11	20	15	18	11	25	19	13	24	16	23	21
Y	2.89	4.20	4.17	3.69	3.78	3.56	3.35	2.69	3.58	3.06	2.41	3.30	3.05	4.18	4.08

6-35. Using the principle of least squares, derive the normal equations for the following model:

$$\hat{Y} = b_1 X_1 + b_2 X_2 + b_3 X_3$$

6-36. The cost of producing power (P) in mills per kilowatt hour is a function of the load factor (L) in percent and the cost of coal (C) in cents per million Btu. Perform a regression analysis to develop a regression equation for predicting P using the following data:

L	C	P
85	15	4.1
80	17	4.5
70	27	5.6
74	23	5.1
67	20	5.0
87	29	5.2
78	25	5.3
73	14	4.3
72	26	5.8
69	29	5.7
82	24	4.9
89	23	4.8

6-37. Visitation to lakes for recreation purposes varies directly with the population of the nearest city or town and inversely with the distance between the lake and the city. Develop a regression equation to predict visitation (V) in person-days per year, using as predictors the population (P) of the nearby town and the distance (D) in miles between the lake and the city. Use the following data:

D	P	V
10	10,500	22,000
22	27,500	98,000
31	18,000	24,000
18	9,000	33,000
28	31,000	41,000
12	34,000	140,000
21	22,500	78,000
13	19,000	110,000
33	12,000	13,000

6-38. Using the method of least squares, derive the normal equations for the model $\hat{Y} = b_1 X + b_2 X^2$. Evaluate the coefficients using the data of Problem 6-9.

6-39. Using the data of Problem 6-33, fit a linear model and a quadratic polynomial and make a complete assessment of each. Which model is best for prediction? Support your choice.

6-40. Using the data of Problem 6-34, fit a linear model and a quadratic polynomial and make a complete assessment of each. Which model is best for prediction? Support your choice.

6-41. Using the method of least squares, derive the normal equations for the model $\hat{Y} = ax^b$. Evaluate the coefficients using the data of Problem 6-9. (*Hint*: Convert the equation to linear form by making a logarithmic transformation.)

6-42. Using the data of Problem 6-33, fit a linear model and a power model. Make a complete assessment of each model. Which model is best for prediction? Support your choice.

6-43. Using the data of Problem 6-34, fit a linear model and a power model. Make a complete assessment of each. Which model is best for prediction? Support your choice.

6-44. Using the data of Problem 6-36, fit a multiple power model. Make a complete assessment of the model.

6-45. Using the method of least squares, derive the normal equations for the model $\hat{Y} = b_0 + b_1 \sqrt{X}$. Evaluate the coefficients using the data of Problem 6-9.

6-46. Use simulation to verify the relationships for computing the standard errors of the two coefficients of the linear bivariate regression equation.

6-47. Use simulation to find the standard error and the critical values for 10 and 5% of the slope coefficient of the linear zero-intercept bivariate regression model ($\hat{Y} = bX$) for a sample size of 10. Assume X to be normally distributed, $X \sim N(\text{mean} = 20, \text{standard deviation} = 4)$, and generate values of Y based on the generated values of X using the following equation:

$$Y = \beta_1 + Z\sigma_Y\sqrt{1-\rho^2}$$

where β_1 = population value of b_1, Z = random normal variate, σ_Y = population standard deviation of Y, and ρ = population value of correlation between X and Y. Use $\rho = 0.6$, $n = 10$, $\mu_X = 20$, $\sigma_X = 4$, $\beta_1 = 3$, and $\sigma_Y = 2$.

Chapter 7
Simulation

TABLE OF CONTENTS

7.1. INTRODUCTION

7.1.1. Engineering Decision Making

Assume that you are a county transportation engineer and that you have found from studying traffic accident reports that traffic fatalities at a particular intersection in the county are especially high. The intersection is presently uncontrolled except for stop signs for the north-south street; there are no stop signs on the east-west street. As a possible measure for reducing traffic accidents, you propose installing a four-way traffic light. However, you want the light to operate in a pattern that will cause the least impedance on traffic. Therefore, you must decide the length of the red-green cycle in each direction and the proportion of time that it will be green in each direction. For example, will the light be green in the north-south direction less time, equal time, or more time than in the east-west direction? Also, will the entire cycle be 1 min? 2 min? It is necessary for you, the transportation engineer, to establish the traffic-light operation policy that will reduce traffic accidents with minimal impedance of traffic.

One possible solution to your dilemma would be to install the traffic light, assume some light-operation sequence, and then observe its effectiveness over 2 weeks or a month. Then a second sequence could be used and traffic observed for another 2 weeks. This process could be repeated until the optimum light-operation sequence is found. Obviously, this is an unrealistic option. But an arbitrary setting of the light-operation sequence is also an undesirable option. A method of simulating the actual process in a very short period of time is needed.

Consider a second case. This involves the task of a water-supply engineer who must establish a policy for public water use during periods of drought. Let's assume here that drought is defined as an extended period of time when the normal demand for water exceeds the available supply. It is the task of the engineer to set a policy that sets a limit on the daily use of water. Common limitations on water use include bans on car washing and lawn watering, and limiting home use to some percentage of normal use. In order to ensure that the policy is enforceable, it must be applied consistently over a relatively long period of time; it cannot be changed daily. But what policy should the water-supply engineer recommend to the water commissioner? The engineer wants to minimize inconvenience to the public, but at the same time water must be available for consumption, fire safety, public health, and important commercial and industrial uses. The policy must take into consideration the engineer's expectation for rain over the next month or so, but this is an unknown, yet important, variable. The engineer would have some knowledge of the depth of rain that has occurred in the past, but past sequences of rain are unlikely to occur exactly in the future. How can the water supply engineer combine knowledge of existing water supplies, public demand for water, potential amounts of rainfall, and the effects of alternative policies for limiting water use? Just as in the traffic-light problem, the engineer needs a method of combining knowledge of each part of the process in a way that simulates the parts and their interaction so that estimates of the effects of alternative water policies can be assessed.

7.1.2. Sampling Variation

The topic of sampling variation has been introduced previously. The standard error of the sample mean was given as $S/\sqrt{n}$. Confidence intervals are a statistical tool that is used to quantify the variation that can be expected in a statistic. The subject of the standard error of estimate was introduced as a measure of the expected variation of the residuals from a regression model. Numerous other examples of sampling variation have been introduced. A slightly different approach is considered here.

Consider the following experiment. A die is rolled three times, with the number of dots from the three rolls summed. We will denote this random variable as S. What is the sample space of S? Obviously, it cannot take on a value smaller than 3 or larger than 18. So the sample space consists of the integer values from 3 to 18. There is only one way to get S equal to 3, namely, three occurrences of a 1: (1, 1, 1). There are three ways of getting S equal to 4: (1, 1, 2), (1, 2, 1), and (2, 1, 1). If we continued to identify all possible outcomes, we would see that there are a total of 216 possibilities. The number of possible outcomes can be converted to a probability by dividing the number of outcomes by 216; these are the population probabilities and are given in Table 7-1 as p. An experiment is to roll the die 150 times, with each group of three rolls used to compute a value of S; thus, the 150 rolls produced 50 values of the random variable. Each group of the 50 will be referred to as a simulation run; the number of values for each S and for each of 17 simulation runs (or cycles) are shown in Table 7-1. The point of this exercise is to show that each simulation run produces a different result. In fact, some runs are quite different; for example, runs 4 and 13 differ in that run 4 has 29 occurrences of S between 3 and 10, while run 13 has 21 occurrences. But each simulation run cannot be the same because run-to-run variation is expected when dealing with a random variable.

This variation occurs in engineering systems. We don't have the same flood patterns every year. Structures are not subjected to the same wind loads every year. Traffic patterns and accidents differ on a day-to-day basis. Construction delays are not systematic. Pollution loads in rivers vary continuously and the level of pollution at any time cannot be predicted with certainty. In problems where uncertainty is significant and important to the decisions that engineers must make, simulation can be an important modeling tool for the engineers.

7.1.3. Coin-Flipping Simulation

Problems like the traffic-light and water-rationing examples can be studied on a computer without the need of conducting exhaustive real-world studies. The essence of the computer approach will be illustrated using an example. The reader is encouraged to perform a similar study of the problem while following the steps shown. The only requirement is a coin and pen and paper to record the results.

First, flip a coin 72 times and record the sequence of heads and tails (H and T, respectively). The following is an example result:

HTT HTH TTH THT TTH HTT HHT HTT TTH TTT HHT HHH

TTH HHT THT THH TTH HTH TTT HHH THH HHH HTH THT

TABLE 7-1 Simulation of the Sampling Variation of a Random Variable

S	n	p	Simulation runs																	Sum	$\hat{p}$
			1	2	3	4	5	6	7	8	9	10	11	12	13	14	15	16	17		
3	1	0.005	0	0	1	0	0	0	1	0	0	1	1	0	0	0	0	0	0	4	0.005
4	3	0.014	0	2	1	1	0	0	1	1	1	1	0	0	0	0	0	0	1	9	0.010
5	6	0.028	2	2	1	1	1	0	0	1	2	2	0	0	0	0	0	2	3	17	0.020
6	10	0.046	0	1	0	4	3	0	2	3	1	1	2	1	2	1	1	3	2	27	0.032
7	15	0.069	3	7	2	4	6	7	5	2	4	3	1	3	4	3	3	2	4	63	0.074
8	21	0.097	6	5	6	6	9	6	5	7	7	6	4	2	3	9	4	6	1	92	0.108
9	25	0.116	10	4	8	9	2	6	4	10	6	4	6	8	8	5	11	6	7	114	0.134
10	27	0.125	3	3	9	4	5	4	5	4	4	10	8	9	4	8	3	7	5	95	0.112
11	27	0.125	8	4	7	7	8	10	5	6	6	5	10	6	4	7	10	5	9	117	0.137
12	25	0.116	5	6	5	2	7	3	7	5	9	8	6	3	5	3	2	5	5	86	0.101
13	21	0.097	4	7	3	5	2	7	6	6	5	2	2	6	6	8	7	5	5	86	0.101
14	15	0.069	3	4	4	5	2	2	3	1	2	3	4	5	5	3	4	4	3	57	0.067
15	10	0.046	2	1	1	1	1	1	3	1	3	1	1	3	3	2	2	3	2	32	0.038
16	6	0.028	2	2	1	0	3	2	2	1	0	2	2	2	3	0	1	2	3	27	0.032
17	3	0.014	2	2	0	1	1	1	1	2	0	0	3	0	3	1	2	0	0	20	0.024
18	1	0.005	0	0	0	0	0	0	1	0	0	1	0	2	0	0	0	0	0	4	0.005
Sum		1.000	50	50	50	50	50	50	50	50	50	50	50	50	50	50	50	50	50	850	1.000

Note: S = sample-space outcome; n = number of ways of getting S; p = probability of getting S; $\hat{p}$ = sample probability of getting S.

Next, divide the sequence into 24 groups of 3 as shown above. This sequence is used to simulate the operation of a manufacturing plant and the selection of a piece of equipment based on a linear cost function. Assume that each group of three coin tosses represents the output from an assembly line. Also, assume that the machine randomly produces defective units, with a unit from machine A being defective for any of the following sequences: (TTT), (TTH), or (HTT); all other combinations indicate a nondefective unit. For machine A, the sequence of defective (D) and nondefective (N) units from the above coin tosses can be, therefore, expressed as

$$DNDNDDNDDDNNDNNNDNDNNNN$$

The manufacturing engineer is considering the installation of a more reliable and more costly machine, which is called machine B. For machine B, the following sequences indicate a defective unit: (TTT) or (TTH); thus, the sequence of defective and nondefective units from the above coin tosses is

$$NNDNDNNNDDNNDNNNDNDNNNN$$

Machine A produced ten defective units (41.7%), while machine B produced only seven defective units (29.2%). Lets assume that there is a $5 profit for a nondefective unit and a $2 loss for each defective unit. Machine A costs $12,000, while machine B costs $15,000. Based on a useful life of 1 year for the machines during which 10,000 units would be produced, the expected profit for machine A is

$$\text{Profit for A} = \$5\left(\frac{14}{24}\right)(10,000) - \$2\left(\frac{10}{24}\right)(10,000) - \$12,000 = \$8,833$$

The expected profit for machine B is

$$\text{Profit for B} = \$5\left(\frac{17}{24}\right)(10,000) - \$2\left(\frac{7}{24}\right)(10,000) - \$15,000 = \$14,583$$

Therefore, based on the sample estimates obtained from the simulated sequences of defective and nondefective items, the profit expected from machine B exceeds the profit expected from machine A. Thus, an engineer would recommend the purchase of machine B.

This example demonstrates the point that it was not necessary to actually produce 10,000 units with each machine in order to estimate the profit for the two alternatives. The random nature of the production of defective units was represented by a simple experiment that had the same statistical characteristics as the real system.

7.1.4. Definitions

Before defining what is meant by engineering simulation, it is necessary to give a few other definitions. First, a system is defined herein as a set of processes or components that are interdependent with each other. This could be a natural system such as a watershed, a geographic system such as a road network, or a structural system such as a high-rise building. An accident on one road can lead to traffic congestion on a nearby road, so the roads are interdependent.

Second, a model is defined herein as a representation of a real system as was discussed in Chapter 1. The model could be either a physical model, such as those used in laboratories, or a mathematical model. Models can be developed from either theoretical laws or empirical analyses. The model includes components that reflect the processes and provides for interaction between components.

Third, an experiment is defined for our purposes here as the process of observing the system or the model. Where possible, it is usually preferable to observe the real system. However, the lack of control may make this option impossible or unrealistic.

Given these three definitions, we can now provide a preliminary definition for simulation. Specifically, simulation is the process of conducting experiments on a model when we cannot experiment directly on the system. The uncertainty or randomness inherent in model elements is incorporated into the model and the experiments are designed to account for this uncertainty. To illustrate this aspect, consider the coin-tossing example. The occurrence of defective items was considered to be random, so the model incorporated the randomness of coin tossing to represent the randomness of the real system. Both sequences of defective and nondefective items appear to be random. The $5 profit and $2 loss, as well as the cost of the machine, were not subject to uncertainty, so they were included in the model as constants.

The term "simulation run or cycle" is defined as an execution of the model through all operations for a length of simulated time. The 72 coin tosses shown above could represent one simulation run, with the 24 items being the length of the simulation run. Using the above model, a simulation run that consisted of 30,000 flips can be used to yield 10,000 units.

Some additional terms that need to be defined, and used in this chapter, are as follows:

1. A model parameter is a value that is held constant over a simulation run but can be changed from run to run.
2. A variable is a model element whose value can vary during a simulation run.
3. Input variables are those model variables that require values to be input prior to the simulation run.
4. Output variables are model variables that reflect the end state of the system and can consist of single values or a vector of values.
5. Initial conditions are the values of model variables and parameters that establish the initial state of the model at the beginning of a simulation run.

7.1.5. Benefit of Simulation

Simulation is widely used in our everyday lives. Flight simulators are widely reported in the space and aircraft industries. Activities at the leading amusement parks simulate exciting space travel. Even video games use simulation to mimic life-threatening activities.

Simulation is widely used in engineering decision making. It is a popular modeling tool because it enables a representation of the system to be manipulated when manipulating the real system is impossible or too costly. Simulation allows the time or space framework of the problem to be changed to a more convenient framework. Simulation enables the representation of the system to be changed in order to better understand the real system; of course, this requires the model to be a realistic representation of the system. Simulation enables an analyst to control any of the model parameters, variables, or initial conditions, something that is not possible with the real system. Simulation can be used to evaluate the response of the system for conditions that have not occurred in the past.

While simulation is extremely useful, there are a few problems. First, it is quite possible to develop several different, but realistic, models of the system. The different models could lead to different decisions. Second, the data used to calibrate the model may be limited, so extrapolations beyond the range of the measured data may be especially inaccurate.

7.2. MONTE CARLO SIMULATION

The interest in simulation methods started in the early 1940s for the purpose of developing inexpensive techniques for testing engineering systems by imitating their real behavior. These methods are commonly called Monte Carlo simulation techniques. The principle behind the methods is to develop an analytical model, which is computer based, that predicts the behavior of a system. Then the model is evaluated using data measured from a system, and then the model predicts the behavior of the system, usually for many simulation runs. Each evaluation (or simulation cycle) is based on a certain randomly selected set of conditions for the input parameters of the system. Certain analytical tools are used to assure the random selection of the input parameters according to their respective probability distributions for each evaluation. As a result, several predictions of the behavior are obtained. Then statistical methods are used to evaluate the moments and distribution type for the system's behavior.

The analytical and computational steps that are needed for performing Monte Carlo simulation are (1) definition of the system using a model, (2) generation of random numbers, (3) generation of random variables, (4) evaluation of the model, (5) statistical analysis of the resulting behavior, and (6) study of simulation efficiency and convergence. The definition of the system should include its boundaries, input parameters, output (or behavior) measures, architecture, and models that relate the input parameters and architecture to the output parameters. The accuracy of the results of simulation are highly dependent on having an accurate definition for the system. All critical parameters should be included in the model. The definition of the input parameters should include their statistical or probabilistic characteristics, i.e., knowledge of their moments and distribution types. It is common to assume the architecture of the system in Monte Carlo simulation to be nonrandom. However, modeling uncertainty can be incorporated in the analysis in the form of bias factors and additional variability, e.g., coefficients of variation. The result of these generations are a set of certain values for the input parameters. These values should then be substituted in the model to obtain an output measure. By repeating the procedure N times (for N simulation cycles), N response measures are obtained. Statistical methods can now be used to obtain, for example, the mean value, variance, or distribution type for the response. The accuracy of the resulting measures for the behavior are expected to increase by increasing the number of simulation cycles. The convergence of the simulation methods can be investigated by studying their limiting behavior as N is increased. Also, the efficiency of simulation methods can be increased by using variance reduction techniques. These variance reduction techniques are not described in this chapter; they are described in Chapter 8 through their use in reliability analysis.

7.3. RANDOM NUMBERS

Random numbers are real values; if they are normalized by dividing each value by the largest possible value, the resulting numbers are real values in the range [0,1]. Also, the normalized numbers have a uniform distribution on the range [0,1]. Therefore, it can be stated that a random number can be modeled as a random variable U that has a uniform distribution on the continuous range [0,1]. A set of random numbers should also satisfy the condition of no serial correlation for the purpose of simulation use.

The importance of uniform random numbers is that they can be transformed into real values that follow any distribution of interest. Therefore, they are the basis for most engineering simulation.

In the early years of simulation, mechanical random-number generators were used, e.g., drawing numbered balls, throwing dice, or dealing out cards. Many lotteries are still operated this way. Then, after several stages of development, arithmetic random-number generators were developed that are computer-based and use some analytical models. In these generators, a random number is obtained based on a previous value (or values) and fixed mathematical equations. Therefore, a seed is needed to start the process of random-number generation, resulting in a stream of random values. The main advantages of arithmetic random-number generators over mechanical generators are speed, that they do not require memory for storage of numbers, and repeatability. The conditions of a uniform distribution and the absence of serial correlation should also be satisfied. Due to the nature of arithmetic generation of random numbers, a given seed should result in the same stream of random values all the times. This property of repeatability is important for debugging purposes of the simulation algorithm and comparative studies of design alternatives of a system. In this section, arithmetic random-number generators are briefly described.

7.3.1. Arithmetic Generators

There are many arithmetic random-number generators, for example, midsquare, linear congruential generator, mixed generators, multiplicative generators, general congruences, composite generators, and Tausworthe generators. The midsquare generator was used in Chapters 1 to 6 to illustrate the use of simulation. All these generators are based on the same principle of starting with a seed and having fixed mathematical equations for obtaining the random value, then the resulting value is used in the same equations to obtain another random value. By repeating this recursive process N times, N random numbers in the range [0,1] are obtained. However, these methods differ in the mathematical equations that are used as the recursive model. In all recursive models, the period for the generator is of concern. The period is defined as the number of generated random values before the stream of values starts to repeat itself. It is always desirable to have random-number generators with large periods, for example, much larger than the number of simulation cycles needed in a simulation study of a system.

In almost all modern computers, a random-number generator is provided as a standard function. Modern generators are sometimes based on the direct use of bits and binary digits. However, the great majority of random-number generators are linear congruential generators. In these generators, a sequence of integers I_1, I_2, I_3, ... is defined by the following recursive equation:

$$I_i = \left(aI_{i-1} + b\right) - \text{Int}\left(\frac{aI_{i-1} + b}{c}\right)c \quad \text{for } i = 1, 2, 3, \ldots \tag{7-1}$$

where $\text{Int}\left(\dfrac{aI_{i-1} + b}{c}\right)$ = the integer of the result of the division, a = multiplier, b = increment, and c = modulus. These model constants a, b, and c are nonnegative integers. The starting value I_0 is called the seed, which should be provided by the user of the model. The value I_i is obtained by dividing ($a\,I_{i-1} + b$) by c and letting I_i be the remainder of the division. The random number U_i is defined as

$$U_i = \frac{I_i}{c} \tag{7-2}$$

The I_i value is normalized by dividing by c, since $0 \leq I_i \leq c$. The parameters of this recursive model should also satisfy the following conditions: $0 < c$, $a < c$, $b < c$, and $I_0 < c$. It is evident from this recursive model that the process is not random, since it can be repeated with the same results all the time. For this reason, this process is commonly called *pseudo* random-number generation. The period of this generator is less than or equal to c. For this reason and others, the value of c in real generators should be very large, e.g., $c \geq 10^9$. In addition to some computers and calculators which have random-number generators, tables of random numbers have also been published (for example, Table A-6). A special case of the linear congruential generators is where $b = 0$, called the multiplicative generators. If $b > 0$, they are called mixed generators.

Example 7-1: Random-Number Generators

For a recursive multiplicative generator that is defined by the following equation:

$$I_i = (2I_{i-1}) - \text{Int}\left(\frac{2I_{i-1}}{5}\right)5 \quad \text{for } i = 1, 2, 3, \ldots \quad (7-3)$$

and a seed $I_0 = 2$, the random numbers shown in Table 7-2 are obtained. The period for this generator is 3. This is not a good generator since it has a small period. The period is less than or equal to c, which has a value of 5 in this case. In real generators, c should be very large. Also, the resulting random numbers from the generator are discrete, i.e., 0, $1/c$, $2/c$, ..., $(c-1)/c$ (in this example, 0, 0.2, 0.4, and 0.8). By selecting a large value for c, the number of discrete values becomes large, resulting in a better approximation of a continuous uniform distribution.

TABLE 7-2 Random Numbers Generated Using a Recursive Relation

i	I_i	U_i	i	I_i	U_i	i	I_i	U_i
0	2	seed	5	4	0.8	10	3	0.6
1	4	0.8	6	3	0.6	11	1	0.2
2	3	0.6	7	1	0.2	12	2	0.4
3	1	0.2	8	2	0.4	13	4	0.8
4	2	0.4	9	4	0.8	14	3	0.6

7.3.2. Testing of Generators

Before using a random-number generator, two important tests should be performed on the generator. These tests are (1) test of uniformity and (2) test of serial correlation. These tests can be performed either theoretically or empirically. A theoretical test is defined as an evaluation of the recursive model itself of a random-number generator. The theoretical tests include an assessment of the suitability of the parameters of the model without performing any generation of random numbers. An empirical test is a statistical evaluation of streams of random numbers resulting from a random-number generator. The empirical tests start by generating a stream of random numbers, i.e.,

N random values in the range [0,1]. Then, statistical tests for distribution types, i.e., goodness-of-fit tests such as the chi-square test, are used to assess the uniformity of the random values. Therefore, the objective in the uniformity test is to make sure that the resulting random numbers follow a uniform continuous probability distribution. This test is discussed in Chapter 5. The serial test of the runs are used to assess the serial correlation of the resulting random vector, where each value in the stream is considered to come from a different, however, identical, uniform distribution. The serial correlation tests are beyond the scope of this book.

7.4. GENERATION OF RANDOM VARIABLES

The process of generating a random variable according to its probability distribution can be viewed as a sampling procedure with a sample size N, where N = the number of simulation cycles. It is desirable, and in some cases necessary, that a random number generator possesses the following properties: (1) exactness, (2) efficiency, (3) simplicity, (4) robustness, and (5) synchronism. The exactness property is necessary to assure that the resulting values for a generated random variable exactly follow the assumed probability distribution. The efficiency property is measured in terms of needed computer memory, and setup and execution time for generating a random variable. It is desirable to have generation methods that are efficient. It is also desirable to have a mathematically and computationally simple procedure for generating a random variable. The methods must also be robust, i.e., they can provide reasonable random values for wide ranges of values for the parameters of the distribution. In certain applications, it might be necessary to have synchronized random values of the input random variables. In such cases, the synchronism property must be satisfied. In developing or selecting a random-variable generator, some of these properties must be satisfied and others need to be satisfied in different levels, using a trade-off approach. For example, increasing the simplicity might reduce exactness, whereas increasing exactness might increase complexity and reduce efficiency.

7.4.1. Inverse Transformation Method

This method is simple and direct. A random number u is first generated in the range [0,1], i.e., $u \in U[0,1]$, where $U[0,1]$ is a continuous uniform probability distribution. Then a generated continuous random variable, X, is determined as follows:

$$x = F_X^{-1}(u) \tag{7-4}$$

where F_X^{-1} = the inverse of the cumulative distribution function of the random variable X. Since $F_X(x)$ is in the range [0,1], a unique value for x is obtained all the time in each simulation cycle.

For a discrete random variable, a random number is generated, $u \in U[0,1]$. Then the value of the generated random variable, X, is determined as follows:

$$x_i \text{ such that } i \text{ is the smallest integer with } u \leq F_X(x_i) \qquad (7\text{-}5)$$

where x_i, $i = 1, 2, 3, ..., m$, are m discrete values of the random variable X with a cumulative mass distribution function $F_X(x)$.

Example 7-2: Inverse Transformation for a Discrete Distribution

In this case, we have a uniform random number generator, $U(0,1)$, and we want to transform U values so that they have a specified discrete mass function. The discrete function of interest could be either some known function such as a binomial or Poisson function or a custom-made mass function. The transformation process is the same for both cases.

Consider the case where we have collected a sample of data on a discrete random variable X and the data yields a sample estimate of the mass function as shown in Figure 7-1a. The cumulative function for values of X from 0 to 5 is $F_X(x) = \{0.10, 0.30, 0.60, 0.85, 0.95, 1.00\}$, for $x = 0, 1, 2, 3, 4,$ and 5, respectively. This is shown in Figure 7-1b. For any value of X greater than 5, the cumulative function equals 1, and for values of X less than 0, the cumulative function equals 0.

To transform continuous uniform variates U to discrete variates having $f_X(x)$ as a mass function, the value of U is entered as the ordinate on the cumulative function of Figure 7-1b, with the point projected horizontally until it intersects the cumulative function. The point of intersection identifies the value of X.

Consider the following set of uniform variates: $U = \{0.47, 0.91, 0.08\}$. These are transformed to X variates as shown in Figure 7-1c. The generated values of X are $X = \{2, 4, 0\}$.

Example 7-3: Inverse Transformation for a Continuous Distribution

The process for generating continuous variates follows the same steps of the discrete case. We have a generator that gives values with a uniform distribution $U(0,1)$. A density function must be identified, usually from empirical analysis, but possibly from theoretical considerations. The cumulative distribution for the assumed density function is formed.

Let's assume that the objective is to transform uniform variates of Figure 7-2a to variates that have a triangular density function as shown in Figure 7-2b. The following is the density function, which can be determined using the constraint $\int_{-\infty}^{\infty} f_X(x)\, dx = 1$ and the equation for the function as $f_X(x) = k(x - x_L)$:

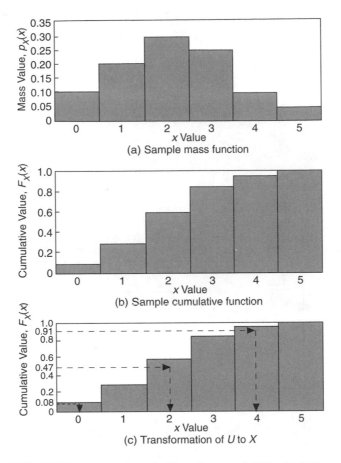

FIGURE 7-1 Inverse transformation for a discrete probability distribution.
(a) Sample mass function; (b) Sample cumulative function; (c) Transformation of U to X.

$$f_X(x) = \begin{cases} \dfrac{2(x - x_L)}{(x_U - x_L)^2} & \text{for } x_L \le x \le x_U \\ 0 & \text{otherwise} \end{cases} \qquad (7\text{-}6)$$

The cumulative function, $F_X(x)$, can be determined by integrating the density function of Equation 7-6 from x_L to any value x as follows:

$$F_X(x) = \begin{cases} 0 & \text{for } x < x_L \\ \dfrac{(x - x_L)^2}{(x_U - x_L)^2} & \text{for } x_L \le x \le x_U \\ 1 & \text{for } x_U < x \end{cases} \qquad (7\text{-}7)$$

The cumulative function is shown in Figure 7-2c.

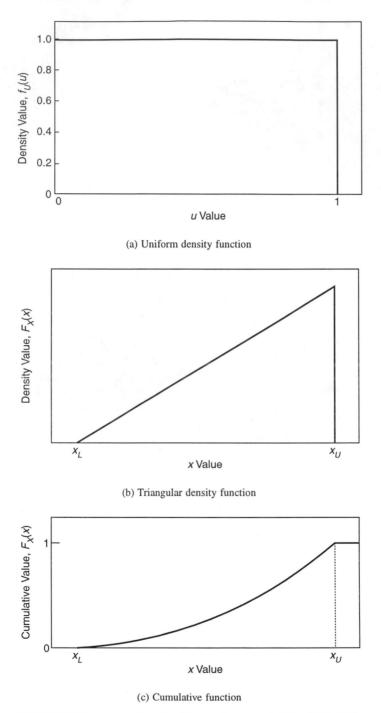

(a) Uniform density function

(b) Triangular density function

(c) Cumulative function

FIGURE 7-2 Inverse transformation for a continuous probability distribution.
(a) Uniform density function; (b) Triangular density function; (c) Cumulative function.

To generate a value of X for a given uniform variate U, the cumulative function $F_X(x)$ is set equal to a generated random number u, and Equation 7-7 is solved for x. Solving the quadratic yields

$$x = x_L + (x_U - x_L)\sqrt{u} \tag{7-8}$$

A graphical solution is also possible using the approach shown for the discrete variate. For the case where $x_L = 1$, $x_U = 3$, and $U = \{0.47, 0.91, 0.08\}$, the values of X are computed from:

$$x = 1 + 2\sqrt{u} \tag{7-9}$$

which gives $X = \{2.371, 2.908, 1.566\}$.

Example 7-4: Transformation to a Normal Distribution

A common need is to transform uniform variates to normal variates. For transformation to a standard normal deviate, the standard normal table can be used. For a normal distribution with mean μ and standard deviation σ, the uniform variate is first transformed to the standard normal deviate z, which is then transformed to normal deviate x by

$$x = \mu + z\sigma \tag{7-10}$$

A uniform deviate is transformed to a standard normal deviate by entering the standard normal table (Table A-1) using u as $\Phi(z)$ and then moving across to the left of the table to find z. That is, the value of u assumes the probability from which deviate z is taken from the table.

Consider the case where the sample consists of the following uniform variates:

$$U = \{0.82, \ 0.43, \ 0.71, \ 0.18, \ 0.66\}$$

To find the corresponding z values, enter the standard normal table for a probability of 0.82, for example, and read the z value ($z = 0.915$). Continuing for each u_i value yields:

$$Z = \{0.915, \ -0.176, \ 0.556, \ -0.915, \ 0.413\}$$

7.4.2. Composition Method

This method can be used to generate a random variable X that has the following composite probability distribution $F_X(x)$:

$$F_X(x) = \sum_{i=1}^{m} w_i \, F_X^i(x) \qquad (7\text{-}11)$$

where $F_X^i(x)$ = ith cumulative distribution function (CDF) for the random variable X, and $w_i = i$th weight factor. The weight factors must satisfy the following condition:

$$\sum_{i=1}^{m} w_i = 1 \qquad (7\text{-}12)$$

The random variable X is generated in the following two stages:

1. A random number u_1 is generated in the range $[0,1]$. This random number is used to select the appropriate CDF, $F_X^i(x)$, $i = 1, 2, ..., m$, for random variable generation. The selection can be based on treating the weight factors to constitute the following discrete probability function:

$$p_W(w) = \begin{cases} w_1 & \text{for } F_X^1(x) \\ w_2 & \text{for } F_X^2(x) \\ \vdots & \vdots \\ w_{m-1} & \text{for } F_X^{m-1}(x) \\ w_m & \text{for } F_X^m(x) \end{cases} \qquad (7\text{-}13)$$

2. A second random number u_2 is then used to generate the random variable according to the selected distribution in step 1. The inverse transformation method or other methods can be used for this purpose.

7.4.3. Function-Based Generation

Some random variables can be expressed in terms of other random variables that can be easily generated. Such functions can be used for generation purposes. For example, the addition of m random variables $X_1, X_2, ..., X_m$ that have independent exponential distributions with a parameter λ is a gamma distribution with parameters (m, λ). Therefore, a random variable Y that has a gamma distribution with parameters (m, λ) can be expressed as

$$Y = X_1 + X_2 + X_3 + ... + X_m \qquad (7\text{-}14)$$

where m = an integer. This random variable can be generated by randomly generating m independent random variables using an exponential distribution with a parameter λ. The result is $x_1, x_2, ..., x_m$. Therefore, a generated value for Y is given by

$$y = x_1 + x_2 + x_3 + \ldots + x_m \tag{7-15}$$

The function-based method can be used to generate random variables in a simple manner, but it can be inefficient. Direct generation of random variables should be used wherever possible.

7.4.4. Acceptance-Rejection Method

This method can be considered as an indirect generation of random variables, especially if it is compared with the inverse transformation, composition, or the function-based methods. For a random variable X that has a probability density function $f_X(x)$, the acceptance-rejection method for generating X can be summarized in the following steps:

1. A random variable Y is selected to facilitate the generation of X with the following density function $f_Y(y)$:

$$f_Y(y) = \frac{g(y)}{\int\limits_{-\infty}^{\infty} g(y) \, dy} \tag{7-16}$$

where $g(y) =$ a generally simple arbitrary function such $g(x) \geq f_X(x)$ for any x, and

$$\int\limits_{-\infty}^{\infty} g(y) \, dy < \infty \tag{7-17}$$

2. A random number u should be generated independent of Y. The random number $u \in [0,1]$.
3. A random value y should be generated based on the density function of Equation 7-16.
4. A generated value for the random variable X is taken as

$$x = y \quad \text{if } u \leq \frac{f_X(y)}{g(y)} \tag{7-18}$$

Otherwise, i.e., if $u > \dfrac{f_Y(y)}{g(y)}$, the value of y is rejected and a new pair of u and y values needs to be generated repeatedly until the condition of Equation 7-18 is satisfied and a value for x is obtained.

7.4.5. Generation Based on Special Properties

In this method, special properties or relations between distributions are used for the purpose of generating random variables. For example, if X has a standard normal distribution, i.e., $X \sim N(0,1)$, and utilizing the property that $Y = X^2$ has a chi-square distribution with one degree of freedom, Y can be generated by (1) generating X as a standard normal variate, then (2) using $y = x^2$. To generate a random variable Z that has a chi-square distribution with v degrees of freedom, the following property can be used

$$Z = Y_1 + Y_2 + Y_3 + \ldots + Y_v \qquad (7\text{-}19)$$

where Y_i, $i = 1, 2, \ldots, v$, have independent chi-square distribution with 1 degree of freedom each. Each Y_i can be generated as X^2, where X has a standard normal. Then the results are added to obtain the distribution of Z. This method is not efficient. There are other more efficient methods for generating random variables of chi-square distributions with v degrees of freedom that are not discussed herein.

7.5. GENERATION OF SELECTED DISCRETE RANDOM VARIABLES

7.5.1. Bernoulli Distribution

The Bernoulli distribution can be generated using the inverse transformation method. Starting with a generated random number $u \in [0,1]$, a random variable X can be generated according to a Bernoulli distribution as follows:

$$x = \begin{cases} 1 & \text{if } u \leq p \\ 0 & \text{otherwise} \end{cases} \qquad (7\text{-}20)$$

where p = a specified parameter for the Bernoulli distribution.

7.5.2. Binomial Distribution

A discrete random variable that has two possible outcomes can follow a binomial mass function. There are four assumptions that underlie a binomial process: (1) there are only two possible outcomes for each trial; (2) there are n trials in the experiment; (3) the n trials are independent; and (4) the probability of each outcome remains constant from trial to trial. The probability of exactly X occurrences in n trials is given by:

$$p_X(x) = \binom{n}{x} p^x (1-p)^{n-x} \qquad \text{for } x = 0, 1, 2, \dots, n \qquad (7\text{-}21)$$

where $\binom{n}{x}$ is the binomial coefficient and equals $\dfrac{n!}{x!(n-x)!}$. In this section, the notation $b(X;n,p)$ is used for the binomial distribution.

A random variable X of a binomial distribution with the parameters n and p can be generated by generating n independent random variables, $X_1, X_2, \dots, X_n$, of a Bernoulli distribution, each with a parameter p. Each X_i can be generated using the inverse transformation method for the Bernoulli distribution as discussed in Section 7.5.1. Then a generated value of y is given by

$$x = x_1 + x_2 + x_3 + \dots + x_n \qquad (7\text{-}22)$$

where x_i, $i = 1, 2, \dots, n$, are the respective generated values of the random variables $X_1, X_2, \dots, X_n$.

This method becomes inefficient for large values of n. In such cases, direct inverse transformation can be utilized.

Binomial experiments can be simulated using uniform variates. The procedure is as follows for N experiments, with each having n trials and an outcome A with an occurrence probability of p and an outcome B with an occurrence probability of $1 - p$:

1. For each trial in the experiment, generate a uniform variate, u_i.
2. If $u_i < p$, then assume outcome A has occurred; otherwise, assume outcome B has occurred.
3. Determine the number of A outcomes, X, that occurred in the n trials.
4. After completing the N experiments, with each having n trials, compute the relative frequency of X outcomes.

Example 7-5: Generation of a Binomial Distribution

This problem has three trials per experiment where the probability (p) of outcome A is 0.3. For example, if we have an unbalanced coin with a probability of 0.3 of getting a head, then we would be interested in the probabilities of getting 0, 1, 2, or 3 heads in three flips of the coin. From the binomial mass function of Equation 7-21, we can compute the following population probabilities:

$$b(0;\ 3,\ 0.3) = \binom{3}{0} 0.3^0\, 0.7^3 = 0.343$$

$$b(1;\ 3,\ 0.3) = \binom{3}{1} 0.3^1\, 0.7^2 = 0.441$$

$$b(2; 3, 0.3) = \binom{3}{2} 0.3^2 \, 0.7^1 = 0.189$$

$$b(3; 3, 0.3) = \binom{3}{3} 0.3^3 \, 0.7^0 = 0.027$$

The sum of these probabilities is 1.

To illustrate the simulation of binomial probabilities, let $N = 15$. For each of the 15 experiments, we generated three uniform variates and counted the number of values less than 0.3, which is then the sample estimate of X. One possible simulation is shown in Table 7-3.

TABLE 7-3 Generation of a Binomial Distribution

Experiment	u_i variates	x
1	0.78, 0.51, 0.43	0
2	0.85, 0.95, 0.22	1
3	0.88, 0.36, 0.20	1
4	0.79, 0.55, 0.71	0
5	0.11, 0.09, 0.70	2
6	0.11, 0.70, 0.98	1
7	0.79, 0.45, 0.86	0
8	0.62, 0.11, 0.77	1
9	0.45, 0.40, 0.13	1
10	0.37, 0.31, 0.12	1
11	0.54, 0.99, 0.95	0
12	0.79, 0.36, 0.37	0
13	0.73, 0.67, 0.60	0
14	0.56, 0.40, 0.71	0
15	0.24, 0.21, 0.23	3

From the 15 experimental estimates of X, there are seven 0's, six 1's, one 2, and one 3. This yields sample estimates of the probabilities of $7/15 = 0.467$, $6/15 = 0.400$, $1/15 = 0.067$, and $1/15 = 0.066$, respectively. As N becomes larger and larger, the sample estimates of the probabilities would more closely approximate the population probabilities.

Example 7-6: Life of a Project

Consider the case of a project that has a design life of 5 years, with the probability of failure in any 1 year equal to 0.1. A binomial process such as this case can be simulated using $n = 5$ and $p = 0.1$, with values of $X = 0, 1, 2, 3, 4,$ and 5. To estimate the probability of no failures, we would be interested in the case of $X = 0$. Ten experiments were made and the sample estimate of the probability of a project not failing were computed as shown in Table 7-4a. The sample ($\hat{p}$) and population (p) probabilities are shown in Table 7-4b. For a sample of 10, the sample estimates are

TABLE 7-4a Simulation Results for a Binomial Distribution

Project	u_i variates	x
1	0.05, 0.37, 0.23, 0.42, 0.05	2
2	0.60, 0.32, 0.79, 0.45, 0.20	0
3	0.67, 0.86, 0.66, 0.55, 0.23	0
4	0.41, 0.03, 0.90, 0.98, 0.74	1
5	0.94, 0.58, 0.31, 0.45, 0.31	0
6	0.69, 0.93, 0.77, 0.37, 0.72	0
7	0.55, 0.69, 0.01, 0.51, 0.58	1
8	0.33, 0.58, 0.72, 0.22, 0.97	0
9	0.09, 0.29, 0.81, 0.44, 0.68	1
10	0.29, 0.54, 0.75, 0.36, 0.29	0

TABLE 7-4b Sample (p) and Population ($\hat{p}$) Probabilities

X	Sample probability ($\hat{p}$)	Population probability (p)
0	0.60	0.5905
1	0.30	0.3280
2	0.10	0.0729
3	0.00	0.0081
4	0.00	0.0005
5	0.00	0.0000
Column summations	1.00	1.0000

in good agreement with the population values. The sample estimate of the probability of a project not failing in 5 years is 0.60.

7.5.3. Geometric Distribution

The geometric distribution can be generated using the inverse transformation method. An alternate method can be used by setting $i = 1$ and then performing the following steps:

1. Generate a random number $u \in [0,1]$.
2. If $u \leq p$, where p = a specified parameter for the geometric distribution, return $x = i$; otherwise replace i by $i + 1$ and go back to step 1.

A third method of random generation for the geometric distribution starts with generating a random number $u \in [0,1]$, and then x is calculated as

$$x = \frac{\ln(u)}{\ln(1-p)} \tag{7-23}$$

The result from Equation 7-23 is a real value; therefore, the integer of this real value should be used as the generated value and should be assigned to x.

7.5.4. Poisson Distribution

Discrete random variables that can take on integer values may be represented by a Poisson mass function, which has the following form:

$$P_{X_t}(x) = \frac{e^{-\lambda t}(\lambda t)^x}{x!} \qquad \text{for } x = 0, 1, 2, \ldots \qquad (7\text{-}24)$$

where X_t is the number of occurrences in a reference period t, λ is a parameter of the function, and 0! is defined as 1. The cumulative function is

$$F_{X_t}(x) = \sum_{i=0}^{x} \frac{e^{-\lambda t}(\lambda t)^i}{i!} \qquad \text{for } x = 0, 1, 2, \ldots \qquad (7\text{-}25)$$

The mean and variance of the function both equal λt, so the sample mean can be used as an estimator of λt.

The Poisson distribution can be generated using the inverse transformation method based on Eq. 7-25 or using its relationship to the exponential distribution.

According to the first method, uniform random variates can be used to generate random variates that have a Poisson distribution for any λ. The procedure for any given λ is as follows:

1. Calculate the cumulative function $F_{X_t}(x_i)$ for $x_i = 0, 1, 2, \ldots$.
2. Generate a unit variate u_i in the range [0,1].
3. If $F_{X_t}(x_{i-1}) \le u_i < F_{X_t}(x_i)$, then the Poisson variate is x_i.

The procedure in the second case is based on the special relation between the Poisson distribution and the exponential distribution. In order to generate a random variable X that has a Poisson distribution with a parameter λ, the following needs to be set:

$$a = \exp(-\lambda t), b = 1, \text{ and } i = 0 \qquad (7\text{-}26)$$

In the $(i + 1)$ random generation cycle, generate a random number $u \in [0,1]$ and assign a value to X as follows:

$$x = \begin{cases} i & \text{if } bu < a \\ \text{increment } i \text{ by 1 and repeat the process} & \text{otherwise} \end{cases} \qquad (7\text{-}27)$$

Example 7-7: Floods in Northern Idaho

The number of floods per year (n) on a river in northern Idaho was obtained from streamflow records over a 38-year period. The sample analysis is shown in Table 7-5. There were a total of 120 floods in the 38 years, which yields a mean

TABLE 7-5 Frequency of Flows above 28,000 cfs on Clearwater River at Kamiah, ID

Year	n	Year	n	Year	n	Year	n	Year	n
1911	4	1919	2	1927	5	1935	3	1943	6
1912	3	1920	2	1928	2	1936	5	1944	1
1913	4	1921	2	1929	3	1937	2	1945	2
1914	3	1922	3	1930	1	1938	4	1946	7
1915	1	1923	3	1931	2	1939	2	1947	3
1916	5	1924	2	1932	4	1940	2	1948	6
1917	5	1925	3	1933	4	1941	1		
1918	3	1926	3	1934	4	1942	3		

TABLE 7-6 Poisson Approximation of Flood Counts

	Sample		Population	
			Mass function	Cumulative function
X	N_X	N_X/N	p_{X_t}	$F_{X_t}(x)$
0	0	0	0.043	0.043
1	4	0.105	0.134	0.177
2	10	0.263	0.212	0.389
3	11	0.290	0.223	0.612
4	6	0.158	0.176	0.788
5	4	0.105	0.111	0.899
6	2	0.053	0.059	0.958
7	1	0.026	0.026	0.984
≥8	0	0	0.016	1.000

annual flood count of 3.158; thus, the sample estimate of the Poisson parameter λ is 3.158, with $t = 1$ year. The assumed population mass function is

$$p_{X_1}(x) = \frac{e^{-3.158}(3.158)^x}{x!} \qquad \text{for } x = 0, 1, 2, \ldots \qquad (7\text{-}28)$$

The population probabilities are shown in Table 7-6.

Example 7-8: Hurricanes Hitting the U.S. Coasts

Records indicate that the annual number of hurricanes hitting the U.S. coast has an average of 4. The discrete variable can be represented by a Poisson mass function with $\lambda = 4$. Thus, the probability of the coast not being hit by a hurricane in any year is

$$p_{X_1}(0) = \frac{e^{-4}(4)^0}{0!} = 0.0183 \qquad (7\text{-}29)$$

where 0! is defined as 1.

Example 7-9: Annual Number of Floods

In Example 7-7 the mean annual number of floods on a particular river was found to be 3.158. Assume that it is necessary to simulate the number of floods in a 10-year period. The cumulative function is given in Table 7-6. For a given vector of uniform variates u_i, the generated number of floods per year is shown in Table 7-7. Based on this sample of ten simulated values, the mean number of floods is $26/10 = 2.6$. As the sample size increases, the sample mean would be expected to be closer to the assumed population mean of 3.158.

TABLE 7-7 Synthetic Series of the Frequency of Floods

Year	u_i	x_i	$F_{x_i}(x_{i-1})$	$F_{x_i}(x_i)$
1	0.13	1	0.043	0.177
2	0.04	0	0.000	0.043
3	0.66	4	0.612	0.788
4	0.80	5	0.788	0.899
5	0.77	4	0.612	0.788
6	0.52	3	0.389	0.612
7	0.30	2	0.177	0.389
8	0.14	1	0.043	0.177
9	0.37	2	0.177	0.389
10	0.65	4	0.612	0.788
Column summation		26		

7.6. GENERATION OF SELECTED CONTINUOUS RANDOM VARIABLES

7.6.1. Uniform Distribution

Variates with a uniform distribution that has parameters other than 0 and 1 can be generated using the inverse transformation method. Starting with a generated random number $u \in [0,1]$, a random variable X can be generated according to a uniform distribution as follows:

$$x = a + (b - a)u \qquad (7\text{-}30)$$

where a and b = specified parameters for the uniform distribution that correspond to the lower and upper limits of the random variable, respectively.

7.6.2. Normal Distribution

A random variable X can be generated according to a normal distribution $N(\mu, \sigma^2)$ by generating the first variate Z with a standard normal distribution, and then using

$$x = z\sigma + \mu \qquad (7\text{-}31)$$

The normal distribution can be generated by several methods. A direct, however, approximate, method is to develop regression equations for segments of the CDF distribution of the standard normal. Then inverse transformation can be used to generate a variate Z according to a standard normal distribution using the following relation

$$z = \Phi^{-1}(u) \qquad (7\text{-}32)$$

The polar method can also be used to generate standard normal variates. The polar method starts by generating two independent random numbers u_1 and u_2 from the interval $[0,1]$. Then, the following variables y_1, y_2, and z are defined:

$$y_1 = 2u_1 - 1 \qquad (7\text{-}33)$$

$$y_2 = 2u_2 - 1 \qquad (7\text{-}34)$$

$$w = y_1^2 + y_2^2 \qquad (7\text{-}35)$$

If $w > 1$, then generate new random numbers u_1 and u_2 and repeat the evaluation of y_1, y_2, and w. If $w \le 1$, then evaluate a pair of independent standard normal variates as follows:

$$z_1 = \sqrt{\frac{-2\ln(w)}{w}} y_1 \qquad (7\text{-}36)$$

$$z_2 = \sqrt{\frac{-2\ln(w)}{w}} y_2 \qquad (7\text{-}37)$$

7.6.3. Lognormal Distribution

The generation of lognormal distributions can be based on generating standard normal variates and the special relationship between the lognormal and the standard normal distributions. In order to generate a variate X with given mean value and variance according to a lognormal distribution, the parameters of the lognormal distribution, μ_Y and σ_Y, should first be determined from the given mean and variance. The relations between the parameters of a distribution and its mean and variance were previously discussed in Chapter 3. Then a normal variate Y with the parameters μ_Y and σ_Y is generated resulting in a normal value y. The generated lognormal value is taken as

$$x = \exp(y) \qquad (7\text{-}38)$$

The generation of normal variates was discussed Section 7.6.2.

7.6.4. Exponential Distribution

The exponential distribution can be generated using the inverse transformation method. Starting with a generated random number $u \in [0,1]$, a random variable X can be generated according to a exponential distribution as follows:

$$x = \frac{-\ln(u)}{\lambda} \qquad (7\text{-}39)$$

where λ = a specified parameter for the exponential distribution.

7.7. APPLICATIONS

7.7.1. Simulation of a Queuing System

A classical problem in engineering is establishing policies for problems involving queues. The following problem illustrates the use of simulation in the evaluation of policies on the operation of a checkout station in a retail store. The manager of the store is interested in the operation of the checkout station because she is aware that customers who spend excessive time waiting to check out will be less likely to return to the store. Although the total time that a customer spends in the checkout process is of interest, the manager is also interested in both the time spent waiting to be served and the time being served; the total time is the sum of the two. The manager also recognizes that customers are discouraged by long lines; therefore, the number of customers waiting to be served is also a criterion of interest. The manager would be interested in examining alternatives that would reduce both the total time and the length of the queue. For example, the manager could install a second checkout counter, which would require increases in costs associated with installation, operation, and maintenance of the added counter. As an alternative, the manager might decide to install a new checkout system that is more efficient; this system would reduce the time required for checkout without the costs associated with the installation of a new counter and the added labor. Of course, there would be a cost associated with the purchase of the new equipment.

Since the manager is most concerned with peak demand hours, she collects data on the current operation of the checkout counter. Each evening of 1 week, the manager measures the time between arrivals of customers and the time required for a customer to be serviced. The manager finds that the service time can be approximated by a uniform probability function with a mean of 2 min; based on measure-

TABLE 7-8 Computation of Cumulative Probability Function of Interarrival Times

Interarrival time (min)	Number of arrivals	Probability function	Cumulative function
0.0–0.5	20	0.031	0.031
0.5–1.0	50	0.077	0.108
1.0–1.5	90	0.138	0.246
1.5–2.0	140	0.215	0.461
2.0–2.5	120	0.185	0.646
2.5–3.0	100	0.154	0.800
3.0–3.5	70	0.108	0.908
3.5–4.0	30	0.046	0.954
4.0–4.5	20	0.031	0.985
4.5–5.0	10	0.015	1.000
Column Summations	650	1.000	

ments the manager believes that the service time can be approximated with probabilities of 0.25 for service times of 1.25, 1.75, 2.25, and 2.75 min; these are the center points of the computational interval of 0.5 min. The time between arrivals is shown in Table 7-8 for 0.5-min intervals. During the time of peak demand all interarrival times were 5 min or less, with an average of 2.2 min. Because the interarrival times did not follow a known probability function, the empirical function was used for simulations. The discrete cumulative function for the interarrival times is shown in Figure 7-3. The discrete function is used because the data were collected on a 0.5-min interval; the center point of each interval is used to represent the interval. The derivation of the cumulative function for the interarrival times is given in Table. 7-8.

To demonstrate the simulation technique, a 1-h period was simulated. The interarrival times were simulated by generating a sequence of random numbers from a uniform distribution having location and scale parameters of 0 and 1, respectively. Each of the numbers in the sequence was entered into Figure 7-3 and the interarrival times determined. The interarrival times are given in Table 7-9, with values ranging between 0.25 and 4.75 on a 0.5-min interval. The second customer arrived at the checkout counter 1.25 min after the first customer. The third customer arrived at a time of 3 min, which is 1.75 min after the second customer. The time of arrival of each customer is also given in Table 7-9.

A second sequence of random numbers having a uniform probability function ($a = 0$, $b = 1$) is generated. Since these values are used for simulating the time required for service, there are four service times (i.e., 1.25, 1.75, 2.25, and 2.75) with an equal probability of occurrence. The random numbers were used to generate the service times, which are shown in Table 7-9, for the existing situation; this is called policy 1. The time of departure equals the sum of the service time and either the time that the previous customer leaves the checkout counter or the time that the person arrives at the counter when no one is being served. The waiting time for a customer, which is also given in Table 7-9, equals the difference between the time

FIGURE 7-3 Comparison of the distributions of the interarrival times and output from the system.

of arrival and the time that the customer starts service. The total time equals the sum of the waiting time and servicing time. The number of customers in the queue is determined by comparing the time of departure of a person and the time of arrival of customers who follow the person being served. For example, the 11th customer departs at time 21.75. At that time, customers 12, 13, 14, and 15 have already arrived at the counter (at times 17.75, 19.50, 20.75, and 21.00), so there are four customers waiting in the queue to be served when customer 11 departs.

The manager is interested in determining the effect of a new means of checking out a customer. The equipment is expected to decrease the service time. The simulation based on this service time distribution is referred to as policy 2. The probabilities for service times of 1.25, 1.75, 2.25, and 2.75 min are 0.35, 0.30, 0.20, and 0.15, respectively. The same sequence of arrivals was used to simulate the effect of service policy 2. The service times and times of departure were computed and are given in Table 7-9. The total time, waiting time, and number in the queue were computed in the same way as for policy 1. The distribution of waiting times is given in Table 7-10. It is quite evident that the waiting time is considerably less with policy 2 than with policy 1. The mean waiting times for policies 1 and 2 are 3.27 and 1.08 min, respectively. The mean service times were 1.78 and 1.59 min, respectively. It is evident from both the mean values and the distributions of service times shown in Table 7-11 that the sample values for the simulated period of 1 hr are less than what would be expected. As the length of the period of simulation is increased, one can expect the sample values to be closer approximations to the true values.

TABLE 7-9 Simulations of Interarrival and Servicing Times for Alternative Policies

Customer	Random number	Interarrival time (min)	Time of arrival (min)	Random number	Policy 1					Policy 2				
					Servicing time (min)	Time of departure (min)	Waiting time (min)	Total time (min)	Number waiting in queue	Servicing time (min)	Time of departure (min)	Waiting time (min)	Total time (min)	Number waiting in queue
1	0.19	1.25	0	0.57	2.25	2.25	0.00	2.25	1	1.75	1.75	0.00	1.75	1
2	0.37	1.75	1.25	0.84	2.75	5.00	1.00	3.75	2	2.25	4.00	0.50	2.75	2
3	0.09	0.75	3.00	0.54	2.25	7.25	2.00	4.25	2	1.75	5.75	1.00	2.75	2
4	0.30	1.75	3.75	0.25	1.25	8.50	3.50	4.75	2	1.25	7.00	2.00	3.25	1
5	0.76	2.75	5.50	0.95	2.75	11.25	3.00	5.75	2	2.75	9.75	1.50	4.25	1
6	0.53	2.25	8.25	0.14	1.25	12.50	3.00	4.25	2	1.25	11.00	1.50	2.75	1
7	0.09	0.75	10.50	0.21	1.25	13.75	2.00	3.25	2	1.25	12.25	0.50	1.75	1
8	0.26	1.75	11.25	0.51	2.25	16.00	2.50	4.75	2	1.75	14.00	1.00	2.75	1
9	0.43	1.75	13.00	0.16	1.25	17.25	3.00	4.25	2	1.25	15.25	1.00	2.25	1
10	0.13	1.25	14.75	0.39	1.75	19.00	2.50	4.25	2	1.75	17.00	0.50	2.25	1
11	0.25	1.75	16.00	0.85	2.75	21.75	3.00	5.75	4	2.25	19.25	1.00	3.25	1
12	0.42	1.75	17.75	0.82	2.75	24.50	4.00	6.75	4	2.25	21.50	1.50	3.75	3
13	0.31	1.75	19.50	0.16	1.25	25.75	5.00	6.25	4	1.25	22.75	2.00	3.25	2
14	0.15	1.25	20.75	0.14	1.25	27.00	5.00	6.25	3	1.25	24.00	2.00	3.25	2
15	0.01	0.25	21.00	0.80	2.75	29.75	6.00	8.75	4	2.25	26.25	3.00	5.25	2
16	0.54	2.25	23.25	0.24	1.25	31.00	6.50	7.75	3	1.25	27.50	3.00	4.25	2
17	0.28	1.75	25.00	0.12	1.25	32.25	6.00	7.25	3	1.25	28.75	2.50	3.75	2
18	0.50	2.25	27.25	0.87	2.75	35.00	5.00	7.75	4	2.75	31.50	1.50	4.25	2
19	0.12	1.25	28.50	0.06	1.25	36.25	6.50	7.75	3	1.25	32.75	3.00	4.25	1
20	0.67	2.75	31.25	0.33	1.75	38.00	5.00	6.75	3	1.25	34.00	1.50	2.75	1

TABLE 7-9 Simulations of Interarrival and Servicing Times for Alternative Policies (continued)

					Policy 1					Policy 2				
Customer	Random number	Interarrival time (min)	Time of arrival (min)	Random number	Servicing time (min)	Time of departure (min)	Waiting time (min)	Total time (min)	Number waiting in queue	Servicing time (min)	Time of departure (min)	Waiting time (min)	Total time (min)	Number waiting in queue
21	0.22	1.25	32.50	0.44	1.75	39.75	5.50	7.25	3	1.75	35.75	1.50	3.25	1
22	0.54	2.25	34.75	0.49	1.75	41.50	5.00	6.75	3	1.75	37.50	1.00	2.75	1
23	0.58	2.25	37.00	0.22	1.25	42.75	4.50	5.75	2	1.25	38.75	0.50	1.75	1
24	0.18	1.25	38.25	0.13	1.25	44.00	4.50	5.75	2	1.25	40.00	0.50	1.75	0
25	0.53	2.25	40.50	0.03	1.25	45.25	3.50	4.75	1	1.25	41.75	0.00	1.25	0
26	0.75	2.75	43.25	0.79	2.75	48.00	2.00	4.75	2	2.25	45.50	0.00	2.25	0
27	0.76	2.75	46.00	0.01	1.25	49.25	2.00	3.25	2	1.25	47.25	0.00	1.25	1
28	0.05	0.75	46.75	0.27	1.75	51.00	2.50	4.25	1	1.25	48.50	0.50	1.75	0
29	0.57	2.25	49.00	0.18	1.25	52.25	2.00	3.25	1	1.25	50.25	0.00	1.25	0
30	0.69	2.75	51.75	0.28	1.75	54.00	0.50	2.25	0	1.25	53.00	0.00	1.25	0
31	0.61	2.25	54.00	0.14	1.25	55.25	0.00	1.25	0	1.25	55.25	0.00	1.25	0
32	0.83	3.25	57.25	0.06	1.25	58.50	0.00	1.25	1	1.25	58.50	0.00	1.25	1
33	0.00	0.25	57.50	0.65	2.25	60.75	1.00	3.25	0	1.75	60.25	1.00	2.75	0
34	0.81	3.25	60.75	0.34	1.75	62.50	0.00	1.75	—	1.25	62.00	0.00	1.25	—

TABLE 7-10 Computation of Cumulative Probability Function of Waiting Times

Waiting time (min)	Policy 1			Policy 2		
	Frequency	Probability function	Cumulative function	Frequency	Probability function	Cumulative function
0.0	3	0.091	0.091	8	0.242	0.242
0.5	1	0.030	0.121	6	0.182	0.424
1.0	2	0.061	0.182	6	0.182	0.606
1.5	0	0.000	0.182	6	0.182	0.788
2.0	5	0.152	0.333	3	0.091	0.879
2.5	3	0.091	0.424	1	0.030	0.909
3.0	4	0.121	0.545	3	0.091	1.000
3.5	2	0.061	0.606			
4.0	1	0.030	0.636			
4.5	2	0.061	0.697			
5.0	5	0.152	0.848			
5.5	1	0.030	0.878			
6.0	2	0.061	0.939			
6.5	2	0.061	1.000			
Column summations	33	1.000		33	1.000	

TABLE 7-11 Distributions of Total Time in Systems and Service Time for Two Policies

Total time (min)	Frequency for policy 1	Frequency for policy 2	Service time (min)	Frequency for policy 1	Frequency for policy 2
1.25	2	7	1.25	16	20
1.75	1	5	1.75	7	7
2.25	2	3	2.25	4	5
2.75	0	7	2.75	7	2
3.25	4	5			
3.75	1	2			
4.25	5	4			
4.75	4	0			
5.25	0	1			
5.75	4				
6.25	2				
6.75	3				
7.25	2				
7.75	3				
8.25	0				
8.75	1				
Column Summations	34	34		34	34

7.7.2. Warehouse Construction

A warehouse is to be constructed from precast concrete elements that are produced by a nearby precast factory. The following construction tasks are identified for building the warehouse:

A: excavation of foundations
B: construction of foundations
C: construction of precast elements at factory
D: transportation of precast elements to construction site
E: assembly of elements at site
F: construction of roof
G: exterior and interior finishing

FIGURE 7-4. Construction network.

Figure 7-4 shows the logical network for performing these activities. The figure indicates that tasks C and D can be performed in parallel to tasks A and B, i.e., as excavation and construction of the footings are being performed at the site, the precast elements can be constructed at the factory and then transported to the construction site. Table 7-12 shows the means and standard deviations for the completion times of these tasks. Normal probability distributions and statistical noncorrelation are assumed for these times. This example was used in Section 4.4.5 to illustrate the estimation of the mean and variance of the completion time of the project using approximate methods. In this section, simulation is used to compute these moments.

TABLE 7-12 Moments of Completion Times

Task	Name	Mean (days)	Standard deviation (days)
A	Foundation excavation	3.0	1.0
B	Foundation construction	2.0	0.5
C	Precast-elements construction	4.0	1.0
D	Transportation of elements	0.5	0.5
E	Assembly of elements	4.0	1.0
F	Roofing	2.0	1.0
G	Finishing	3.0	1.0

The project completion time, T, is a dependent random variable that is given by

$$T = \max\{A + B, \ C + D\} + E + F + G \qquad (7\text{-}40)$$

where A, B, C, D, E, F, and G are random variables that model the times for completing the corresponding tasks. The random numbers used for generating the completion times for tasks A, B, C, D, E, F, and G are shown in Table 7-13. In this example, 20 simulation cycles were used. The random numbers were used to generate the random times as shown in Table 7-14 utilizing the inverse transformation method. Then Equation 7-40 was used to compute the completion time for the project for each simulation cycle as shown in Table 7-14. The mean value and variance of the completion time of the project were computed using values in the last column of Table 7-14. The resulting statistics are

$$\text{Mean value} = 14.59 \text{ days} \qquad (7\text{-}41a)$$

$$\text{Variance} = 5.55 \ (\text{days})^2 \qquad (7\text{-}41b)$$

The approximate mean duration of the project, $\overline{T}$, was computed in Section 4.4.5 as

$$\overline{T} = \max\{3 + 2, \ 4 + 0.5\} + 4 + 2 + 3 = 14 \text{ days} \qquad (7\text{-}42)$$

The approximate variance of the project duration, $\text{Var}(T)$, was computed as

$$\text{Var}(T) = \text{Var}(A) + \text{Var}(B) + \text{Var}(E) + \text{Var}(F) + \text{Var}(G) \qquad (7\text{-}43a)$$

$$= (1)^2 + (0.5)^2 + (1)^2 + (1)^2 + (1)^2 = 4.25 \ (\text{days})^2 \qquad (7\text{-}43b)$$

These values differ from the computed values using simulation. The simulation results are also approximate because of the small number of cycles used to compute the results. By increasing the number of simulation cycles, the accuracy of the results is expected to increase. The assumed probability distributions for the task durations resulted in negative durations that are not realistic as shown in Table 7-14. Other distributions for finite durations might be appropriate in this case. Also, the simulation results in the case of a small number of cycles are dependent on the random numbers used. For example, by using a different set of random numbers as those used in Table 7-15, the results shown in Table 7-16 are obtained. The statistics in this case are

$$\text{Mean value} = 13.98 \text{ days} \qquad (7\text{-}44a)$$

$$\text{Variance} = 3.67 \ (\text{days})^2 \qquad (7\text{-}44b)$$

TABLE 7-13 Random Numbers Used for Completion Time of Tasks

Task A	Task B	Task C	Task D	Task E	Task F	Task G
0.642707	0.758002	0.547225	0.510713	0.924981	0.444910	0.671304
0.240297	0.092418	0.847150	0.071252	0.981120	0.793358	0.780596
0.169051	0.446979	0.990008	0.079644	0.391058	0.793205	0.276989
0.457609	0.521270	0.606333	0.006137	0.479270	0.121284	0.343670
0.386325	0.395759	0.956544	0.432595	0.723067	0.448813	0.008538
0.313708	0.061922	0.343042	0.230356	0.538481	0.636290	0.211676
0.137571	0.078837	0.471558	0.383158	0.203166	0.500447	0.101354
0.296782	0.610994	0.785467	0.282056	0.282186	0.560465	0.539651
0.908314	0.124274	0.709123	0.508328	0.496352	0.886927	0.720611
0.763968	0.327695	0.506164	0.246872	0.743617	0.275227	0.218178
0.139498	0.935402	0.789508	0.966422	0.440431	0.682035	0.476614
0.220256	0.040641	0.347426	0.282962	0.178687	0.092735	0.964860
0.344963	0.100168	0.963482	0.569873	0.933351	0.646640	0.858627
0.095613	0.791418	0.726318	0.376506	0.872995	0.895403	0.962331
0.225540	0.262949	0.632760	0.550859	0.198235	0.077169	0.086730
0.239485	0.985236	0.212528	0.445724	0.662470	0.325610	0.025242
0.191603	0.108613	0.897544	0.990706	0.933851	0.557361	0.050711
0.946010	0.241317	0.187334	0.015071	0.228146	0.832563	0.816427
0.973859	0.343243	0.197940	0.177672	0.125638	0.099943	0.747989
0.484109	0.214928	0.020997	0.424466	0.893968	0.866459	0.706856

TABLE 7-14 Generated Random Values for Completion Time of Tasks

Task A	Task B	Task C	Task D	Task E	Task F	Task G	Project completion time
3.3652600	2.3498042	4.1183849	0.5133909	5.4396627	1.8617675	3.4430860	16.459580
2.2949296	1.3368915	5.0242746	−0.2334048	6.0778971	2.8179440	3.7739874	17.460698
2.0421427	1.9334970	6.3270846	−0.2038625	3.7238635	2.8174073	2.4085497	15.073043
2.8937885	2.0266006	4.2693511	−0.7523029	3.9481516	0.8313093	2.5979742	12.297824
2.7115217	1.8680395	5.7122992	0.4152843	4.5916169	1.8716254	0.6144956	13.205321
2.5150499	1.2304214	3.5962670	0.1312882	4.0963777	2.3481148	2.1995787	12.389543
1.9086613	1.2934043	3.9288268	0.3516179	3.1698067	2.0011170	1.7259547	11.177323
2.4667156	2.1407416	4.7905834	0.2118119	3.4240093	2.1518299	3.0993185	13.677553
4.3306524	1.4230146	4.5504403	0.5104081	3.9908831	3.2104754	3.5842925	16.539318
3.7188585	1.7770724	4.0154067	0.1579639	4.6542217	1.4032765	2.2218555	13.775285
1.9173874	2.7587914	4.8045186	1.4155181	3.8504386	2.4729750	2.9415001	15.484950
2.2288943	1.1281585	3.6081638	0.2131524	3.0797160	0.6756985	4.8105053	12.387236
2.6014871	1.3596149	5.7930090	0.5878512	5.5015181	2.3758197	4.0742012	18.332399
1.6928452	2.4055804	4.6013610	0.3428837	5.1407426	3.2559441	4.7787973	18.119729
2.2466222	1.6830249	4.3387303	0.5637742	3.1522166	0.5753632	1.6386032	10.268687
2.2923139	3.0884019	3.2025201	0.4319117	4.4187744	1.5483608	1.0437219	12.391573
2.1281350	1.3829576	5.2678487	1.6770614	5.5053989	2.1439730	1.3616434	15.955925
4.6076817	1.6491037	3.1123607	−0.5843253	3.2552784	2.9642836	3.9017196	16.378067
4.9412389	1.7984060	3.1511551	0.0379101	2.8526594	0.7179441	3.6678706	13.978119
2.9602643	1.6053855	1.9659854	0.4049415	5.2480626	3.1098640	3.5438339	16.467410

TABLE 7-15 Random Numbers Used for Completion Time of Tasks

Task A	Task B	Task C	Task D	Task E	Task F	Task G
0.606883	0.093464	0.125703	0.736216	0.585157	0.033755	0.719628
0.277315	0.682777	0.759930	0.485396	0.288004	0.697372	0.101427
0.725585	0.326737	0.091488	0.718726	0.819744	0.912300	0.910932
0.179915	0.471119	0.072710	0.293896	0.559946	0.441863	0.749723
0.152424	0.240208	0.294833	0.769227	0.786163	0.121520	0.663357
0.168486	0.035771	0.513560	0.880006	0.748794	0.115441	0.953369
0.915682	0.243600	0.610186	0.848375	0.102922	0.009326	0.801494
0.124135	0.682049	0.610019	0.203327	0.081627	0.866440	0.514767
0.342101	0.739733	0.131999	0.569512	0.388688	0.518582	0.204704
0.985961	0.613146	0.914132	0.898415	0.543517	0.091718	0.970920
0.336867	0.616759	0.402409	0.268781	0.913337	0.098700	0.545388
0.583809	0.471045	0.343964	0.278476	0.128413	0.359243	0.341192
0.798033	0.053788	0.467997	0.405734	0.923671	0.587813	0.126547
0.688703	0.028898	0.021365	0.039026	0.483284	0.546590	0.267746
0.959589	0.749079	0.914929	0.729020	0.917082	0.870119	0.652013
0.331024	0.626462	0.697033	0.771629	0.382801	0.702866	0.060994
0.201754	0.233297	0.417021	0.770881	0.034672	0.724181	0.395496
0.633503	0.380850	0.538246	0.326588	0.633842	0.176778	0.346776
0.840578	0.895108	0.071531	0.714916	0.400981	0.243865	0.211002
0.531249	0.463470	0.952944	0.073020	0.345216	0.578557	0.214954

TABLE 7-16 Generated Random Values for Completion Time of Tasks

Task A	Task B	Task C	Task D	Task E	Task F	Task G	Project completion time
3.2707799	1.3400400	2.8529732	0.8156949	4.2147164	0.1713280	3.5813727	12.578237
2.4095244	2.2375281	4.7058017	0.4817430	3.4411537	2.5164544	1.7263661	12.871519
3.5991597	1.7757440	2.6681462	0.7893469	4.9142851	3.3552784	4.3467318	17.991199
2.0844111	1.9638631	2.5438200	0.2291748	4.1505155	1.8540637	3.6733168	13.726170
1.9739182	1.6473216	3.4610715	0.8680251	4.7929741	0.8324813	3.4212035	13.375756
2.0399005	1.0987845	4.0339016	1.0875602	4.6703954	0.8017872	4.6788003	15.272445
4.3768346	1.6527585	4.2793768	1.0147408	2.7347579	−0.3528514	3.8468091	12.258309
1.8453495	2.2365066	4.2789404	0.0851884	2.6055421	3.1097732	3.0369229	13.116367
2.5937050	2.3210996	2.8829455	0.5873924	3.7176888	2.0464701	2.1752441	12.854208
5.1966454	2.1435484	5.3668740	1.1363712	4.1090447	0.6695408	4.8949117	17.013691
2.5794094	2.1482732	3.7532931	0.1919199	5.3618234	0.7108243	3.1137570	13.914087
3.2112634	1.9637700	3.5987720	0.2064942	2.8660028	1.6399625	2.5912269	12.272226
3.8344438	1.1952375	3.9198922	0.3809357	5.4304724	2.2215315	1.8570525	14.538738
3.4917621	1.0511701	1.9732268	−0.3812494	3.9582012	2.1167830	2.3806975	12.998614
4.7463211	2.3356456	5.3719798	0.8047524	5.3859459	3.1270236	3.3903168	18.985253
2.5633476	2.1610288	4.5154829	0.8719888	3.7023022	2.5322679	1.4531975	13.075239
2.1647959	1.6361127	3.7908605	0.8707520	2.1834205	2.5949514	2.7353984	12.175383
3.3407015	1.8485931	4.0957866	0.2755382	4.3416040	1.0723773	2.6064022	13.209678
3.9968036	2.6271607	2.5352335	0.7837149	3.7496031	1.3063624	2.1972483	13.877178
3.0782144	1.9542634	5.6744610	−0.2269675	3.6021733	2.1978306	2.2108603	13.458358

7.8. PROBLEMS

7-1. Using the linear congruential generator given by Equation 7-1, generate a stream of random numbers based on the following model parameters: $a = 7$, $b = 5$, $c = 26$, and $I_0 = 3$.

7-2. Using the linear congruential generator given by Equation 7-1, generate a stream of random numbers based on the following model parameters: $a = 7$, $b = 5$, $c = 345$, and $I_0 = 3$.

7-3. Using the linear congruential generator given by Equation 7-1, generate a stream of random numbers based on the following model parameters: $a = 77$, $b = 345$, $c = 26$, and $I_0 = 3$.

7-4. Perform a parametric analysis of the linear congruential generator given by Equation 7-1 by varying the values of a, b, c, and I_0 to investigate their importance. Based on your results, discuss the importance of these parameters.

7-5. The change in the length of a rod due to axial force P is given by

$$\Delta L = \frac{PL}{AE}$$

where L = length of rod, P = applied axial force, A = cross-sectional area of rod, and E = modulus of elasticity. Using 20 simulation cycles, determine the mean and variance of ΔL assuming noncorrelated random variables with the following probabilistic characteristics:

Random variable	Mean value	Coefficient of variation	Distribution type
P	100 kips	0.35	Lognormal
L	20 in.	0.05	Normal
E	30,000 ksi	0.10	Lognormal
A	1 in.^{2}	0.05	Normal

7-6. For the rod in Problem 7-5, study the effect of increasing the number of simulation cycles on the estimated mean and variance of the deformation. Use the following numbers of simulation cycles: 20, 100, 500, 1000, 2000, and 10,000. Provide your results in the form of plots of estimated statistics as a function of the number of simulation cycles.

7-7. For the rod in Problems 7-5 and 7-6, plot two frequency histograms for the deformation of rod based on 20 and 100 simulation cycles. Suggest a distribution type for the deformation.

7-8. For the rod in Problems 7-5 and 7-6, plot a frequency histogram for the deformation of rod based on 1000 simulation cycles. Suggest a distribution type for the deformation.

7-9. For the rod in Problems 7-5 and 7-6, compare the results of Problems 7-7 and 7-8. Provide a discussion.

7-10. The ultimate moment capacity, M, of an underreinforced concrete rectangular section is given by

$$M = A_s f_y \left(d - \frac{a}{2} \right)$$

where

$$a = \frac{A_s f_y}{0.85 b f'_c}$$

in which the following are random variables: A_s is the cross-sectional area of the reinforcing steel, f_y is the yield stress (strength) of the steel, d is the distance from the reinforcing steel to the top fibers of the beam, b is the width of the beam, and f'_c is the ultimate stress (strength) of the concrete. If the random variables are assumed to be statistically noncorrelated, determine the mean and variance of the moment capacity using the following information:

Random variable	Mean value	Coefficient of variation	Distribution type
A_s	0.25 in.^{2}	0.10	Lognormal
f_y	40,000 psi	0.10	Normal
d	20 in.	0.05	Lognormal
b	12 in.	0.05	Normal
f'_c	4000 psi	0.20	Lognormal

Use 100 simulation cycles. Is this a sufficient number of cycles? Why? Provide a discussion.

7-11. A random variable is Poisson distributed with $\lambda = 1.5$. Using the following random variates from a uniform distribution, generate Poisson variates. Compute the sample mean and standard deviation of the Poisson variates and compare them to the population values. Comment on the differences.

$$u_i = \begin{cases} 0.43, & 0.53, & 0.65, & 0.66, & 0.48, & 0.87, & 0.67, & 0.66, & 0.85, & 0.81, \\ 0.44, & 0.37, & 0.54, & 0.71, & 0.27, & 0.15, & 0.34, & 0.89, & 0.52, & 0.64 \end{cases}$$

7-12. A discrete random variable (N) has a geometric distribution with parameter $p = 0.5$. Using the first ten values of u_i in Problem 7-11, generate values of N using the inverse transformation method. Compare these to values generated with Equation 7-23.

7-13. A discrete random variable X has a Bernoulli distribution with parameter $p = 0.3$. Using the following uniform variates, generate Bernoulli variates using Equation 7-20. Determine the sample proportion and compare it to the population value.

$$u_i = \{0.79,\ 0.34,\ 0.27,\ 0.55,\ 0.06,\ 0.13,\ 0.70,\ 0.18,\ 0.16,\ 0.50,\ 0.19,\ 0.63\}$$

7-14. A random variable that can take on values of x and y has $P(x) = 0.2$ and $P(y) = 1 - P(x) = 0.8$. Assume that an experiment was conducted with three trials per experiment. Use the following 30 uniform variates to generate 10 binomial variates. Compare the sample and population probabilities:

$$u_i = \begin{cases} 0.47, 0.06, 0.02, 0.77, 0.44, 0.73, 0.40, 0.01, 0.53, 0.00, 0.83, 0.71, 0.74, \\ 0.34, 0.04, 0.62, 0.82, 0.90, 0.97, 0.29, 0.61, 0.32, 0.01, 0.18, 0.22, 0.95, \\ 0.67, 0.64, 0.68, 0.67 \end{cases}$$

Chapter 8
Reliability and Risk Analysis

TABLE OF CONTENTS

8.1. INTRODUCTION

The process of engineering analysis and design can be systematically performed within a systems framework. Generally, an engineering project can be modeled to include a segment of its environment that interacts significantly with it, defining an engineering system. The limits of the system are drawn based on the nature of the project, class of performances (including failures) under consideration, and the objectives of the analysis.

The first step in solving any engineering problem is to define the architecture of the system. The definition of a system can be based on observations about its source and data elements, interaction among the elements, and behavior of the system. Each level of knowledge which is obtained about an engineering problem can be said to define a system on the problem. As additional levels of knowledge are added to previous ones, higher epistemological levels of system definition and description are generated which, taken together, form a hierarchy of such system descriptions.

Uncertainties in civil engineering systems can be mainly attributed to ambiguity and vagueness in defining the variables and parameters of the systems and their relations. The ambiguity component is generally due to noncognitive sources. These sources include (1) physical randomness; (2) statistical uncertainty due to the use of limited information to estimate the characteristics of the variables and parameters; and (3) model uncertainties which are due to simplifying assumptions in analytical and prediction models, simplified methods, and idealized representations of real performances. The vagueness-related uncertainty is due to cognitive sources that include (1) the definition of certain variables or parameters, e.g., structural performance (failure or survival), quality, deterioration, skill and experience of construction workers and engineers, environmental impact of projects, conditions of existing structures; (2) other human factors; and (3) defining the interrelationships among the parameters of the problems, especially for complex systems. These uncertainty types were discussed in Chapter 1.

Reliability considerations are important in the analysis and design of engineering projects. The reliability of a project is commonly stated in reference to some performance criteria. The need for the reliability analysis stems from acknowledging the presence of uncertainty in the definition, understanding, modeling, and behavior prediction of the system that models the project. The objective of the analysis is the assurance of some level of reliability. There are numerous sources of uncertainties in an engineered system. The absolute safety of the system cannot be guaranteed due to the unpredictability of the demand conditions for the system, our inability to obtain and express the properties and performances of the system accurately, the use of simplified assumptions in predicting the behavior of the system under consideration, the limitations in the numerical methods used, and human factors, e.g., errors or omissions. However, the likelihood of unacceptable performances can be limited to a reasonable level. Estimation of this likelihood, even when used to compare various design alternatives, is an important task for an engineer. The objective of this chapter is to provide an introduction to the area of reliability assessment.

8.2. TIME TO FAILURE

The early developments in reliability were based on assessing the time to failure of an engineered component, e.g., a mechanical or electrical product. In these applications, the reliability of a component was defined as the probability that the component would perform according to a specified performance criterion for at least a specified period under specified conditions. The performance period T can be defined as the time to failure. Distance, number of cycles, or any other convenient and suitable measure can be used instead of time. The performance period is a random variable that has a probability density function $f_T(t)$ and a cumulative distribution function $F_T(t)$. Therefore, the reliability function $R_T(t)$ can be defined as

$$R_T(t) = \int_t^{\infty} f_T(t)\, dt = 1 - F_T(t) \qquad (8\text{-}1)$$

The expected value (mean) for the time to failure, μ_T, can be computed as

$$\mu_T = E(T) = \int_0^\infty t \, f_T(t) \, dt \qquad (8\text{-}2)$$

The result from Equation 8-2 can be viewed as the expected life. It can be shown that the expected life can be computed as

$$\mu_T = E(T) = \int_0^\infty R_T(t) \, dt \qquad (8\text{-}3)$$

The variance of the time to failure (or life), σ_T^2, can be computed as

$$\sigma_T^2 = \text{Var}(T) = \int_0^\infty (t - \mu_T)^2 f_T(t) \, dt \qquad (8\text{-}4)$$

It can be shown that the variance of life can be computed as

$$\sigma_T^2 = \text{Var}(T) = \int_0^\infty 2t \, R_T(t) \, dt - \left(\int_0^\infty R_T(t) \, dt \right)^2 \qquad (8\text{-}5)$$

Another aspect to the reliability problem in the context of the time to failure is the assessment of *failure rate*. In general, the failure rate of a component can be time variant as the component ages. In reliability analysis, it is common to measure this failure rate with the *hazard function*. The hazard function $h_T(t)$ can be viewed as the probability that the component will fail in the next time increment Δt, given that it survived up to some time t. This conditional probability can be expressed as

$$P(t < T \le t + \Delta T \mid T > t) = \frac{P(t < T \le t + \Delta T)}{P(T > t)} \qquad (8\text{-}6)$$

Dividing both sides of Equation 8-6 with ΔT and taking the limiting case as $\Delta T \to 0$, the hazard function results as

$$h_T(t) = \frac{f_T(t)}{R_T(t)} \qquad (8\text{-}7)$$

Therefore, based on the definition of the hazard function, as the failure rate increases, the hazard function increases.

FIGURE 8-1 The hazard function.

Experimentally developed hazard functions for many components can be modeled as a *bathtub* distribution as shown schematically in Figure 8-1. In this typical bathtub distribution, three regions can be identified, the warranty-failures region, the chance-failures region, and the aging-failures region. The high failure rate (hazard function) in the first region is generally attributed to several factors that are related to manufacturing errors, imperfections, deviations from standards, poor quality, and human factors. This region represents the break-in period and is commonly covered by a warranty by the manufacturer. The hazard function in this region can be reduced by increasing the quality control at the production level. The hazard level in the second region represents the hazard during the service life of the component. In the third region, the hazard level starts to increase again due to aging. During this region, a decision needs to be made concerning replacement, life extension, or repair of the component.

Example 8-1: Time to Failure for a Construction Crane

A major construction company has 100 construction cranes that are within the same size and capacity class, and level of use. The records of the company indicate that the average time between failures for these cranes is 2 years. Assuming that the failure rate is constant, therefore, the hazard function is given by

$$h_T(t) = \lambda = \frac{1}{2} = 0.5 \tag{8-8}$$

where λ = the failure rate. Using Equation 8-2, with $\mu_T = 2$ years, the following $f_T(t)$ results:

$$f_T(t) = 0.5\exp(-0.5t) \tag{8-9}$$

Therefore, the cumulative distribution function is

$$F_T(t) = 1 - \exp(-0.5t) \tag{8-10}$$

FIGURE 8-2 Density, cumulative, and hazard functions.

Equations 8-9 and 8-10 are the density and the cumulative functions for the exponential distribution as described in Section 3.7.4. The reliability function is given by

$$R_T(t) = \exp(-0.5t) \tag{8-11}$$

The density, cumulative, reliability, and hazard functions are shown in Figure 8-2. In this example, the reliability of a crane decreases exponentially as a function of time, with a constant hazard level.

8.3. RELIABILITY OF COMPONENTS

The reliability of a component can be defined as the probability that the component meets some specified demands. For example, the reliability of a structural component, e.g., a beam, can be defined as the probability that the structural strength, e.g., ultimate moment capacity, exceeds the applied loads, e.g., the moment due to the total combined loads. This definition of the reliability R_e can be expressed mathematically as

$$R_e = P(R > L) \tag{8-12}$$

where R = the structural resistance or strength of the component, and L = the load effect expressed in the same units as the resistance R. In general, R can be the resistance, supply, capacity, or input for the component; and L can be the load effect, demand, or needs. Although most of the presented methods in this section were used for structural reliability assessment, they can be used for nonstructural problems.

In general, R and L in Equation 8-12 are functions of several random variables that are called the basic random variables. For example, the moment capacity of a beam is a function of the stress-strain characteristics of the materials that make up the beam, the dimensions of the cross section, and the definition of capacity for the beam. In this case, the variables, such as yield stress, width of the cross section of the beam, and height of the cross section, are basic random variables. In reliability analysis, it is common to define a performance function that describes the performance of the component in meeting the demand. The performance function Z is expressed as

$$Z = R - L \tag{8-13a}$$

or

$$Z = g(X_1, X_2, \ldots, X_n) \tag{8-13b}$$

where X_1, X_2, ..., and X_n are n basic random variables for R and L; and $g(.)$ is a function that describes the relationship between R-L and the basic random variables. The performance function in Equations 8-13 is expressed such that failure of the component results in a negative sign for the function (i.e., $Z < 0$), survival of the component results in a positive sign for the function (i.e., $Z > 0$), and limit-state for the beam results in $Z = 0$.

The probability of failure P_f for the component is defined as

$$P_f = P(R < L) \tag{8-14a}$$

$$= 1 - R_e \tag{8-14b}$$

In terms of the performance function, the probability of failure is expressed as

$$P_f = P(Z < 0) \tag{8-15a}$$

$$= P\left[g(X_1, X_2, \ldots, X_n) < 0 \right] \tag{8-15b}$$

For the general case, where the basic random variables can be correlated, the probability of failure can be determined by solving the following integral:

$$P_f = \int \cdots \int_{\text{over } Z \le 0} f_{\underline{X}}(x_1, x_2, \ldots, x_n) \, dx_1 \, dx_2 \ldots dx_n \tag{8-16}$$

where $f_{\underline{X}}$ = joint probability density function (PDF) of the random vector $\underline{X}$ = $\{X_1, X_2, \ldots, X_n\}$; and the integration is performed over the region where $Z = g(.) < 0$.

The determination of the probability of failure by evaluating the integral of Equation 8-16 can be a difficult task.

In this section, methods for assessing the reliability of components are presented and discussed. These methods include the first-order reliability method (FORM), the advanced second-moment (ASM) method, direct Monte Carlo simulation (MCS) method, and variance reduction techniques (VRT) in simulation.

8.3.1. First-Order Reliability Method

In order to avoid some computational difficulties of Equation 8-16 in the evaluation of the probability of failure, the FORM is used in structural reliability analysis. The approximate mean and standard deviation of Z in Equation 8-13 (μ_Z and σ_Z, respectively) are estimated using the information on means, μ_{X_i}, and standard deviations, σ_{X_i}, or coefficient of variations (δ_{X_i}) of the basic random variables, X_i. Expanding $g(.)$ in Equation 8-13 using a Taylor series expansion about the mean values of the basic random variables and truncating the series at the first-order terms, the first-order approximate mean, and variance of Z are, respectively, given by

$$\mu_Z \cong g\left(\mu_{X_1}, \mu_{X_2}, \ldots, \mu_{X_n}\right) \tag{8-17}$$

and

$$\sigma_Z^2 \cong \sum_{i=1}^{n} \sum_{j=1}^{n} \left(\frac{\partial Z}{\partial X_i}\right)_\mu \left(\frac{\partial Z}{\partial X_j}\right)_\mu \mathrm{Cov}\left(X_i, X_j\right) \tag{8-18}$$

where the partial derivatives of Z are evaluated at the mean values of the basic random variables; $\mathrm{Cov}(X_i, X_j)$ is the covariance of X_i and X_j. For statistically uncorrelated random variables, Equation 8-18 can be expressed as

$$\sigma_Z^2 \cong \sum_{i=1}^{n} \left(\frac{\partial Z}{\partial X_i}\right)_\mu^2 \sigma_{X_i}^2 \tag{8-19}$$

The second-order mean (considering the square term in the Taylor series expansion) can be used to improve the accuracy of the estimation of the mean as follows:

$$\mu_Z \cong g\left(\mu_{X_1}, \mu_{X_2}, \ldots, \mu_{X_n}\right) + \frac{1}{2} \sum_{i=1}^{n} \sum_{j=1}^{n} \left(\frac{\partial^2 Z}{\partial X_i \, \partial X_j}\right)_\mu \mathrm{Cov}\left(X_i, X_j\right) \tag{8-20}$$

Again the partial derivatives are evaluated at the mean values of the basic random variables. The second-order variance can be similarly computed. For practical purposes, the use of the second-order mean and the first-order variance can be adequate for some engineering applications.

A measure of reliability can be defined by introducing a parameter β, called the reliability index, as

$$\beta = \frac{\mu_Z}{\sigma_Z} \tag{8-21}$$

If Z is assumed to have a normal probability distribution, the probability of failure P_f is given by

$$P_f = 1 - \Phi(\beta) \tag{8-22}$$

in which Φ = cumulative probability distribution function of the standard normal variate (Appendix A-1) and β is called the reliability index.

An illustrative special case for the structural reliability problem is given for the following performance function:

$$Z = R - L \tag{8-23}$$

This two-random variable case is shown in Figure 8-3. The performance region (sample space) in this figure is divided into two mutually exclusive and collectively exhaustive events, i.e., failure and survival. Assuming normal probability distributions for R and L results in Z being normally distributed with the following mean and variance, respectively:

$$\mu_Z = \mu_R - \mu_L \tag{8-24}$$

and

$$\sigma_Z^2 = \sigma_R^2 + \sigma_L^2 \tag{8-26}$$

where μ = mean value and σ = standard deviation. The random variables R, L, and Z are shown in Figure 8-4. The probability of failure according to this figure can be computed as the area under $f_Z(z)$ for $z < 0$. This area equals the cumulative distribution function of Z evaluated at $z = 0$, i.e.,

$$P_f = P(Z < 0) \tag{8-26}$$

or

$$P_f = F_Z(0) \tag{8-27}$$

FIGURE 8-3 Performance space for special case.

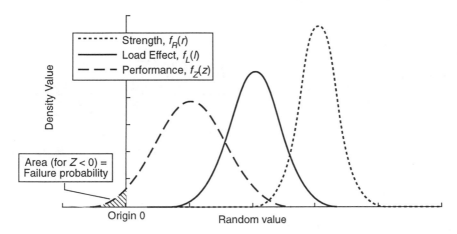

FIGURE 8-4 Random variables for special case.

Since Z is normally distributed, the probability of failure can be computed as

$$P_f = \Phi\left(\frac{0 - \mu_Z}{\sigma_Z}\right) \qquad (8\text{-}28a)$$

$$= \Phi(-\beta) = 1 - \Phi(\beta) = \Phi\left(-\frac{\mu_R - \mu_L}{\sqrt{\sigma_R^2 + \sigma_L^2}}\right) \qquad (8\text{-}28b)$$

The reliability index according to Equation 8-28b is given by

$$\beta = \frac{\mu_R - \mu_L}{\sqrt{\sigma_R^2 + \sigma_L^2}} \qquad (8\text{-}29)$$

Another illustrative special case for the structural reliability problem is given for the case where R and L have lognormal probability distributions. In this case, the performance function should be expressed as

$$Z = \frac{R}{L} \tag{8-30}$$

where failure is defined as $Z < 1$, survival is $Z > 1$, and the limit state is $Z = 1$. In this case, Z is lognormally distributed. It can be shown that the probability of failure can be computed as

$$P_f = P(Z < 1) \tag{8-31a}$$

resulting into the following expression:

$$P_f = \Phi(-\beta) = \Phi\left(\frac{-\ln\left(\frac{\mu_L}{\mu_R}\sqrt{\frac{\delta_R^2 + 1}{\delta_L^2 + 1}}\right)}{\sqrt{\ln\left[\left(\delta_R^2 + 1\right)\left(\delta_L^2 + 1\right)\right]}}\right) \tag{8-31b}$$

where μ = mean value and δ = coefficient of variation. The reliability index in this case is given by

$$\beta = \frac{\ln\left(\frac{\mu_L}{\mu_R}\sqrt{\frac{\delta_R^2 + 1}{\delta_L^2 + 1}}\right)}{\sqrt{\ln\left(\left(\delta_R^2 + 1\right)\left(\delta_L^2 + 1\right)\right)}} \tag{8-32}$$

The first-order reliability method has some shortcomings. According to this method, the function $g(.)$ in Equation 8-13b is linearized at the mean values of the basic random variables. Where $g(.)$ is nonlinear, significant error may be introduced by neglecting higher-order terms. This approximate method also may yield different results for different mechanically equivalent formulations of the same problem, e.g., stress formulation vs. strength formulation. Moreover, the method completely ignores the information on distributions of the basic random variables. In other words, the method results in accurate reliability assessments only where the basic random variables are normally distributed and the function $g(.)$ is linear.

Example 8-2: Tensile Reliability of a Truss Member

A tension member in a truss has an ultimate tensile strength T with a mean value of 120 kips and a standard deviation of 10 kips. The tension load L in this member has a mean value of 80 kips and standard deviation of 20 kips. Normal probability distributions are assumed for T and L. The reliability index β for the member can be computed using Equation 8-29 as

$$\beta = \frac{120 - 80}{\sqrt{10^2 + 20^2}} = 1.79 \qquad (8\text{-}33)$$

Therefore, the probability of failure according to Equation 8-28b is

$$P_f = \Phi(-1.79) = 1 - \Phi(-1.79) = 0.03673 \qquad (8\text{-}34)$$

Example 8-3: Flexural Reliability of a Beam

The following performance function describes the flexural behavior of a simply supported beam of a span length $\sqrt{2}\,L$ that carries a uniform load W:

$$Z = YS - \frac{WL^2}{4} \qquad (8\text{-}35)$$

where Y = yield stress of the material of the beam and S = elastic section modulus. In this example, failure is defined as yielding at the extreme material fibers of the cross section of the beam. The mean values and standard deviations of the variables are given in Table 8-1.

TABLE 8-1 Mean Values and Standard Deviations for Random Variables Used in Example 8-3

Random variable	Mean value	Standard deviation	Coefficient of variation
Y	38 ksi	1.9 ksi	0.05
S	100 in.^{3}	5 in.^{3}	0.05
W	0.3 kips/in.	0.075 kips/in.	0.25
L	180 in.	9 in.	0.05

The following first-order mean of Z can be computed by substituting the mean values of the basic random variables in Equation 8-35:

$$\mu_Z \cong (38)(100) - \frac{(0.3)(180)^2}{4} = 1370 \text{ kip-in.} \qquad (8\text{-}36)$$

Assuming the random variables to be statistically uncorrelated, the first-order variance is computed using Equation 8-19 as

$$\sigma_Z^2 \cong (100)(1.9)^2 + (38)(5)^2 + \left(\frac{(180)^2}{4}\right)^2 (0.075)^2 + \left(\frac{(0.3)(180)}{4}\right)^2 (9)^2 \qquad (8\text{-}37)$$

$$\cong 429,416.25 \text{ (kip-in.)}^2$$

Therefore,

$$\sigma_Z \cong 655.30 \text{ kip-in.}$$

The coefficient of variation is $655.30/1370 = 0.478$. A measure of reliability can be computed using Equation 8-21 as

$$\beta = \frac{\mu_Z}{\sigma_Z} = 2.091 \tag{8-38}$$

If Z is assumed to have a normal probability distribution, the probability of failure P_f is given by

$$P_f = 1 - \Phi(2.091) = 0.01835 \tag{8-39}$$

8.3.2. Advanced Second-Moment Method

According to this method, the basic random variables are transformed into a reduced (or normalized) coordinate system $\underline{X}' = \{X'_1, X'_2, ..., X'_n\}$ according to the following transformation:

$$X'_i = \frac{X_i - \mu_{X_i}}{\sigma_{X_i}} \qquad \text{for } i = 1, 2, ..., n \tag{8-40}$$

It should be noted that the resulting reduced variables have the following mean and standard deviation:

$$\mu_{X'_i} = 0$$

and

$$\sigma_{X'_i} = 1$$

For the following fundamental performance function:

$$Z = R - L \tag{8-41}$$

The transformation results in the following expression for the performance function:

$$Z = R'\sigma_R + \mu_R - \left(L'\sigma_L + \mu_L\right) \tag{8-42}$$

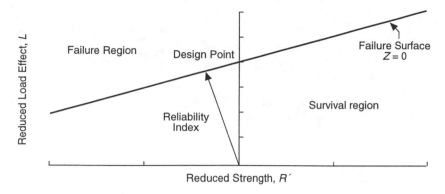

FIGURE 8-5 Performance space in reduced coordinates.

The limit state (i.e., $Z = 0$) can be expressed as

$$L' = R'\frac{\sigma_R}{\sigma_L} + \frac{\mu_R - \mu_L}{\sigma_L} \qquad (8\text{-}43)$$

The resulting performance function in the reduced coordinate system is shown in Figure 8-5. By comparing Figures 8-3 and 8-5, it is evident that the limit state in the reduced coordinates does not pass through the origin. It can be shown that in the reduced coordinates, the shortest distance from the origin to the limit state is given by

$$\beta = \frac{\mu_R - \mu_L}{\sqrt{\sigma_R^2 + \sigma_L^2}} \qquad (8\text{-}44)$$

The resulting distance is the same as the reliability index given by Equation 8-29. Therefore, it can be stated that a measure of safety in the form of the reliability index β is defined by the advanced second-moment method as the shortest distance from the origin to the failure surface (limit state) in the reduced coordinates. The reliability index according to this definition is commonly referred to as the Hasofer-and-Lind index. This definition of β is valid for a nonlinear performance function of the type

$$Z = g\left(X_1, X_2\right) \qquad (8\text{-}45)$$

such that the failure region is for $Z < 0$, survival region is for $Z > 0$, and the failure surface (limit state) is $Z = 0$. The nonlinear case of two random variables is shown in Figure 8-6. The point on the failure surface that corresponds to the shortest distance is called the design (or failure) point. It can be shown that it is

FIGURE 8-6 Performance space in reduced coordinates — nonlinear case.

the most likely failure point. This point has some significance in defining partial safety factors in reliability-based design codes. For the nonlinear case, it is possible to obtain the β value by approximating the failure surface by a linear tangent at the design point. Numerically, the approximation can be achieved by a first-order Taylor series approximation of the failure surface at the design point. The procedure is similar to the Taylor series expansion performed in Equations 8-17 and 8-18, with the expansion point being the design point instead of the mean values. The resulting reliability index based on this approximation of the failure surface should be exactly the same as the result from the case without approximation (i.e., nonlinear failure surface).

For the general case where the nonlinear performance function is

$$Z = g(X_1, X_2, \ldots, X_n) \tag{8-46}$$

the reliability index has the same definition as the shortest distance from the origin to the failure surface in the reduced coordinates. Mathematically, this definition can be expressed as

$$\beta = \min_{\underline{x}'} \sqrt{\sum_{i=1}^{n} x_i'^2} \quad \text{such that } g(\underline{x}') = 0 \tag{8-47}$$

where the square root term in this equation is the distance in the reduced coordinates; and the point $\underline{x}'$ corresponds to the design point and results in the least (shortest) distance. It is common to use the notation $\underline{x}^*$ for the design point in the regular coordinate and $\underline{x}'^*$ in the reduced coordinates. Equation 8-47 can be solved using one of several nonlinear optimization methods. For example, the solution can be achieved iteratively by simultaneously solving the following set of $n + 1$ equations:

$$\alpha_i^* = \frac{\left(\dfrac{\partial g}{\partial X_i'}\right)_*}{\sqrt{\displaystyle\sum_{i=1}^{n}\left(\dfrac{\partial g}{\partial X_i'}\right)_*^2}} \qquad \text{for } i = 1, 2, \ldots, n \qquad (8\text{-}48a)$$

$$x_i'^* = -\alpha_i^* \beta \qquad \text{for } i = 1, 2, \ldots, n \qquad (8\text{-}48b)$$

$$Z = g\left(x_1'^*, x_2'^*, \ldots, x_n'^*\right) = 0 \qquad (8\text{-}48c)$$

Using the following relation between a random variable and its reduced form:

$$x_i^* = \mu_{X_i} + \sigma_{X_i} x_i'^* = \mu_{X_i} - \alpha_i^* \sigma_{X_i} \beta \qquad (8\text{-}49)$$

Equation 8-48c can be expressed as

$$Z = g\left[\left(\mu_{X_1} - \alpha_1^* \sigma_{X_1} \beta\right), \left(\mu_{X_2} - \alpha_2^* \sigma_{X_2} \beta\right), \ldots, \left(\mu_{X_n} - \alpha_n^* \sigma_{X_n} \beta\right)\right] = 0 \qquad (8\text{-}50)$$

A subscript or superscript of * means the evaluation of the noted quantity (or term) at the design point. The set of equations represents a first-order (linear) approxima-tion of the failure surface. The α_i' values are the directional cosines of the tangent hyperplane in the reduced coordinates. The directional cosines are indicators of the sensitivity of the performance function to variations in the corresponding basic random variables. The solution of Equations 8-48 and 8-50 can be achieved by performing the following computational steps:

1. Assume an initial value for the design point. It is common to start with the mean values of the basic random variables. The design point in the reduced coordinates should then be computed using

$$x_i'^* = \frac{x_i - \mu_{X_i}}{\sigma_{X_i}} \qquad \text{for } i = 1, 2, \ldots, n \qquad (8\text{-}51)$$

2. Evaluate the directional cosines at the failure point. The partial derivatives that are needed for computing the directional cosines (Equation 8-48a) can be obtained as

$$\left(\frac{\partial g}{\partial X_i'}\right)_* = \left(\frac{\partial g}{\partial X_i} \frac{\partial X_i}{\partial X_i'}\right)_* = \left(\frac{\partial g}{\partial X_i}\right)_* \sigma_{X_i} \qquad \text{for } i = 1, 2, \ldots, n \qquad (8\text{-}52)$$

3. Solve the following equation for the root β:

$$g\left[\left(\mu_{X_1} - \alpha_1^*\sigma_{X_1}\beta\right), \left(\mu_{X_2} - \alpha_2^*\sigma_{X_2}\beta\right), ..., \left(\mu_{X_n} - \alpha_n^*\sigma_{X_n}\beta\right)\right] = 0 \qquad (8\text{-}53)$$

4. Using the β obtained from step 3, evaluate a new design point using the following equation:

$$x_i^* = \mu_{X_i} - \alpha_i^*\sigma_{X_i}\beta \qquad (8\text{-}54)$$

5. Repeat steps 1 to 4 until convergence of β is obtained.

The important relation between the probability of failure and the reliability index can be maintained as

$$P_f = 1 - \Phi(\beta) \qquad (8\text{-}55)$$

Equation 8-55 results in the exact probability if the performance function is linear and the basic random variables are normally distributed. For the case of a nonlinear performance function, the error depends on the level of nonlinearity of the performance function. This error is due to approximating the performance function by a tangent hyperplane at the design point. Although the reliability index does not have any error, P_f is approximate. In reliability problems, it is common to deal with nonlinear performance functions with a relatively small level of nonlinearity, resulting in very small errors in the estimated P_f. Therefore, in these cases, for all practical purposes, P_f is of sufficient accuracy.

Example 8-4: Flexural Reliability of a Beam — Nonlinear Case

The performance function describing the flexural behavior of a simply supported beam of a span length $\sqrt{2}\,L$ that carries a uniform load W was discussed in Example 8-3. It is given by

$$Z = YS - \frac{WL^2}{4}$$

where Y = yield stress of the material of the beam and S = elastic section modulus. In this example, failure is defined as yielding at the extreme material fibers of the cross section of the beam. The probabilistic characteristics of the variables are given in Table 8-2.

TABLE 8-2 Probabilistic Characteristics of Random Variables in Example 8-4

Random variable	Mean value	Standard deviation	Coefficient of variation	Distribution type
Y	38 ksi	1.9 ksi	0.05	Normal
S	100 in.^{3}	5 in.^{3}	0.05	Normal
W	0.3 kips/in.	0.075 kips/in.	0.25	Normal
L	180 in.	9 in.	0.05	Normal

In this case, the random variables are assumed to have normal probability distributions. The partial derivatives of the performance function with respect to the reduced variables are

$$\left(\frac{\partial g}{\partial Y'}\right)_* = s^* \sigma_Y \tag{8-56a}$$

$$\left(\frac{\partial g}{\partial S'}\right)_* = y^* \sigma_S \tag{8-56b}$$

$$\left(\frac{\partial g}{\partial w'}\right)_* = \frac{-l^{*2}}{4} \sigma_w \tag{8-56c}$$

$$\left(\frac{\partial g}{\partial L'}\right)_* = \frac{-w^* l^*}{2} \sigma_L \tag{8-56d}$$

Now, the previously described iterative steps can be performed as shown in Table 8-3. The solution in this case converges in two iterations. Therefore, the probability of failure is

$$P_f = 1 - \Phi(1.83653) = 0.0341 \tag{8-56e}$$

The FORM solution in Example 8-3 produced a probability of failure of 0.01835. Therefore, the relative error of the FORM solution is –46%.

Nonnormal probability distributions — The advanced second-moment method deals with nonnormal probability distributions for the basic random variables by determining an equivalent normal distribution at the design point. In the iterative solution, this is performed at each iteration. The equivalent normal distribution at the design point x^* for a nonnormal basic random variable X with a density function $f_X(x)$ and cumulative distribution function $F_X(x)$ is defined by the following mean μ_X^N and standard deviation σ_X^N:

TABLE 8-3 Iterations Using the Advanced Second-Moment Method for Example 8-4

Iteration number	Random variable	Design point	$\left(\dfrac{\partial g}{\partial X'}\right)_*$	Directional cosine, α_i^*	New design point that satisfies, $g(.) = 0$, according to Equation 8-53
1	Y	38	190	0.26862	37.05645020
	S	100	190	0.26862	97.51697430
	W	0.3	−607.5	−0.85890	0.41908722
	L	180	−243	−0.34350	185.71618600
					0.00078733 = sum
				Therefore,	β = 1.84874
2	Y	37.056450	185.28	0.23732	37.17189830
	S	97.516970	185.28	0.23732	97.82078500
	W	0.419087	−646.70	−0.82830	0.41409282
	L	185.716200	−350.24	−0.44860	187.41490300
					−0.00056764 = sum
				Therefore,	β = 1.83653

$$\mu_X^N = x^* - \Phi^{-1}\left(F_X\left(x^*\right)\right)\sigma_X^N \tag{8-57}$$

and

$$\sigma_X^N = \frac{\phi\left[\Phi^{-1}\left(F_X\left(x^*\right)\right)\right]}{f_X\left(x^*\right)} \tag{8-58}$$

where $\Phi(z)$ is the cumulative distribution function for the standard normal evaluated at z, and $\phi(z)$ is the density function of the standard normal. The computational steps in this case are revised as follows:

1. Assume an initial value for the design point. It is common to start with the mean values of the basic random variables. The design point in the reduced coordinates should then be computed using

$$x_i'^* = \frac{x_i - \mu_{X_i}}{\sigma_{X_i}} \qquad \text{for } i = 1, 2, \ldots, n \tag{8-59}$$

2. Evaluate the equivalent normal distributions for the nonnormal basic random variables at the design point using Equations 8-57 and 8-58.
3. Evaluate the directional cosines at the design point. The partial derivatives that are needed for computing the directional cosines (Equation 8-48a) can be obtained in this case as

$$\left(\frac{\partial g}{\partial X_i'}\right)_* = \left(\frac{\partial g}{\partial X_i}\frac{\partial X_i}{\partial X_i'}\right)_* = \left(\frac{\partial g}{\partial X_i}\right)_* \sigma_{X_i}^N \qquad \text{for } i = 1, 2, \ldots, n \tag{8-60}$$

4. Solve the following equation for the root β:

$$Z = g\left[\left(\mu_{X_1}^N - \alpha_1^*\sigma_{X_1}^N\beta\right), \left(\mu_{X_2}^N - \alpha_2^*\sigma_{X_2}^N\beta\right), ..., \left(\mu_{X_n}^N - \alpha_n^*\sigma_{X_n}^N\beta\right)\right] = 0 \qquad (8\text{-}61)$$

5. Using the β obtained from step 4, evaluate a new design point using the following equation:

$$x_i^* = \mu_{X_i}^N - \alpha_i^*\sigma_{X_i}^N\beta \qquad (8\text{-}62)$$

6. Repeat steps 1 to 5 until convergence of β is obtained.

The important relation between the probability of failure and the reliability index can be maintained as

$$P_f = 1 - \Phi(\beta) \qquad (8\text{-}63)$$

Example 8-5: Flexural Reliability of a Beam — Nonlinear Nonnormal Case

The performance function for the flexural behavior of a simply supported beam of a span length $\sqrt{2}\,L$ that carries a uniform load W was discussed in Examples 8-3 and 8-4. It is given by

$$Z = YS - \frac{WL^2}{4}$$

where Y = yield stress of the material of the beam and S = elastic section modulus. Failure was defined as yielding at the extreme material fibers of the cross section of the beam. The mean values, standard deviations, and distribution types of the variables are given in Table 8-4.

TABLE 8-4 Probabilistic Characteristics of Random Variables Used in Example 8-5

Random variable	Mean value	Standard deviation	Coefficient of variation	Distribution type
Y	38 ksi	1.9 ksi	0.05	Lognormal
S	100 in.^{3}	5 in.^{3}	0.05	Lognormal
W	0.3 kips/in.	0.075 kips/in.	0.25	Lognormal
L	180 in.	9 in.	0.05	Lognormal

Now, the previously described iterative steps can be performed as shown in Table 8-5 to obtain the reliability index β. The parameters of the lognormal distributions for Y, S, W, and L are given by

Random variable	Mean value	Standard deviation	First parameter (μ_Y)	Second parameter (σ_Y)
Y	38 ksi	1.9 ksi	3.634	0.04997
S	100 in.³	5 in.³	4.604	0.04997
W	0.3 kips/in.	0.075 kips/in.	−1.234	0.2462
L	180 in.	9 in.	5.192	0.04997

Now the equivalent normal can be computed for these nonnormal random variables as shown in Table 8-5. The starting design point is assumed to be the mean values of the random variables. Two iterations are needed in this case to obtain the reliability index. Therefore, the probability of failure is

$$P_f = 1 - \Phi(\beta) = 1 - \Phi(1.7526) = 0.0398$$

The assumption of normal probability distributions for all random variables in Example 8-4 produced a probability of failure of 0.0341. Therefore, in this case the relative difference due to this assumption is −14%.

TABLE 8-5 Iterations Using the Advanced Second-Moment Method for Example 8-5

Iteration number	Random variable	Mean for equivalent normal	Standard deviation for equivalent normal	Directional cosine, α_i^*	New design point
1	Y	37.9526	1.8988	0.27275	36.930
	S	99.8752	4.9969	0.27275	97.190
	W	0.2909	0.0739	−0.85840	0.416
	L	179.7753	8.9944	−0.33830	185.780
				Therefore,	$\beta = 1.97252$
2	Y	37.9390	1.8450	0.20587	37.27290
	S	99.8390	4.8560	0.20587	98.08650
	W	0.2674	0.1024	−0.92370	0.43321
	L	179.6760	9.2830	−0.24920	183.73050
				Therefore,	$\beta = 1.7526$

Example 8-6: Flexural Reliability of a Beam — Nonlinear Nonnormal Case

The performance function described in Example 8-5 is used herein. In this example, the mean values, standard deviations, and distribution types of the variables are given in Table 8-6. The difference between this example and Example 8-5 is in the distribution types. In this example, a mixture of distribution types is used. The extreme value (largest) type I distribution is a continuous probability distribution with the following density and cumulative functions:

TABLE 8-6 Probabilistic Characteristics of Random Variables Used in Example 8-6

Random variable	Mean value	Standard deviation	Coefficient of variation	Distribution type
Y	38 ksi	1.9 ksi	0.05	Normal
S	100 in.^{2}	5 in.^{2}	0.05	Lognormal
W	0.3 kips/in.	0.075 kips/in.	0.25	Extreme value Type I — largest
L	180 in.	9 in.	0.05	Normal

$$f_X(x) = \alpha e^{-\alpha(x-u)} \exp\left[-e^{-\alpha(x-u)}\right] \qquad (8\text{-}64a)$$

$$F_X(x) = \exp\left[-e^{-\alpha(x-u)}\right] \qquad (8\text{-}64b)$$

where α and u are the parameters of the distribution. The mean and variance of X in terms of these parameters are given, respectively, by

$$\mu_X = u + \frac{0.577216}{\alpha} \qquad (8\text{-}65a)$$

$$\sigma_X^2 = \frac{\pi^2}{6\alpha^2} \qquad (8\text{-}65b)$$

The parameters as functions of the mean and variance are

$$u = \mu_X - \frac{0.577216}{\alpha} = 0.2662 \qquad (8\text{-}65c)$$

$$\alpha = \frac{\pi}{\sqrt{6}\,\sigma_X} = 17.10 \qquad (8\text{-}65d)$$

The iterative steps can now be performed as shown in Table 8-7 to obtain the reliability index β. Therefore, the probability of failure is

$$P_f = 1 - \Phi(\beta) = 1 - \Phi(1.7155) = 0.043128$$

8.3.3. Simulation Methods

Monte Carlo simulation techniques can be used to estimate the probabilistic characteristics of the functional relationship Z in Equation 8-13. Monte Carlo, or

TABLE 8-7 Iterations Using the Advanced Second-Moment Method for Example 8-6

Iteration number	Random variable	Mean for equivalent normal	Standard deviation for equivalent normal	Directional cosine, α_i^*	New design point
1	Y	38.	1.9	0.2776	36.92
	S	99.8752	4.9969	0.2774	97.04
	W	0.2873	0.0717	-0.8485	0.4118
	L	180.	9.	-0.3550	186.5
				Therefore,	$\beta = 2.04694$
2	Y	38.	1.9	0.1712	37.44
	S	99.8340	4.849	0.1663	98.45
	W	0.2521	0.1134	-0.9165	0.4309
	L	180.	9.	-0.3210	185.0
				Therefore,	$\beta = 1.7198$
3	Y	38.	1.9	0.1667	37.46
	S	99.86	4.919	0.1641	98.48
	W	0.2416	0.1205	-0.9182	0.4314
	L	180.0	9.	-0.3196	184.9
				Therefore,	$\beta = 1.7155$

direct simulation, consists of drawing samples of the basic variables according to their probabilistic characteristics and then feeding them into the performance function. It is known that failure occurs where $g(.) < 0$, then an estimate of the probability of failure, P_f, can be found by

$$P_f = \frac{N_f}{N} \qquad (8\text{-}66)$$

where N_f is the number of simulation cycles where $g(.) < 0$ and N is the total number of simulation cycles. As N approaches infinity, P_f approaches the true probability of failure. The accuracy of Equation 8-66 can be evaluated in terms of its bias and variance. The estimator of the failure probability according to Equation 8-66 is unbiased. For a small probability of failure and/or a small number of simulation cycles, the variance of P_f can be quite large. Consequently, it may take a large number of simulation cycles to achieve a specified accuracy, especially since the probability of failure is unknown. The variance of the estimated probability of failure can be computed by assuming each simulation cycle to constitute a Bernoulli trial. Therefore, the number of failures in N trials can be considered to follow a binomial distribution. Then the variance of the estimated probability of failure can be approximately computed as

$$\mathrm{Var}\left(P_f\right) \cong \frac{\left(1 - P_f\right)P_f}{N} \qquad (8\text{-}67)$$

It is recommended to measure the statistical accuracy of the estimated probability of failure by computing its coefficient of variation as

$$\text{VOV}\left(P_f\right) \cong \frac{\sqrt{\dfrac{\left(1 - P_f\right)P_f}{N}}}{P_f} \qquad (8\text{-}68)$$

It is evident from Equations 8-67 and 8-68 that as N approaches infinity, $\text{Var}(P_f)$ and $\text{COV}(P_f)$ approach zero.

Example 8-7: Flexural Reliability of a Beam — Nonlinear Normal Case

The performance function describing the flexural behavior of a simply supported beam of a span length $\sqrt{2}\ L$ that carries a uniform load W was discussed in Examples 8-3 and 8-4. It is given by

$$Z = YS - \frac{WL^2}{4}$$

where Y = yield stress of the material of the beam and S = elastic section modulus. In this example, failure is defined as yielding at the extreme material fibers of the cross section of the beam. The probabilistic characteristics of the random variables are given in Table 8-2.

Using direct Monte Carlo simulation, the random variables Y, S, W, and L were randomly generated and substituted in the performance function. Failures were then counted by monitoring the sign of the resulting evaluations of the performance function (negative means failure). Then the probability of failure was estimated as N_f/N. Also, the $\text{COV}(P_f)$ was computed using Equation 8-68. The number of simulation cycles N was varied to illustrate convergence of the simulation process. The results are shown in Figure 8-7 using nonarithmetic scale for the number of simulation cycles (N) in order to show the effect of N on P_f and $\text{COV}(P_f)$ at small as well as large N values.

In the classical use of simulation-based methods, all the basic random variables are randomly generated and Equation 8-13 is evaluated. Failures are then counted depending on the resulting sign of Equation 8-13. The probability of failure is estimated as the ratio of the number of failures to the total number of simulation cycles. Therefore, for smaller probabilities of failure, larger numbers of simulation cycles are needed to estimate the probability of failure within an acceptable level of statistical error. The amount of computer time needed for this method is relatively large, whereas the computational effort is relatively small.

FIGURE 8-7 Results for Example 8-7.

The efficiency of simulation can be improved considerably by using variance reduction techniques. However, the level of computational difficulty increases. Importance sampling, and conditional expectation combined with antithetic variates variance reduction techniques are commonly used for structural reliability assessment. These methods were determined to be highly efficient and converge to the correct probability of failure in a relatively small number of simulation cycles.

Importance sampling (IS) method — The probability of failure of a structure according to the performance function of Equation 8-13 is provided by the integral of Equation 8-16. In evaluating this integral with direct simulation, the efficiency of the simulation process depends on the magnitude of the probability of failure, i.e., the location of the most likely failure point or design point as defined in Section 8.5.2. The deeper the location of the design point in the failure domain, the larger the needed simulation effort to obtain failures; in other words, smaller failure probabilities require larger numbers of simulation cycles. This deficiency can be addressed by using importance sampling. In this method, the basic random variables are generated according to some carefully selected probability distributions with mean values that are closer to the design point than their original (actual) probability distributions. It should be noted that the design point is not known in advance; the analyst can only guess. Therefore, failures are obtained more frequently and the simulation efficiency is increased. To compensate for the change in the probability distributions, the results of the simulation cycles should be corrected. The fundamental equation for this method is given by

$$P_f = \frac{1}{N} \sum_{i=1}^{N} I_f \frac{f_{\underline{X}}\left(x_{1i}, x_{2i}, \ldots, x_{ni}\right)}{h_{\underline{X}}\left(x_{1i}, x_{2i}, \ldots, x_{ni}\right)} \qquad (8\text{-}69)$$

where N = number of simulation cycles, $f_{\underline{X}}(x_{1i}, x_{2i}, ..., x_{ni})$ = the original joint density function of the basic random variables evaluated at the ith generated values of the basic random variables, $h_{\underline{X}}(x_{1i}, x_{2i}, ..., x_{ni})$ = the selected joint density function of the basic random variables evaluated at the ith generated values of the basic random variables, and I_f = failure indicator function that takes values of either 1 for failure and 0 for survival. In Equation 8-69, $h_{\underline{X}}(\underline{x})$ is called the sampling density function or the importance function.

Efficiency (and thus the required number of simulation cycles) depends on the choice of the sampling density function. A number of procedures for selecting the sampling density functions have been suggested and are beyond the scope of this chapter.

Conditional expectation (CE) method — The performance function for a fundamental structural reliability assessment case is given by

$$Z = R - L \tag{8-70a}$$

where R = function of structural strength or resistance and L = function of the corresponding load effect. Therefore, the probability of failure, P_f, is given by

$$P_f = P(Z < 0) = P(R < L) \tag{8-70b}$$

For a randomly generated value of L (or R), say, l_i (or r_i), the probability of failure is given by, respectively,

$$P_{f_i} = \text{Prob}(R < l_i) = F_R(l_i) \tag{8-71a}$$

or

$$P_{f_i} = \text{Prob}(L > r_i) = 1 - F_L(r_i) \tag{8-71b}$$

where F_R and F_L = cumulative distribution functions of R and L, respectively. In this formulation R and L are assumed to be statistically uncorrelated random variables. Thus, for N simulation cycles, the mean value of the probability of failure is given by the following equation:

$$\overline{P}_f = \frac{\displaystyle\sum_{i=1}^{N} P_{f_i}}{N} \tag{8-72}$$

The variance (Var) and the coefficient of variation (COV) of the estimated (i.e., sample) probability of failure are given, respectively, by

$$\text{Var}\left(\overline{P}_f\right) = \frac{\sum_{i=1}^{N}\left(P_{f_i} - \overline{P}_f\right)^2}{N(N-1)} \qquad \text{(8-73a)}$$

and

$$\text{COV}\left(\overline{P}_f\right) = \frac{\sqrt{\text{Var}\left(\overline{P}_f\right)}}{\overline{P}_f} \qquad \text{(8-73b)}$$

For simplicity of notation, $\text{COV}(\overline{P}_f)$ sometimes is referred to as $\text{COV}(P_f)$.

For the general performance function given by Equation 8-13, the conditional expectation method can be utilized by randomly generating all the basic random variables except one variable, called the control variable X_k. The randomly generated variables should be selected as the ones of least variability, and the resulting conditional expectation can be evaluated by some known expression, e.g., the cumulative distribution function of the control random variable that was not randomly generated. This method can be used for any performance function with any probability distributions for the random variables. The only limitation is that the control random variable, X_k, must be statistically uncorrelated to the $g_k(.)$ as shown in the following equation:

$$P_f = E_{X_i: i=1,2,\dots,n \& i \neq k}\left[F_{X_k}\left(g_k(X_i: i=1,\ 2,\ \dots,\ n \& i \neq k)\right)\right] \qquad \text{(8-74a)}$$

where $g_k(X_i: i = 1, 2, \dots, n \& i \neq k)$ is defined as follows:

$$P_f = \text{Prob}\left[X_k < g_k(X_i: i=1,\ 2,\ \dots,\ n \& i \neq k)\right] \qquad \text{(8-74b)}$$

According to this method, the variance of the estimated quantity is reduced by removing the variability of the control variable from the simulation process. In addition, the method converges to the correct probability of failure in a relatively small number of simulation cycles.

Example 8-8: Flexural Reliability of a Beam — Nonlinear Normal Case

The performance function for the flexural behavior of a simply supported beam of a span length $\sqrt{2}\,L$ that carries a uniform load W was discussed in Examples 8-3, 8-4, and 8-7. It is given by

$$Z = YS - \frac{WL^2}{4}$$

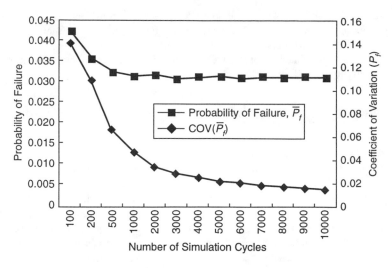

FIGURE 8-8 Results for Example 8-8.

where W = uniform load, L = span length, Y = yield stress of the material of the beam, and S = elastic section modulus. In this example, failure is defined as yielding at the extreme material fibers of the cross section of the beam. The probabilistic characteristics of the random variables are given in Table 8-2.

Using conditional expectation with W as the control variable, Y, S, and L were randomly generated, and the cumulative distribution function of W was used to compute the probability of failure at each simulation cycle. Then the probability of failure was estimated as the average probability of failure based on all the simulation cycles. Also, the $COV(\overline{P}_f)$ was computed. The number of simulation cycles N was varied to illustrate convergence of the simulation process. The results are shown in Figure 8-8 using a nonarithmetic scale for N.

Generalized conditional expectation (GCE) method — The conditional expectation method can be generalized by allowing the number of the control variables to be larger than one. Equation 8-73 can be generalized as follows:

$$P_f = E_{X_i:i=1,2,...,n\&i\neq k}\left[P_f\left(\underline{X}_k\right)\right] \tag{8-75}$$

where $\underline{X}_k$ = a vector of control random variables, X_{k1}, X_{k2}, ..., X_{km}; and $P_f(\underline{X}_k)$ = the probability of failure evaluated in the dimensions of X_{k1}, X_{k2}, ..., X_{km}. This probability can be evaluated using any method, e.g., moment methods, importance sampling, or conditional expectation.

The computational steps according to this generalized approach are summarized in the following:

1. The performance function should be defined according to Equation 8-13.
2. The control random variables, $\underline{X}_k = (X_{k1}, X_{k2}, ..., X_{km})$, are selected on the basis of reducing the dimensionality of the problem to the space of the control random

variables. All other random variables, X_i: $i \notin \underline{k}$, are considered the conditional random variables, and they are generated randomly according to their respective probability distributions.

3. Therefore, the reliability assessment problem is reduced to N evaluations of the probability term (Prob) in the following expression:

$$P_f = E_{X_i:i=1,2,...,n \& i \notin \underline{k}} \left\{ \text{Prob} \left[g\left(X_1, X_2, ..., X_{k1}, X_{k2}, ..., X_{km}, ..., X_n \right) < 0 \right] \right\} \quad (8\text{-}76)$$

where N = number of simulation cycles.

4. In the nongeneralized conditional expectation method, the expression "Prob[$g(x_1$, x_2, ..., X_{k1}, X_{k2}, ..., X_{km}, ..., x_n) < 0]" with $m = 1$ is evaluated using the cumulative distribution function (CDF) of X_k as given in Equation 8-74. It should be noted that the lower cases for x_1, x_2, ..., x_n in this expression indicate generated values of the random variables X_1, X_2, ..., X_n. In the generalized approach, the value of m is larger than one. Therefore, the probability expression can be evaluated using any suitable (or convenient) method. For example, the FORM, advanced second-moment method, importance sampling, conditional expectation (CE), or other structural reliability method can be used for this purpose. The choice of the m random variables of $\underline{X}_k$ should be based on the intended method for the evaluation of the probability expression. Care should be exercised in selecting the m variables that result in simplifying the reliability evaluation, in addition to reducing the dimensionality of the problem. The simplification can be, for example, in the form of (1) reducing a nonlinear performance function into a linear function that is used in the probability expression, (2) using a closed-form expression for evaluating the probability expression, and (3) removing random variables with nonnormal probability distributions from the expression. These concepts are best explained using examples 8-9 and 8-10.

Antithetic variates (AV) method — According to the antithetic variates variance reduction technique, negative correlation between different cycles of simulation is induced in order to decrease the variance of the estimated mean value. If U is a random number uniformly distributed in the range [0,1] and is used in a computer run to determine the probability of failure $P_{f_i}^{(1)}$, the $1 - U$ can be used in another run to determine the probability of failure $P_{f_i}^{(2)}$. Therefore, the probability of failure in the ith simulation cycle is given by

$$P_{f_i} = \frac{P_{f_i}^{(1)} + P_{f_i}^{(2)}}{2} \quad (8\text{-}77)$$

Then, the mean value of probability of failure can be calculated by Equation 8-72. The AV variance reduction technique can be used in combination with the nongeneralized or generalized conditional expectation VRT. The negative correlation can be achieved by using, for example, the inverse transformation method for generating the random variables as discussed in Chapter 7. In the random generation process for each simulation cycle, say the ith cycle, a set of uncorrelated random numbers based on the uniform random variable U is used in the first stage of the ith cycle to determine the probability of failure $P_{f_i}^{(1)}$. In the second stage of the ith cycle, a complementary set of uncorrelated random numbers based on the random variable $(1 - U)$ is used to determine the probability of failure $P_{f_i}^{(2)}$. Therefore, the probability of failure in the ith simulation cycle is given by Equation 8-77. The result is simply additional reduction in the variance of the estimated quantity and expedited convergence.

Example 8-9: Flexural Reliability of a Beam — Nonlinear Normal and Nonnormal Cases

Consider the first-yield failure mode of a structural steel section subjected to a bending moment loading. The performance function is given by

$$Z = YS - M \tag{8-78}$$

where Y = yield stress of material, S = elastic section modulus, and M = moment effect due to applied loading. The statistical characteristics of the variables are shown in Table 8-8. These variables are assumed to be statistically uncorrelated.

The probability of failure of the structural component according to the first-yield failure mode can expressed as

$$P_f = \text{Prob}(Z < 0) = \text{Prob}(M > YS) \tag{8-79}$$

The control random variable, in this case, is selected as the random variable M, because it has the largest coefficient of variation (COV). Therefore, conditioning on Y and S, the cumulative distribution function of M was used to evaluate P_f in each simulation cycle. The variables Y and S were randomly generated using the inverse transformation method. Then the CE variance reduction technique was used to estimate the probability of failure. To improve the performance of the method, antithetic variates (AV) variance reduction technique was combined with CE to estimate the probability of failure.

In this example, two cases were considered, normal random variables (case 1) and nonnormal random variables (case 2), as shown in Table 8-8. For the nonnormal case, the probability of failure according to the specified performance function and the distribution types for the ith simulation cycle is given by

$$P_{f_i} = 1 - F_M(y_i s_i) = 1 - \exp\left\{-\exp\left[-\alpha(y_i s_i - u)\right]\right\} \tag{8-80}$$

in which α and u are the parameters of type I-largest extreme value distribution for the assumed probabilistic characteristics of M, and F_M is the cumulative distribution function of M. The sample mean and COV of the failure probability were then determined using Equations 8-72 and 8-73b, respectively, and are shown in Table 8-9 for the normal and nonnormal probability distributions.

TABLE 8-8 Statistical Characteristics of Variables for Example 8-9

Random variable	Mean value	COV	Case 1	Case 2
Y	275.52 MPa	0.125	Normal	Lognormal
S	8.19×10^{-4} m^3	0.050	Normal	Lognormal
M	1.13×10^5 N-m	0.200	Normal	Type I — largest

TABLE 8-9 Results of Conditional Expectation Method in Example 8-9

Simulation method	Number of cycles	Estimated failure probability $\bar{P}_f$	COV($\bar{P}_f$)
Case 1: normal			
Direct Monte Carlo	200,000	0.00128	0.0625
Conditional expectation (CE)	40,000	0.00118	0.0460
Generalized CE	500	0.00118	0.0380
Case 2: nonnormal			
Direct Monte Carlo	100,000	0.00325	0.0560
Conditional expectation (CE)	2,000	0.00319	0.0460
Generalized CE	500	0.00300	0.0240

For the purpose of comparison, P_f was recalculated using the ASM method. The results are 1.1×10^{-3} and 3×10^{-3} for the normal and nonnormal probability distributions, respectively. For the ASM method, P_f was determined as

$$P_f = 1 - \Phi(\beta) \tag{8-81}$$

where Φ = the CDF of the standard normal probability distribution and β = reliability or reliability index.

Using the generalized conditional expectation (GCE) method, M and Y were selected as the control variables, and S was randomly generated. For the ith simulation cycle, the probability expression as described in Equation 8-76 is given by

$$P_{f_i} = \text{Prob}[Ys_i - M < 0] \tag{8-82}$$

where s_i = randomly generated value of S. The probability expression was then evaluated, for the normal probability distributions (case 1), as follows:

$$P_{f_i} = 1 - \Phi\left[\frac{\mu_Y s_i - \mu_M}{\sqrt{s_i^2 \sigma_Y^2 + \sigma_M^2}}\right] \tag{8-83}$$

where μ = mean value and σ = standard deviation. For the nonnormal probability distributions (case 2), the ASM method was used to determine P_{f_i}. Then the mean value and COV of the failure probability were determined using Equations 8-72 and 8-73b, respectively, for N simulation cycles. The resulting statistical characteristics of P_f are shown in Table 8-9 for the normal and nonnormal probability distributions, respectively.

Example 8-10: Flexural Reliability of a Portal Frame — Linear Nonnormal Case

The probability of occurrence of the most critical mode of a one-bay, one-story portal frame subjected to horizontal and vertical loads is of interest in this example. The performance function for the most critical failure mode is given by

$$Z = 4M_1 + 2M_2 - 1500k - 1000 \qquad (8\text{-}84)$$

where M_1 and M_2 = plastic capacities for the two columns and k = wind load factor. The columns are assumed to be weaker than the beam. The assumed probabilistic characteristics of the basic random variables are summarized in Table 8-10. Since the columns have different plastic moment capacities from the beam, the plastic hinge at the right joint in the most critical failure mechanism is on the column. In this example, combinations of the ASM method, the CE method, the GCE, and the AV method of the variance reduction techniques were used to estimate the probability of failure of the frame. The control variable was selected as k for the CE method, and M_1 and k for the GCE method. The results are shown in Table 8-11. The ASM method resulted in a probability of failure of 4.8×10^{-3}.

TABLE 8-10 Probabilistic Characteristics of Random Variables for Example 8-10

Variable	Mean value	Coefficient of variation	Distribution type
M_1	4.068×10^5 N-m (300 kip-ft)	0.15	Normal
M_2	6.102×10^5 N-m (450 kip-ft)	0.10	Normal
k	0.3	0.33	Normal
L *(nonrandom)*	6.10 m (20 ft)	N/A	N/A
H *(nonrandom)*	4.57 m (15 ft)	N/A	N/A
W *(nonrandom)*	445 kN (100 kips)	N/A	N/A

Note: N/A = not applicable.

TABLE 8-11 Simulation Results of Example 8-10

Simulation method	Number of cycles	Estimated failure probability $\bar{P}_f$	$\text{COV}(\bar{P}_f)$
Direct Monte Carlo	10,000	0.00475	0.046
Conditional expectation (CE)	5,000	0.00467	0.078
Generalized CE	500	0.00481	0.057

It is evident from Examples 8-9 and 8-10 that the assessment of failure probability based on nonlinear performance function using the GCE method can be reduced to averaging N evaluations of the probability of structural failure according to a linear expression of the performance function. This transformation can be achieved by carefully selecting the control random variables. The probabilistic evaluation of the linear expression was performed, in the two examples, using the ASM method. Other methods could have been used to achieve this objective. The choice of the ASM method was for the purpose of illustrating merging moment reliability methods with conditional expectation in Monte Carlo simulation. This concept can be greatly utilized in complex performance functions to transform them into computationally manageable formats.

Another unique aspect of the GCE method is the reduction of the structural reliability assessment problem that has a performance function with nonnormal probability distributions to N probabilistic evaluations based on a performance expression with only normal probability distributions. These evaluations can be determined under certain conditions using closed-form equations. A combination between the above two aspects of the GCE method can be utilized in solving structural reliability problems.

8.4. RELIABILITY OF SYSTEMS

In the previous section, the reliability assessment discussion was limited to the failure of one component according to one failure mode that is expressed by the performance function of Equation 8-13. In general, a component can fail in one of several failure modes. The treatment of multiple failure modes requires modeling the component behavior as a system. Also, a system can be defined as an assemblage of several components that serves some function or purpose. Again, a multicomponent system can fail according to several failure modes. In this case, one or more component failures are needed to fail the system. Modeling multimode components or multicomponent systems can be based on similar concepts. The objective of this section is to introduce the fundamental concepts of system analysis for the purpose of reliability assessment at the system level.

System analysis requires the recognition and modeling of some system characteristics that include (1) postfailure behavior of a component, (2) the contribution of a component failure or failure-mode occurrence to the system's failure, (3) the statistical correlation among failure modes and components' failure, and (4) the definition of failure at the system level. The postfailure behavior of a component is needed in order to determine the remaining effect or contribution of the component after its failure to the system response or failure. For example, in structural engineering, components can be ideally classified into brittle or ductile components according to their potential failure modes. Components lose their strength completely after failure according to a brittle failure mode. Therefore, these components can be removed from the structural analysis of a system upon failure. On the other hand, components that fail according to a ductile failure mode maintain complete or substantial partial force resistance at increasing levels of deformation after failure. Therefore, they continue to contribute to the behavior of the system after failure, and their contribution needs to be considered in the analysis of the system.

The contribution of a component failure or failure-mode occurrence to the system's failure depends on the level of redundancy in the system. The failure of some components can lead to the failure of the system, whereas the failure of other components does not result in system failure, but it can weaken the system.

The statistical correlation among failure modes and components' failure can have a large effect on the reliability of the system. The difficulty in assessing this correlation can result in approximate assessment of system reliability, or interval assessment of the reliability by considering the extreme conditions of statistically uncorrelated and fully correlated failure modes and components' failure.

Finally, the reliability assessment of a system requires defining failure at the system level. For example, in a structural system, the failure definition of a system can be that the remaining (or surviving) elements of a system become structurally unstable.

8.4.1. Systems in Series

A system in series can be represented as shown in Figure 8-9. The failure of one of the n components that are shown in the figure can result in the failure of the system. This type of system is also referred to as a weakest-link system. The connectivity of the components as shown in Figure 8-9 is a logical connectivity in terms of the contribution of component failure to system failure. The logical connectivity can be different than the physical connectivity of the components in the real system. For example, the statistically determinate truss shown in Figure 8-10 is a weakest-link system, because the failure of a component in the system results in the failure of the system. The system model for the truss is shown in Figure 8-11. The difference between the physical connectivity (Figure 8-10) and the logical connectivity (Figure 8-11) can be observed by comparing the two figures.

The reliability of a system in series depends on the level of dependence among the underlying failure events for the components (called E_i, $i = 1, 2, .., n$). The survival of the system ($\bar{E}$) can be modeled as

$$\bar{E} = \bar{E}_1 \cap \bar{E}_2 \cap ... \cap \bar{E}_{n-1} \cap \bar{E}_n \tag{8-85}$$

where $\bar{E}_i$ = the survival event of component i. For independent failure events for the components, the probability of failure of the system can, therefore, be computed as

$$P(E) = 1 - P(\bar{E}) \tag{8-86a}$$

or

$$P(E) = 1 - [1 - P(E_1)][1 - P(E_2)]...[1 - P(E_{n-1})][1 - P(E_n)] \tag{8-86b}$$

FIGURE 8-9 A system of n components in series.

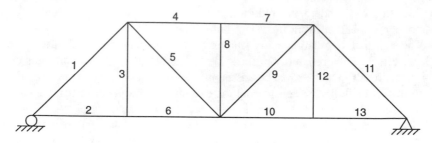

FIGURE 8-10 A truss structural system.

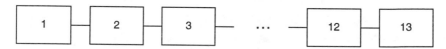

FIGURE 8-11 A system of 13 components in series for the truss.

For fully dependent events, the following relations hold:

$$E_i \subseteq \left(E_i \cap E_j \right) \tag{8-87a}$$

and

$$E_j \subseteq \left(E_i \cap E_j \right) \tag{8-87b}$$

Therefore,

$$P\left(E_i \right) \le P\left(E_i \cap E_j \right) \tag{8-88a}$$

and

$$P\left(E_j \right) \le P\left(E_i \cap E_j \right) \tag{8-88b}$$

For a system of n components with fully dependent failure events, the probability of survival of the system can be determined from Equation 8-85 as

$$P\left(\overline{E} \right) = P\left(\overline{E}_1 \cap \overline{E}_2 \cap \ldots \cap \overline{E}_{n-1} \cap \overline{E}_n \right) \tag{8-89}$$

Therefore, by extension of Equations 8-88a and 8-88b, and using Equation 8-89 results into

$$P(\overline{E}) = \min_{i=1}^{n}\left[P(\overline{E}_i)\right]$$ (8-90)

or

$$P(E) = \max_{i=1}^{n}\left[P(E_i)\right]$$ (8-91)

Equations 8-89 to 8-91 are limited to cases where the underlying performance functions for the failure events have positive statistical correlation. The results from Equations 8-86b and 8-91 provide upper and lower limits on the probability of failure of the system. The results are useful in cases where the positive dependence levels between failure events are unknown. Using Equations 8-86b and 8-91, the probability of failure of the system (P_f) can, therefore, be expressed as

$$\max_{i=1}^{n}\left(P_{f_i}\right) \le P_f \le 1 - \left(1 - P_{f_1}\right)\left(1 - P_{f_2}\right)\dots\left(1 - P_{f_{n-1}}\right)\left(1 - P_{f_n}\right)$$ (8-92a)

or

$$\max_{i=1}^{n}\left(P_{f_i}\right) \le P_f \le 1 - \prod_{i=1}^{n}\left(1 - P_{f_i}\right)$$ (8-92b)

where P_{f_i} = the failure probability of the *ith* component, for $i = 1, 2, .., n$.

Example 8-11: Reliability of a Truss

For the truss given in Figure 8-12, the truss members can be either in compression or tension depending on the applied loads and the truss's own weight. Given some loading that is random and using statics, the truss's member forces can be determined. For tension members, the yield strength of the members can be used to determine their failure probabilities. The compression members of the truss can fail either by compressive yielding or buckling. Using the smaller strength, the failure probabilities can be computed for these members. Assuming the failure probabilities of the truss members as shown in Figure 8-12, the failure probability of the truss system can be computed using Equation 8-92 as

$$\max(0.01, 0.02, 0.03, 0.005) \le P_f$$ (8-93a)

$$1 - (1 - 0.01)^6 (1 - 0.02)^4 (1 - 0.03)^2 (1 - 0.005) \ge P_f$$ (8-93b)

or

$$0.03 \le P_f \le 0.1870$$ (8-94)

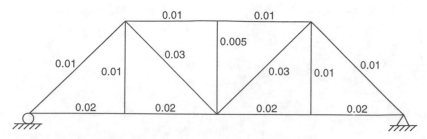

FIGURE 8-12 Example truss structural system with failure probabilities of truss members.

The lower bound in Equation 8-94 is the failure probability of the truss system for failure events that are dependent, whereas the upper bound is the failure probability for the independent case. For real trusses a level of dependence exists which can be related to both the strength and loading aspects of the reliability assessment problem at the member level. The computation of the dependence level can be difficult due to the various ill-defined factors that affect it, such as manufacturing and construction factors for the truss members and the contribution of dead load to the load effect in the members. Therefore, a failure probability at the system level expressed according to Equation 8-94 is commonly used.

8.4.2. Systems in Parallel

A system in parallel can be represented as shown in Figure 8-13. The n components that are shown in the figure need to fail in order to fail the system. This type of system is also referred to as a redundant system. The connectivity of the components as shown in Figure 8-13 is again a logical connectivity in terms of the contribution of the components' failure to the system failure.

The reliability of a system in parallel depends on the level of dependence among the underling failure events for the components (called E_i, $i = 1, 2, .., n$). The failure of the system (E) can be modeled as

$$E = E_1 \cap E_2 \cap ... \cap E_{n-1} \cap E_n \qquad (8\text{-}95)$$

or the survival of the system ($\overline{E}$) can be modeled as

$$\overline{E} = \overline{E}_1 \cup \overline{E}_2 \cup ... \cup \overline{E}_{n-1} \cup \overline{E}_n \qquad (8\text{-}96)$$

where $\overline{E}_i$ = the survival event of component i. Equation 8-95 can be evaluated using conditional probabilities as was discussed in Chapter 3.

The probability of failure of the system (P_f) for a system in parallel can be expressed similar to Equation 8-92 as

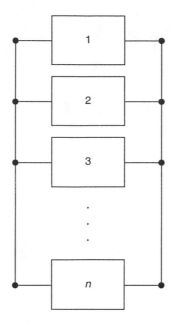

FIGURE 8-13 A system of n components in parallel.

$$\prod_{i=1}^{n} P_{f_i} \leq P_f \leq \min_{i=1}^{n} \left(P_{f_i} \right) \tag{8-97}$$

where P_{f_i} = the failure probability of the ith component, for $i = 1, 2, .., n$. These bounds can be wide. Narrower bounds require the knowledge of correlation levels between the underling failure events or conditional probabilities.

Example 8-12: Reliability of a Piping System in Parallel

The piping system shown in Figure 8-14 has four components with the indicated failure probabilities. The components are in parallel as shown in the figure. The failure of the system is defined as the flow failure from point A to B. Therefore, in order to fail the system, all the components must fail. Assuming the survival events of the pipes to be independent, the survival of the system $(\overline{E})$ can be modeled as

$$\overline{E} = \overline{E}_1 \cup \overline{E}_2 \cup \overline{E}_3 \cup \overline{E}_4 \tag{8-98a}$$

where $\overline{E}_i$ = the survival event of component i. The failure event (E) for the system can be expressed as

$$E = E_1 \cap E_2 \cap E_3 \cap E_4 \tag{8-98b}$$

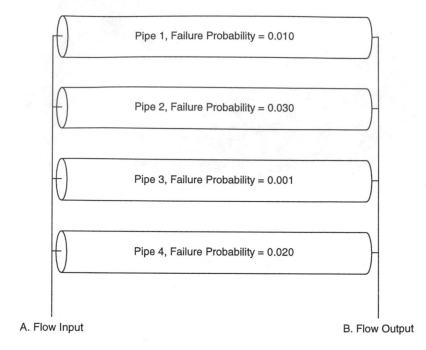

A. Flow Input   B. Flow Output

FIGURE 8-14 A piping system in parallel.

where E_i = the failure event for component i. Therefore, the failure probability of the system can be computed as

$$P_f = (0.01)(0.03)(0.001)(0.02) = 6 \times 10^{-9} \qquad (8\text{-}98c)$$

The resulting small failure probability reflects the high reliability due to redundancy in the piping system.

In this example, the failure of a pipe was assumed to not affect the performance of other pipes that did not fail. If such an effect exists, then fault or event tree analyses as discussed in Sections 8.5.4 and 8.5.5 should be used with conditional probabilities. In this case, the performance of nonfailing pipes, given the failure of one or more pipes, needs to be determined for the purpose of evaluating the system reliability.

8.4.3. Mixed Systems in Series and Parallel

In general, many engineering problems can be modeled using systems with components that are mixed in parallel and series. The representation in this case can be in the form of combined plots using Figure 8-9 for the portions of a system that are in series, and Figure 8-13 for the portions of the system that are in parallel. The computations can also be performed using the mathematical formulations presented

FIGURE 8-15 A piping system.

FIGURE 8-16 A revised piping system.

for the systems in series and parallel. The example that follows illustrates how to model mixed systems and how to compute their probabilities.

Example 8-13: Reliability of a Mixed Piping System

The piping system shown in Figure 8-15 has nine components with the indicated failure probabilities. The components are in series and parallel as shown in the figure. The failure of the system is defined as the flow failure from point A to B.

The pipes 1, 2, 3, and 4 are similar to the pipes discussed in Example 8-12. Therefore, they can represented by a single pipe with a failure probability of 6×10^{-9}. Similarly, pipes 6, 7, 8, and 9 are similar to the pipes discussed in Example 8-12. Therefore, they can be represented by a single pipe with a failure probability of 6×10^{-9}. As a result of these two representations, the system can be modeled as shown in Figure 8-16. The resulting system in series can now be evaluated using Equation 8-86b as follows

$$P_f = 1 - \left[1 - \left(6 \times 10^{-9}\right)\right]\left[1 - \left(6 \times 10^{-9}\right)\right](1 - 0.00001) = 0.0000100 \quad (8\text{-}99)$$

The resulting small failure probability reflects the high reliability due to redundancy in the parallel segments of the piping system.

8.4.4. Fault Tree Analysis

The reliability evaluation of complex engineering systems requires analytical methods that are systematic and methodical. Fault tree analysis (FTA) is a method that can be used for this purpose. The method starts by defining a top event, which is commonly selected as an adverse event. An engineering system can have more than one top event. For example, a building might have the following top events for the purpose of reliability assessment: power failure, sewer failure, water interruption, fire, or structural failure. Then each top event needs to be examined using the following logic: in order for the top event to occur, other events must occur. As a result, a set of lower level events is defined. Also, the form in which these lower level events are logically connected (i.e., in parallel or in series) needs to be defined. The connectivity of these events is expressed using gates. The "AND" or "OR" gates should be used as shown in Figure 8-17. Lower level events are classified into the following types:

1. *Basic events.* These events cannot be decomposed further into lower level events. They are the lowest events that can be obtained. For these events, failure probabilities need be obtained.
2. *Events that can be decomposed further.* These events can be decomposed further to lower levels. Therefore, they should be decomposed until the basic events are obtained.
3. *Undeveloped events.* These events are not basic and can be decomposed further. However, because they are not important, they are not developed further. Usually, the probabilities of these events are very small or the effect of their occurrence on the system is negligible, or can be controlled or mediated.
4. *Switch (or house) events.* These events are not random, and can be turned on or off with full control.

The symbols shown in Figure 8-17 are used for these events. Also, a continuation symbol is shown, which is used to break up a fault tree into several parts for the purpose of fitting it in several pages.

FTA requires the development of a tree-looking diagram for the system. Therefore, the tree shows failure paths and scenarios that can result in the occurrence of a top event. The construction of the tree should be based on the building blocks and the logic gates in Figure 8-17.

The outcome of interest from the FTA is the occurrence probability of the top event. Since the top event was decomposed into basic events, its occurrence can be stated in the form of "AND" and "OR" of the basic events. The resulting statement can be restated by replacing the "AND" with the intersection of the corresponding basic events, and the "OR" with the union of the corresponding basic events. Then the occurrence probability of the top event can be computed by evaluating the probabilities of the unions and intersections of the basic events. The dependence between these events, called sometimes common cause, affects the resulting probability based on the concepts covered in Chapter 3. The computations of fault trees are illustrated using an example at the end of this section.

For large fault trees, the computation of the occurrence probability of the top event can be difficult because of their sizes. In this case, more efficient approaches need to be used for assessing the reliability of a system, such as the minimal cut set approach. According to this approach, each cut set is defined as a set of basic events where the joint occurrence of these basic events results in the occurrence of the top

FIGURE 8-17 Symbols used in fault tree analysis.

event. A minimal cut set is a cut set with the condition that the nonoccurrence of any one basic event from this set results in the nonoccurrence of the top event. Therefore, a minimal cut set can be viewed as a subsystem in parallel. In general, systems have more than one minimal cut sets. The occurrence of the top event of the system can, therefore, be due to any one of these minimal cut sets. As a result, the system can be viewed as the union of all the minimal cut sets for the system. The computations of fault trees using minimal cut sets are illustrated using an example an the end of this section.

Alternatively, for large fault trees the computation of the occurrence probability of the top event can be based on the minimal link set approach. According to this approach, each link set is defined as a set of basic events where the joint nonoccurrence of these basic events results in the nonoccurrence of the top event. Then minimal link sets are defined similar to the minimal cut sets. In general, this approach is more difficult to deal with than the minimal cut set approach.

Example 8-14: Reliability of a Piping System Using Fault Tree Analysis

The piping system that was discussed in Example 8-13 is reexamined herein. The piping system, as shown in Figure 8-15, consists of nine components with the indicated failure probabilities. The components are in series and parallel as shown in the figure. The top event is defined as the flow failure from point A to B. The system is broken down into three piping sections as follows:

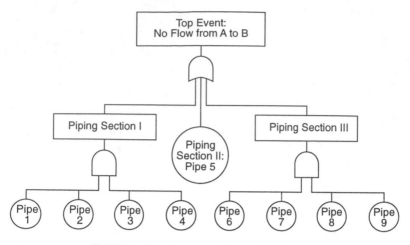

FIGURE 8-18 Fault tree model for a piping system.

1. Section I: pipes 1, 2, 3, and 4
2. Section II: pipe 5
3. Section III: pipes 6, 7, 8, and 9

The fault tree shown in Figure 8-18 was constructed using "AND" and "OR" gates. The events that are shown in the figure are failure events. The occurrence probability of the top event (P_f) can be computed as

$$P_f = P[(P1 \text{ AND } P2 \text{ AND } P3 \text{ AND } P4)$$
$$\text{OR } (P5) \text{ OR } (P6 \text{ AND } P7 \text{ AND } P8 \text{ AND } P9)]$$

(8-100)

where Pi is the failure event of the *ith* pipe. Replacing "AND" and "OR" with intersection and union produces the following:

$$P_f = P[(P1 \cap P2 \cap P3 \cap P4) \cup (P5) \cup (P6 \cap P7 \cap P8 \cap P9)] \quad (8\text{-}101)$$

In Equation 8-101, the failure probabilities of pipe section I (P[S1]) can be computed based on the assumption of independent pipe failures as

$$P(S1) = P[(P1 \cap P2 \cap P3 \cap P4)]$$
$$= P(P1)P(P2)P(P3)P(P4)$$
$$= (0.01)(0.03)(0.001)(0.02)$$
$$= 6 \times 10^{-9}$$

(8-102a)

Similarly, the failure probabilities of sections II and III, respectively, are given by

$$P(S2) = 0.00001 \tag{8-102b}$$

$$P(S3) = P(S1) = 6 \times 10^{-9} \tag{8-102c}$$

Now, the occurrence probability of the top event can be computed as

$$P_f = P[(S1) \cup (S2) \cup (S3)]$$

$$= P(S1) + P(S2) + P(S3) \tag{8-103}$$

$$- P(S1)P(S2) - P(S1)P(S3) - P(S2)P(S3) + P(S1)P(S2)P(S3)$$

Substituting the values of the failure probabilities results into

$$P_f = (6 \times 10^{-9}) + 0.00001 + (6 \times 10^{-9})$$

$$- (6 \times 10^{-9})(0.00001) - (6 \times 10^{-9})(6 \times 10^{-9}) - (0.00001)(6 \times 10^{-9}) \tag{8-104}$$

$$+ (6 \times 10^{-9})(0.00001)(6 \times 10^{-9})$$

$$= 0.0000100$$

Example 8-15: Reliability of a Piping System Using Minimal Cut Sets

The piping system that was discussed in Examples 8-13 and 8-14 is reexamined herein. The piping system, as shown in Figure 8-15, consists of nine components with the indicated failure probabilities. The components are in series and parallel as shown in the figure. The top event is defined as the flow failure from point A to B. In this case, the following are example cut sets:

$$\text{Example cut sets} = \{(P1, P5, P7), \ (P1, P2, P5, P8), \ (P3, P4, P5, P7, P9), \tag{8-105}$$

$$(P2, P3, P5, P7), \ (P1, P2, P3, P4)\}$$

where Pi is the failure event of the ith pipe. The minimal cut sets are

$$\text{Minimal cut sets} = \{(P1, P2, P3, P4), \ (P6, P7, P8, P9), \ (P5)\} \tag{8-106}$$

The occurrence of the top event is, therefore, considered as the union of the minimal cut sets. Each minimal cut set is considered to be the intersection of its basic events. The occurrence probabilities of the minimal cut sets can be computed as the product of the occurrence probabilities of its events. For example, the first minimal cut set (P1,P2,P3,P4) has a probability of

$$P(P1 \cap P2 \cap P3 \cap P4) = (0.01)(0.03)(0.001)(0.02) = 6 \times 10^{-9} \quad (8\text{-}107)$$

Similarly, the probabilities of all the minimal cut sets can be computed. The occurrence probability of the top event can then be evaluated as a system in series of the minimal cut sets. The resulting occurrence probability is the same as the previous example, i.e., 0.0000100. It should be noted that minimal cut sets that include some common basic events are dependent sets. Therefore, the occurrence probability of their union should be based on conditional probabilities as given by Equations 3-10, 3-27, and 3-28.

8.4.5. Event Tree Analysis

Event tree analysis is a method to study the effect of an initiating event on a system. The initiating event can be the failure of a component in the system or the top event as was defined in fault tree analysis. The effect of an initiating event on a system depends on what might happen next and the sequence of occurrence. As a result, several possible scenarios can be developed with possibly very severe impact on the system. In reliability analysis of systems, it is important to identify these scenarios, their occurrence probabilities, and effects on the system. In some systems, an initiating event might appear to have limited consequence on the system, but as its occurrence is combined with the occurrence of other events, the result can have significant impact, event possibly devastating effects, on the system.

Event tree analysis was used in the nuclear industry to investigate the effect of accidents or other initiating events on the integrity of nuclear reactors. The event tree shown in Figure 8-19 was adapted from a reactor safety study performed for the Nuclear Regulatory Commission, called WASH 1400 study. The initiating event was considered for the purpose of illustration to be a pipe break with a probability of P(E). The first subsequent event was selected to be electric power failure with a failure probability P(E1). Therefore, the electric power survival probability is [1 − P(E1)]. The second subsequent event was selected as the failure of the emergency core cooling system with a probability of P(E2). The third subsequent event was selected to be the failure of the fission product removal system with a probability of P(E3). The fourth subsequent event was selected as the failure of the containment (reactor) integrity with a probability of P(E4). The selection of the subsequent events and their order are for the purpose of illustration. In real systems, they should be based on the knowledge of the system, its components, their interaction, functions, and performance. The resulting event tree has 16 branches as shown in Figure 8-19. The subsequent events at any level are not necessarily limited to two. In this example, the treatment was kept simple by considering two subsequent events at each level. In general, it is possible to consider several levels of failure for each subsequent event. However, the tree can grow to a large size. Sometimes, it is possible to identify some unimportant tree branches because of their relatively very small probabilities and small consequences. These branches can be pruned and ignored.

After completing the construction of the tree, the branch probabilities can be computed as the product of all the probabilities of all the events along a branch as shown in Figure 8-19. The assumption behind this mathematical operation is that the underling events are independent, which needs to be carefully examined for its validity. Also, we need to define the consequences for each branch. The consequences can be expressed in different units depending on their types. It is possible, for example, to define the consequences as property damage with estimates in dollars, or potential human injury with estimates in numbers. Also, it is possible to keep the assessed consequences in linguistic terms as shown in Figure 8-19.

Since all the tree branches are possible failure scenarios, they can be considered as failure modes for the system with varying levels of consequence. Branches that produce high probabilities with high consequences are called high risk branches and should be targeted for risk improvement. The meaning of risk in this case is the combination of both occurrence probability and consequences. The combination can be expressed in the form of a product as follows:

$$\text{Risk} = \text{occurrence probability} \times \text{occurrence consequence} \qquad (8\text{-}108)$$

The risk improvement can be either by reducing the occurrence probability through system or component changes, or reducing the potential consequences.

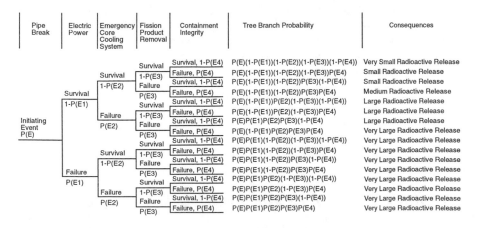

FIGURE 8-19 Event tree model for reactor safety.

Example 8-16: Reliability of a Nuclear Reactor Using Event Tree Analysis

The event tree shown in Figure 8-19 is used in this example to illustrate the computations of branch probabilities. The assumed annual occurrence probabilities for the events E, E1, E2, E3, and E4 are for the purpose of illustration and are given by

$$P(E) = 1E\text{-}04 \tag{8-109a}$$

$$P(E1) = 1E\text{-}07 \tag{8-109b}$$

$$P(E2) = 1E\text{-}05 \tag{8-109c}$$

$$P(E3) = 1E\text{-}06 \tag{8-109d}$$

$$P(E4) = 1E\text{-}05 \tag{8-109e}$$

where 1E-04 = 1×10^{-4}. Table 8-12 shows the computations of the branch probabilities. The table also shows approximate expressions for computing the branch probabilities. Since the event probabilities as given by Equations 8-109a to 8-109e are small, the product of a small number by one minus another small number is approximately equal to the small number, i.e.,

$$P(Ei)[1 - P(Ej)] \approx P(Ei) \tag{8-110}$$

where P(Ei) and P(Ej) are small numbers.

Table 8-12 shows that branch 1 has the largest occurrence probability, but it has a small consequence according to Figure 8-19. According to Figure 8-19, branches 8 to 16 have the highest level of consequence. From these branches, branch 9 has the highest occurrence probability. Therefore, branch 9 can be considered as the highest risk branch. The occurrence probability of this branch can be reduced by improving the failure probability of the pipe breakage (P[E]), or improving the electric power failure probability (P[E1]). Options of system improvement can be identified that affect these two events. Then cost-benefit analysis using decision trees can be utilized to select the best option. Decision analysis is discussed in Section 8.6.

8.5. RISK-BASED DECISION ANALYSIS

The risk of an event was defined in Section 8.4.5 as a combination of both its occurrence probability and its occurrence consequence. The combination can be in

TABLE 8-12 Computation of Branch Probabilities in Event Tree Analysis

Branch no.	Branch probability expression from Figure 8-19	Approximate expression for branch probability	Branch probability
1	$P(E)[1 - P(E1)][1 - P(E2)][1 - P(E3)][1 - P(E4)]$	$P(E)$	1E-04
2	$P(E)[1 - P(E1)][1 - P(E2)][1 - P(E3)]P(E4)$	$P(E)P(E4)$	1E-09
3	$P(E)[1 - P(E1)][1 - P(E2)]P(E3)[1 - P(E4)]$	$P(E)P(E3)$	1E-10
4	$P(E)[1 - P(E1)][1 - P(E2)]P(E3)P(E4)$	$P(E)P(E3)P(E4)$	1E-15
5	$P(E)[1 - P(E1)]P(E2)[1 - P(E3)][1 - P(E4)]$	$P(E)P(E2)$	1E-09
6	$P(E)[1 - P(E1)]P(E2)[1 - P(E3)]P(E4)$	$P(E)P(E2)P(E4)$	1E-14
7	$P(E)[1 - P(E1)]P(E2)P(E3)[1 - P(E4)]$	$P(E)P(E2)P(E3)$	1E-15
8	$P(E)[1 - P(E1)]P(E2)P(E3)P(E4)$	$P(E)P(E2)P(E3)P(E4)$	1E-20
9	$P(E)P(E1)[1 - P(E2)][1 - P(E3)][1 - P(E4)]$	$P(E)P(E1)$	1E-11
10	$P(E)P(E1)[1 - P(E2)][1 - P(E3)]P(E4)$	$P(E)P(E1)P(E4)$	1E-16
11	$P(E)P(E1)[1 - P(E2)]P(E3)[1 - P(E4)]$	$P(E)P(E1)P(E3)$	1E-17
12	$P(E)P(E1)[1 - P(E2)]P(E3)P(E4)$	$P(E)P(E1)P(E3)P(E4)$	1E-22
13	$P(E)P(E1)P(E2)[1 - P(E3)][1 - P(E4)]$	$P(E)P(E1)P(E2)$	1E-16
14	$P(E)P(E1)P(E2)[1 - P(E3)]P(E4)$	$P(E)P(E1)P(E2)P(E4)$	1E-21
15	$P(E)P(E1)P(E2)P(E3)[1 - P(E4)]$	$P(E)P(E1)P(E2)P(E3)$	1E-22
16	$P(E)P(E1)P(E2)P(E3)P(E4)$	None	1E-27

FIGURE 8-20 Example risk plot.

the form of a product as given in Equation 8-108. Alternatively, it can be considered to be an ordered pair of occurrence probability and its occurrence consequence, i.e., probability, consequences. For several events of interest, risk plots can be produced using these ordered pairs for the events and a coordinate system of occurrence probability and consequence as shown in Figure 8-20. The risk plot can be used to easily identify high-risk events, high-probability events, or high-consequence events.

Performing risk analysis requires various decisions that need to be made in order to reduce the risk of a system. Assuming the system to consist of equipment, components, and details, the decisions can include, for example, what and when to inspect components or equipment, which inspection methods to use, assessing the significance of detected damage, and repair/replace actions. These decisions are important in operating, using, and maintaining the system. The risk aspect of the analysis requires obtaining and utilizing information about failure likelihood and consequences. Engineering decisions of these types need to be made using a systematic framework that considers all facets of a decision problem. The decision framework is called the decision model.

The objective of this section is to introduce a decision model (a systematic framework) for decision making in the risk analysis. In order to construct a decision model, the following elements of the decision model need to be defined:

1. Objectives of decision analysis
2. Decision variables
3. Decision outcomes
4. Associated probabilities and consequences

The components of the decision model are described in the following sections.

8.5.1. Objectives of Decision Analysis

Engineering decision problems can be classified into single- and multiple-objective problems. Example objectives are minimizing the total expected cost, maximizing safety, maximizing the total expected utility value, and maximizing the total expected profit. Decision analysis requires the definition of these objectives. For cases of multiple objectives, the objectives need to be stated in the same units, and weight factors that can be used to combine the objectives need to be assigned.

8.5.2. Decision Variables

The decision variables for the decision model also need to be defined. The decision variables are the feasible options or alternatives available to the decision maker at any stage of the decision-making process. Also, ranges of values that can be taken by the decision variables should be defined. Example decision variables can include, for example, what and when to inspect components or equipment, which inspection methods to use, assessing the significance of detected damage, and repair/replace decisions. Therefore, assigning a value to a decision variable means making a decision at that point of a decision-making process. These points within the decision-making process are called decision nodes. The decision nodes are identified in the model using a rectangle or square symbol (■).

8.5.3. Decision Outcomes

The decision outcomes for the decision model also need to be defined. The decision outcomes are the events that can happen as a result of a decision. They are random in nature, and their occurrence cannot be fully controlled by the decision maker. Example decision outcomes can include, for example, the outcomes of an inspection (detection or nondetection of a damage) and the outcomes of a repair (satisfactory or nonsatisfactory repair). Therefore, the decision outcomes with the associated occurrence probabilities need to be defined. The decision outcomes can occur after making a decision at points within the decision-making process, called chance nodes. The chance nodes are identified in the model using the circle symbol (●).

8.5.4. Associated Probabilities and Consequences

The decision variables take values that can have associated costs. These costs can be considered as the direct consequences of making these decisions. The decision outcomes have both consequences and occurrence probabilities. The probabilities are needed due to the random (chance) nature of these outcomes. The consequences can include, for example, the cost of failure due to damage that was not detected by an inspection method.

8.5.5. Decision Trees

The elements of a decision model need to be considered in a systematic form in order to make decisions that meet the objectives of the decision-making process. Decision trees are commonly used to examine the available information for the purpose of decision making. Figure 8-21 shows a general layout for an example decision tree. The decision tree includes the decision and chance nodes. The decision nodes are followed by possible actions (or alternatives, A_i) that can be selected by a decision maker. The chance nodes are followed by outcomes (or chances, O_j) that can happen without the complete control of the decision maker. The actions have costs associated with them (C_{Ai}), whereas the outcomes have both probabilities ($P[O_j]$) and consequences (C_{Oj}). Each line followed from the beginning (left end) of the tree to the end (right end) of the tree is called a tree branch. Each branch represents a possible scenario of decisions and possible outcomes that can happen. The total expected cost for each branch can be computed as shown in Figure 8-21. Then the most suitable decisions can be selected such that the minimum total expected cost is obtained.

In general, utility values can be used instead of cost values. Decision analysis using utility values is not discussed in here. Also, the decision tree shown in Figure 8-21 was developed for the purpose of its use as a computational form, where values for probability and cost can be recorded and computed. It should be noted that in computing the total expected cost of an action, the cost of an action (CA) should not be multiple counted from all the chance-node branches. It should be included once. The figure shows this cost in each branch in order to facilitate the computation of the branch cost.

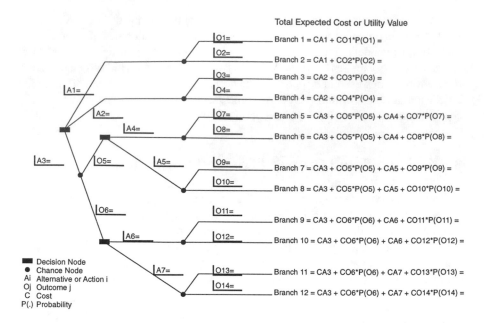

FIGURE 8-21 Example decision tree.

Example 8-17: Decision Analysis for Selection of an Inspection Strategy

The objective herein is to develop an inspection strategy for a selected component of a limited section of a refinery plant. This study is for illustration purposes and is based on hypothetical inspection strategies, probabilities, costs, and consequences. The inspection strategy needs to be selected using decision analysis.

The first step in decision analysis is to select a component for inspection, possibly based on risk analysis, fault trees, and event trees. High-risk components can be selected for this purpose. The shell of a regenerator in a refinery was chosen to illustrate how an inspection strategy could be selected. Then we need to select and define candidate inspection strategies based on previous experience and knowledge of the system and logistics of inspection. For the purpose of illustration, only three candidate inspection strategies are considered. They are internal visual inspection, external visual inspection, and internal and external ultrasonic testing. The probabilities and costs in this example were chosen to provide a range of values to show the scope of the selection procedure.

Figure 8-22 shows the decision tree with the three decision strategies. Each branch provides a portion of the tree and is shown in a separate figure. Figures 8-23a, 8-23b, and 8-23c show the three portions of the decision tree for the three strategies.

In Figure 8-23(a to c), the outcomes of an inspection strategy is either detection or nondetection of a defect. The detection probability is denoted $P1$, and the nondetection probability is $P2$. These outcomes originate from a chance node. The costs of these outcomes are $C2 = 0$ for detection and $C4$ = consequences of nondetection of a defect, respectively. The cost of an inspection strategy is denoted $C1$. At the end of one of these branches, there is a chance node that can lead to damage to the component due to inspection, and no damage to the component due to inspection. The probabilities $P3$ and $P4$ are used to denote the damage probability and the no-damage probability, respectively. The associated cost for damage is $C3$, and no damage is zero. The probability and cost estimates were assumed for each inspection strategy on its portion of the decision tree as shown in Figure 8-23(a to c).

The total expected cost for each branch was computed by summing up the product of the pairs of cost and probability along the branch. Then total expected cost for the inspection strategy was obtained by adding up the total expected costs of the branches of its portion of the decision tree.

The total expected costs of the strategies are summarized in Figure 8-22. Assuming that the decision objective is to minimize the total expected cost, then strategy 3 should be selected as the optimal strategy. Although strategy 3 is the most expensive (its $C1$ value is the highest), its total expected cost ($TC3$) is the smallest. This result can be attributed to the ability of strategy 3 to detect smaller defects and, therefore, reduce the consequences of nondetection.

8.6. APPLICATIONS

8.6.1. Structural Reliability of a Pressure Vessel

Pressure vessels are structures that are designed to support internal pressure, in addition to dead load, wind load, earthquake load, temperature, piping load, impact, and cyclic load. In this example, the reliability of a pressure vessel subjected to wind load is assessed. The example vessel is 120 ft high with 6.0 ft in diameter. The vessel is shown in Figure 8-24 with the assumed wind load distribution. The vessel is assumed to have a uniform shell thickness. The bending moment due to the wind load at the base of the vessel is considered to define the performance function. First yield of the cross section was assumed to be the failure mode of interest.

The wind load (W_1, W_2, and W_3) as shown in Figure 8-24 is given by the following empirical equation according to the American National Standard Institute (1972):

$$W_i = 0.00256 C K_i G D V^2 \qquad (8\text{-}111)$$

where, in this example, W_i = total design wind load per unit length of the vessel in pounds per foot; C = net pressure coefficient; K_i = velocity pressure coefficient, which depends on the exposure type and height above ground level; G = gust factor, which depends on the exposure type and dynamic response characteristics of the structure; V = basic wind speed in miles per hour recorded at 30 ft above ground level; and D = the effective diameter in feet, which considers the effect of ladders,

FIGURE 8-22 Decision tree for inspection.

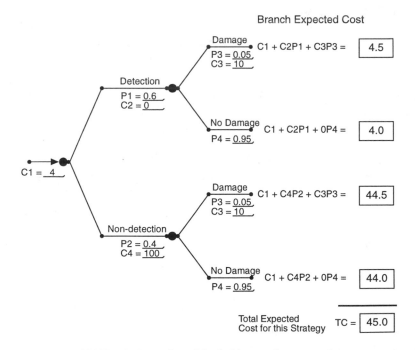

FIGURE 8-23a Portion of the decision tree for strategy 1.

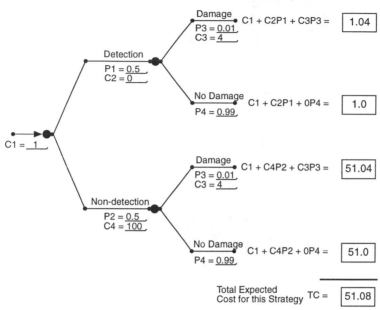

FIGURE 8-23b Portion of the decision tree for strategy 2.

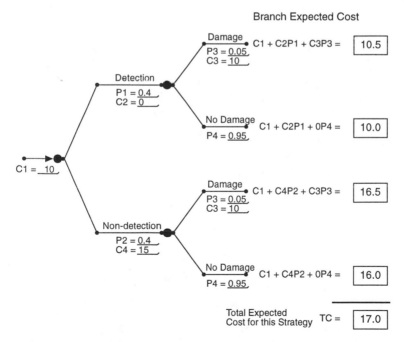

FIGURE 8-23c Portion of the decision tree for strategy 3.

FIGURE 8-24 Wind load distribution on a pressure vessel.

insulation, and pipes attached to the vessel. Therefore, the load effect (L) in the form of a bending moment at the base is

$$L = W_1 \frac{30^2}{2} + W_2(45)\left(30 + \frac{45}{2}\right) + W_3(45)\left(30 + 45 + \frac{45}{2}\right) \qquad (8\text{-}112)$$

The moment capacity (R) of the circular hollow cross section at the base of the vessel is given by

$$R = \pi Y R_d^2 T \qquad (8\text{-}113)$$

where Y = yield strength of the material, which is assumed to be steel; R_d = radius of the vessel; and T = wall thickness.

Using Equations 8-112 and 8-113, the performance function (Z) can be expressed as

$$Z = R - L \qquad (8\text{-}114)$$

or

$$Z = \pi Y R_d^2 T - W_1 \frac{30^2}{2} + W_2(45)\left(30 + \frac{45}{2}\right) + W_3(45)\left(30 + 45 + \frac{45}{2}\right) \qquad (8\text{-}115a)$$

Substituting Equation 8-111 into Equation 8-115a results into

$$Z = \pi Y R_d^2 T - 0.00256 CGDV^2 \left[450K_1 + 2362.5K_2 + 4387.5K_3 \right] \quad (8\text{-}115b)$$

Since K_1, K_2, and K_3 are fully correlated with the assumed mean values 1, 1.2, and 1.4, respectively, they can be normalized with respect to K_1. Therefore, Equation 8-115b can be expressed as

$$Z = \pi Y R_d^2 T - 24.1344 CGDKV^2 \quad (8\text{-}116)$$

where $K = K_1$. Table 8-13 provides the assumed probabilistic characteristics of the random variables. Several thicknesses are examined in this example as shown in Table 8-13. The random variables X_1, X_2, and X are used in the example to reduce the number of the random variables in the problem. The reduction is based on the property that the product or division of random variables with lognormal distribution is a random variable that has a lognormal distribution. The wind speed was assumed to have probability distribution called the extreme value (largest) type II. This distribution has the following probability density and cumulative function, respectively:

$$f_V(v) = \frac{k}{v} \left(\frac{u}{v} \right)^k \exp\left[-(u/v)^{-k} \right] \quad (8\text{-}117)$$

and

$$F_V(v) = \exp\left[-(u/v)^{-k} \right] \quad (8\text{-}118)$$

where u and k are the parameters of the distribution. These parameters are related to the mean and variance, respectively, as follows:

$$E(V) = u\Gamma\left(1 - \frac{1}{k} \right) \quad (8\text{-}119)$$

and

$$\mathrm{Var}(V) = u^2 \left[\Gamma\left(1 - \frac{2}{k} \right) - \Gamma^2\left(1 - \frac{1}{k} \right) \right] \quad (8\text{-}120)$$

where $\Gamma(.)$ is the gamma function as described in Chapter 5 and tabulated in Appendix B. The mean and coefficient of variation for the wind speed V provided in Table 8-13 results into the following parameters:

$$u = 58.369 \tag{8-121a}$$

and

$$k = 8.85 \tag{8-121b}$$

Using the probabilistic characteristics of the random variables as shown in Table 8-13, the advanced second-moment method was used to assess the reliability of the vessel according to the performance function of Equation 8-116. Nonnormal random variables were transformed to normal random variables using Equations 8-57 and 8-58. The random variables were assumed to be uncorrelated. The resulting safety indices and failure probabilities for the four thicknesses in Table 8-13 are shown in Table 8-14.

Direct Monte Carlo simulation was also used to estimate the failure probability. Utilizing the property that the product or division of random variables with lognormal distribution is a random variable that has a lognormal distribution, the performance function was simplified as follows:

$$Z = \pi X_1 Y - 24.1344 C X_2 V^2 \tag{8-122}$$

TABLE 8-13 Probabilistic Characteristics of Random Variables for Pressure Vessel

Random variable	Mean value	Coefficient of variation	Probability distribution
C	0.672	0.10	Normal
G	1.283	0.12	Lognormal
V	63 mph	0.16	Extreme value type II
D	8.3 ft	0.10	Lognormal
K	1.0	0.10	Lognormal
T	1/8, 3/16, 1/4, 5/16 in.	0.05	Lognormal
R_d	3.08 ft	0.05	Lognormal
F	38 ksi	0.10	Normal
$X_1 = TR^2$	0.0988, 0.1482, 0.1976, 0.2470 in.^{3}	0.08666	Lognormal
$X_2 = KGD$	10.6489 ft	0.18652	Lognormal
$X = X_1/X_2$	0.0096, 0.0144, 0.0192, 0.0240	0.20632	Lognormal

TABLE 8-14 Results of the Advanced Second-Moment Method for Pressure Vessel

Thickness (in.)	Safety index, β	Failure probability
1/8	2.18	0.014629
3/16	2.80	0.002555
1/4	3.19	0.000711
5/16	3.47	0.000260

Generating the random variables and substituting in Equation 8-122, failure can be counted based on the sign of Z. Then the probability of failure can be estimated using Equation 8-66. The results of direct simulation for ¼-in.-thick vessel are shown in Table 8-15. The number of simulation cycles was increased from 500 to 20,000 in increments. Also, the reliability of the vessel was estimated using conditional expectation without and with antithetic variates variance reduction techniques. In these methods, the wind speed V was selected as the control variable. The cumulative distribution function of V (Equation 8-118) was used to compute the probability of failure as follows:

$$P_f = P(Z < 0) \tag{8-123a}$$

$$= P(\pi X_1 Y - 24.1344 C X_2 V^2 < 0) \tag{8-123b}$$

$$= P\left(V > \sqrt{0.1301707 \frac{XY}{C}}\right) \tag{8-123c}$$

$$= 1 - F_V\left(\sqrt{0.1301707 \frac{XY}{C}}\right) \tag{8-123d}$$

Then Equations 8-72 and 8-73 were used to estimate the failure probability and its coefficient of variation. The results are also shown in Table 8-15.

Table 8-15 indicates that as the number of simulation cycles is increased, the accuracy of the estimated failure probability increases. Also, the table shows the inability of direct simulation to estimate the failure probability using small numbers of simulation cycles. Based on this table, a value for N of 5000 simulation cycles was selected to study the other thicknesses in Table 8-13. The results are shown in Table 8-16. The table shows that the vessel reliability improves as the thickness is increased. Also, the table indicates that as the failure probability becomes small, the direct simulation method becomes incapable of estimating it

TABLE 8-15 Simulation Results for Pressure Vessel with 1/4 in. Thickness

Number of simulation cycles (N)	Direct simulation (P_f)	Conditional expectation		Conditional expectation with antithetic variates	
		$\overline{P}_f$	$COV(\overline{P}_f)$	$\overline{P}_f$	$COV(\overline{P}_f)$
500	0.000000 [a]	0.0008440	0.0642	0.0007427	0.0370
1,000	0.000000 [a]	0.0008029	0.05085	0.0007472	0.0288
5,000	0.000000 [a]	0.0007415	0.0208	0.0007362	0.0118
20,000	0.0004500	0.0007451	0.0107	0.0007389	0.0063

[a] Number of failures = 0 for these cases.

Table 8-16. Simulation Results for Pressure Vessel for All Thicknesses Using 5000 Simulation Cycles

Mean thickness (in.) $\overline{T}$	Direct simulation P_f	Conditional expectation		Conditional expectation with antithetic variates	
		$\overline{P}_f$	$\mathrm{COV}(\overline{P}_f)$	$\overline{P}_f$	$\mathrm{COV}(\overline{P}_f)$
1/8	0.015000	0.015561	0.0199	0.015459	0.0111
3/16	0.003000	0.002640	0.0206	0.002622	0.0117
1/4	0.000000 [a]	0.000742	0.0208	0.000736	0.0118
5/16	0.000000 [a]	0.000276	0.0208	0.000274	0.0118

[a] Number of failures = 0 for these cases.

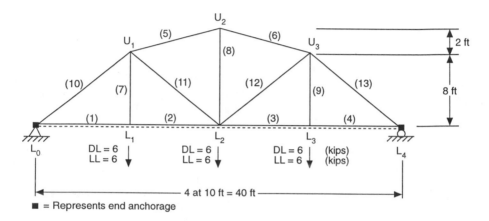

FIGURE 8-25 Posttensioned truss.

8.6.2. System Reliability of a Posttensioned Truss

Trusses are structural systems that are commonly used for roofs or bridges. An example truss is shown in Figure 8-25. This truss has 13 members with loads applied at truss joints L_1, L_2, and L_3 as shown in the figure. The loads are of two types: dead load (DL) and live load (LL). The loads were given the mean values shown in the figure. The coefficient of variation of the load effect (i.e., member forces) was assumed to be 0.25, and its distribution type was assumed to be normal. In order to strengthen this truss, a posttensioning cable was added as shown in Figure 8-25 in the form of a dashed line along the bottom chord of the truss. The cable was anchored at truss joints L_0 and L_4.

The truss members are assumed to have the mean tensile capacity and mean compressive capacity shown in Table 8-17. The coefficient of variation for the tensile and compressive capacities was assumed to be 0.1. A normal distribution was assumed for the capacities. The posttensioning cable was assumed to have mean tensile capacity of 33.75 kips, coefficient of variation of 0.1, and normal probability distribution. The posttensioning force was assumed to be deterministic with a value of 10 kips.

TABLE 8-17 Mean Strength Values for Truss Members

Truss member	Mean tensile capacity (kips)	Mean compressive capacity (kips)
$L_0 L_1$	44	40
$L_1 L_2$	44	40
$U_1 U_2$	37	52
$L_1 U_1$	25	12
$L_2 U_2$	26	13
$L_0 U_1$	27	64
$L_2 U_1$	5	7

TABLE 8-18 Mean Member Forces for the Truss

Truss member	Mean tensile force (mean compressive force) in kips
$L_0 L_1$	10.8
$L_1 L_2$	10.8
$U_1 U_2$	(24.5)
$L_1 U_1$	12
$L_2 U_2$	9.6
$L_0 U_1$	(28.8)
$L_2 U_1$	1.9
Cable	11.7

TABLE 8-19 Failure Probabilities for the Truss Members

Truss member	Failure probability
$L_0 L_1$	1.28×10^{-12}
$L_1 L_2$	1.28×10^{-12}
$U_1 U_2$	3.13×10^{-4}
$L_1 U_1$	4.34×10^{-4}
$L_2 U_2$	1.88×10^{-4}
$L_0 U_1$	1.48×10^{-4}
$L_2 U_1$	4.32×10^{-6}

Stiffness structural analysis of the truss subjected to the mean dead and live loads, and posttensioning force resulted in the member forces that are shown in Table 8-18. Using Equation 8-28b, the failure probabilities of the truss members were computed. The member forces and their strength values are as shown in Tables 8-18 and 8-17, respectively. The resulting failure probabilities are shown in Table 8-19.

Event tree analysis was used to evaluate the reliability of the truss system. Starting with an intact posttensioned truss, the truss elements were failed one by one to develop the branches of the event tree. Then subsequent member failures were introduced, until a system failure was obtained. For each partially failed system, the

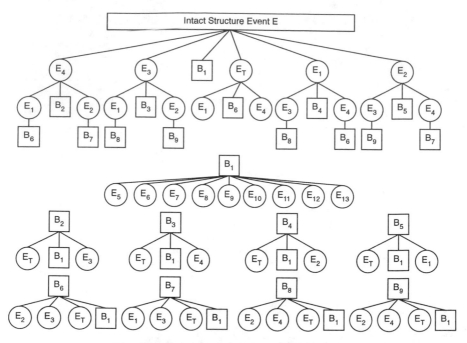

FIGURE 8-26 Event tree model for posttensioned truss.

failure probabilities of the remaining truss members were evaluated. These failure probabilities can be viewed as conditional probabilities for the condition of partial failure. The failure of the truss members was assumed to be brittle, i.e., members without postfailure (or residual) strength. The resulting event tree is shown in Figure 8-26. In the figure E_i means failure of the ith truss member; E_T means cable failure; and B_j is a continuation reference code for the tree. The occurrence probability of each branch of the tree was computed by multiplying the conditional failure probabilities of all the truss members along the branch. The failure probability of the system (P_f) was then computed as a system in series with the branches as the elements of the system, using Equation 8-92 as follows:

$$\max_{i=1}^{n}\left(P_{f_i}\right) \leq P_f \leq 1-\left(1-P_{f_1}\right)\left(1-P_{f_2}\right)\ldots\left(1-P_{f_{n-1}}\right)\left(1-P_{f_n}\right) \qquad (8\text{-}124)$$

where P_{f_i} = the failure probability of the ith branch, for i = 1, 2, .., n branches of the event tree. The lower and upper limits correspond to fully dependent and independent branch occurrence (i.e., failure scenarios). The result of the system failure-probability evaluation is

$$0.434\times10^{-3} \leq P_f \leq 1.990\times10^{-3} \qquad (8\text{-}125)$$

The failure-probability of the original truss without posttensioning is

$$1.350 \times 10^{-3} \leq P_f \leq 7.390 \times 10^{-3} \qquad (8\text{-}126)$$

From Equations 8-125 and 8-126, it can be observed that the failure probability was improved approximately by a factor of 3.5 due to posttensioning.

8.7. PROBLEMS

8-1. For the following performance function, determine the safety index (β) using (a) the first-order reliability method and (b) the advanced second-moment method:

$$Z = X_1 X_2 - \sqrt{X_3}$$

The noncorrelated random variables are assumed to have the following probabilistic characteristics:

Random variable	Mean value	Coefficient of variation	Distribution type
X_1	1	0.25	Normal
X_2	5	0.05	Normal
X_3	4	0.20	Normal

8-2. For the following performance function, determine the safety index (β) using (a) the first-order reliability method and (b) the advanced second-moment method:

$$Z = X_1 - X_2 - \sqrt{X_3}$$

The noncorrelated random variables are assumed to have the following probabilistic characteristics:

Random variable	Mean value	Coefficient of variation	Distribution type
X_1	1	0.25	Lognormal
X_2	5	0.05	Lognormal
X_3	4	0.20	Lognormal

8-3. Redo Problem 8-1 using (a) the direct Monte Carlo simulation method and (b) the conditional expectation method. Use 100 and 2000 simulation cycles to estimate the failure probability and its coefficient of variation.

8-4. Redo Problem 8-2 using (a) the direct Monte Carlo simulation method and (b) the conditional expectation method. Use 100 and 2000 simulation cycles to estimate the failure probability and its coefficient of variation.

8-5. For the following performance function, determine the safety index (β) using (a) the first-order reliability method and (b) the advanced second-moment method:

$$Z = X_1 - X_2 - X_3$$

The noncorrelated random variables are assumed to have the following probabilistic characteristics:

Random variable	Mean value	Coefficient of variation	Distribution type
X_1	10	0.25	Normal
X_2	5	0.05	Normal
X_3	3	0.20	Normal

8-6. For the following performance function, determine the safety index (β) using (a) the first-order reliability method and (b) the advanced second-moment method:

$$Z = X_1 - X_2 - X_3$$

The noncorrelated random variables are assumed to have the following probabilistic characteristics:

Random variable	Mean value	Coefficient of variation	Distribution type
X_1	10	0.25	Normal
X_2	5	0.05	Normal
X_3	3	0.20	Type I largest

8-7. For the following performance function, determine an estimated failure probability and its coefficient of variation using (a) the direct simulation method and (b) the conditional expectation method with 100 and 2000 cycles:

$$Z = X_1 - X_2 - X_3 + I_1 \sin\left(\frac{I_2}{10} Z\right)$$

The last term in this equation introduces noise to the performance function. Assume that the constants $I_1 = 1$ and $I_2 = 1$. The noncorrelated random variables are assumed to have the following probabilistic characteristics, with Z having a standard deviation of 1:

Random variable	Mean value	Coefficient of variation	Distribution type
X_1	10	0.25	Normal
X_2	5	0.05	Normal
X_3	3	0.20	Normal
Z	0	Not applicable	Normal

8-8. In Problem 8-7, study the effect of changing the level of noise in the performance function. Investigate the effect of using $I_1 = 1$, 5, and 10 given that $I_2 = 1$ or 10 using the conditional expectation method, with $N = 10$, 100, 1000, and 2000 cycles. Discuss your results.

8-9. Redo Problem 8-7 using (a) importance sampling and (b) conditional expectation with antithetic variates.

8-10. Redo Problem 8-8 using (a) importance sampling and (b) conditional expectation with antithetic variates.

8-11. The change in the length of a rod due to axial force P is given by

$$\Delta L = \frac{PL}{AE}$$

where L = length of rod, P = applied axial force, A = cross-sectional area of rod, and E = modulus of elasticity. Assume that the change in length of the rod must not exceed 0.15 in. Using 2000 simulation cycles in (a) the direct Monte Carlo simulation method and (b) the conditional expectation method, determine the failure probability of the rod in meeting this limit. Assume the noncorrelated random variables to have the following probabilistic characteristics:

Random variable	Mean value	Coefficient of variation	Distribution type
P	100 kips	0.35	Lognormal
L	20 in.	0.05	Normal
E	30,000 ksi	0.10	Lognormal
A	1 in.^{2}	0.05	Normal

8-12. For the rod in Problem 8-11, study the effect of increasing the number of simulation cycles on the estimated failure probability. Use the following numbers of simulation cycles: 20, 100, 500, 1000, and 2000. Provide your results in the form of plots of estimated failure probability and its coefficient of variation as a function of the number of simulation cycles.

8-13. The ultimate moment capacity, M, of an underreinforced concrete rectangular section is given by

$$M = A_s f_y \left(d - \frac{a}{2} \right)$$

where

$$a = \frac{A_s f_y}{0.85 b f_c'}$$

in which the following are random variables: A_s = the cross-sectional area of the reinforcing steel, f_y = the yield stress (strength) of the steel, d = the distance from the reinforcing steel to the top fibers of the beam, b = the width of the beam, and f_c' = the ultimate stress (strength) of the concrete. Assume the random variables to be statistically noncorrelated with the following characteristics:

Random variable	Mean value	Coefficient of variation	Distribution type
$F = f_y A_s$	40,000 lb	0.15	Lognormal
d	15 in.	0.05	Lognormal
b	12 in.	0.05	Normal
f_c'	3000 psi	0.20	Lognormal

The applied random moment has a coefficient of variation of 0.2 and lognormal distribution type. Using simulation methods, determine the mean value of the applied moment that results in a failure probability of about 0.001 and a coefficient of variation of the estimated failure probability of, at the most, 0.05.

8-14. The following system consists of components with the indicated nonfailure probabilities:

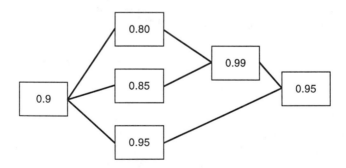

Compute the reliability of the system assuming independent failure events for the components.

8-15. For the system described in Problem 8-14, develop a fault tree model and evaluate the minimal cut set. Determine the reliability of the system.

8-16. The following system consists of components with the indicated nonfailure probabilities:

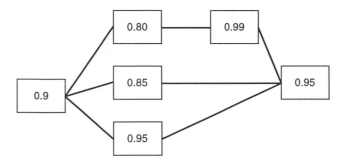

Compute the reliability of the system assuming independent failure events for the components.

8-17. For the system described in Problem 8-16, develop a fault tree model and evaluate the minimal cut set. Determine the reliability of the system.

8-18. Select an engineering system that interests you, define the system, define a system failure criterion, and develop (a) a fault tree model for the system and (b) an event tree model.

8-19. The following system consists of six components with the following reliability values and associated costs:

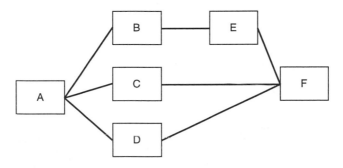

Components	Reliability	Initial (or component) cost	Failure cost
A	0.99	100	1000
B	0.95	150	2000
C	0.85	150	4000
D	0.90	100	2000
E	0.85	120	1000
F	0.99	500	5000

(a) Develop an event tree model for the system and (b) using risk analysis, which of the following two improvement options would you select:

Option	New reliability	New initial (or component) cost	New failure cost
Replace component B with a new design B'	0.99	200	800
Replace component C with a new design C'	0.9	250	3500

Chapter 9
Bayesian Methods

TABLE OF CONTENTS

9.1. INTRODUCTION

Engineers commonly need to solve a problem and make decisions based on limited information about one or more of the parameters of the problem. The types of information available to them can be

1. Objective information based on experimental results or observations
2. Subjective information based on experience, intuition, other previous problems that are similar to the one under consideration, or the physics of the problem

The first type of information can be dealt with using the theories of probability and statistics as was described in the previous chapters. In this type, probability is interpreted as the frequency of occurrence assuming sufficient repetitions of the problem, its outcomes, and parameters, as a basis of the information. The second type of information is subjective and can depend on the engineer or analyst studying the problem. In this type, uncertainty that exists needs to be dealt with using probabilities. However, the definition of probability is not the same as the first type because it is viewed herein as a subjective probability that reflects the state of knowledge of the engineer or the analyst.

9.2. BAYESIAN PROBABILITIES

It is common in engineering to encounter problems with both objective and subjective types of information. In these cases, it is desirable to utilize both types of

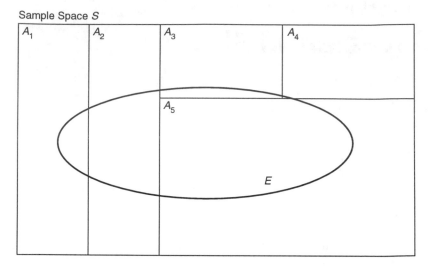

Sample Space S

FIGURE 9-1 Bayes' theorem.

information to obtain solutions or make decisions. The subjective probabilities are assumed to constitute a prior knowledge about a parameter, with gained objective information (or probabilities). Combining the two types produces posterior knowledge. The combination is performed based on Bayes' theorem as described in Chapter 3.

If A_1, A_2, ..., A_n represent the prior (subjective) information, or a partition of a sample space S, and $E \subset S$ represents the objective information (or arbitrary event) as shown in Figure 9-1, the theorem of total probability states that

$$P(E) = P(A_1) P(E|A_1) + P(A_2) P(E|A_2) + ... + P(A_n) P(E|A_n) \qquad (9\text{-}1)$$

This theorem is very important in computing the probability of the event E, especially in practical cases where the probability cannot be computed directly, but the probabilities of the partitioning events and the conditional probabilities can be computed.

Bayes' theorem is based on the same conditions of partitioning and events as the theorem of total probability and is very useful in computing the posterior (or reverse) probability of the type $P(A_i|E)$, for $i = 1, 2, ..., n$. The posterior probability can be computed as follows:

$$P(A_i|E) = \frac{P(A_i) P(E|A_i)}{P(A_1) P(E|A_1) + P(A_2) P(E|A_2) + ... + P(A_n) P(E|A_n)} \qquad (9\text{-}2)$$

The denominator of this equation is $P(E)$, which is based on the theorem of total probability. According to Equation 9-2, the prior knowledge, $P(A_i)$, is updated using the objective information, $P(E)$, to obtain the posterior knowledge, $P(A_i|E)$.

Example 9-1: Defective Products in Manufacturing Lines

Consider line 3 of the three manufacturing lines of Example 3-6. The three lines manufacture 20, 30, and 50% of the components, respectively. The quality assurance department of the producing factory determined that the probability of having defective products from lines 1, 2, and 3 are 0.1, 0.1, and 0.2, respectively. The following events were defined:

$$L_1 = \text{Component produced by line 1} \qquad\qquad (9\text{-}3a)$$

$$L_2 = \text{Component produced by line 2} \qquad\qquad (9\text{-}3b)$$

$$L_3 = \text{Component produced by line 3} \qquad\qquad (9\text{-}3c)$$

$$D = \text{Defective component} \qquad\qquad (9\text{-}3d)$$

Therefore, the following probabilities are given:

$$P(D|L_1) = 0.1 \qquad\qquad (9\text{-}4a)$$

$$P(D|L_2) = 0.1 \qquad\qquad (9\text{-}4b)$$

$$P(D|L_3) = 0.2 \qquad\qquad (9\text{-}4c)$$

Since these events are not independent, the joint probabilities can be determined as follows:

$$P(D \cap L_1) = P(D|L_1)\, P(L_1) = 0.1(0.2) = 0.02 \qquad\qquad (9\text{-}5a)$$

$$P(D \cap L_2) = P(D|L_2)\, P(L_2) = 0.1(0.3) = 0.03 \qquad\qquad (9\text{-}5b)$$

$$P(D \cap L_3) = P(D|L_3)\, P(L_3) = 0.2(0.5) = 0.10 \qquad\qquad (9\text{-}5c)$$

The theorem of total probability can be used to determine the probability of a defective component as follows:

$$P(D) = P(D|L_1)\, P(L_1) + P(D|L_2)\, P(L_2) + P(D|L_3)\, P(L_3)$$

$$= 0.1(0.2) + 0.1(0.3) + 0.2(0.5) = 0.02 + 0.03 + 0.10 \qquad (9\text{-}6)$$

$$= 0.15$$

FIGURE 9-2 Prior probability distribution for defective probability of line 3.

Therefore, on the average, 15% of the components produced by the factory are defective.

Because of the high contribution of line 3 to the defective probability, a quality assurance engineer subjected the line to further analysis. The defective probability for line 3 was assumed to be 0.2. An examination of the source of this probability revealed that it is subjective and is also uncertain. A better description of this probability can be as shown in Figure 9-2 in the form of a prior discrete distribution for the probability. The distribution is denoted as $P_P(p)$. The mean defective component probability $\bar{p}(D)$ based on this distribution is

$$\bar{p}(D) = 0.1(0.45) + 0.2(0.43) + 0.4(0.05) + 0.6(0.04) + 0.8(0.02) + 0.9(0.01)$$

$$= 0.200$$

(9-7)

Now assume that a component from line 3 was tested and found to be defective; the subjective prior distribution of Figure 9-3 needs to be revised to reflect the new (objective) information. The revised distribution is called the posterior distribution $[P'_P(p)]$ and can be computed using Equation 9-2 as follows:

$$P'_P(0.1) = \frac{0.45(0.1)}{0.2} = 0.225$$

(9-8a)

Similarly, the following posterior probabilities can be computed:

$$P'_P(0.2) = \frac{0.43(0.2)}{0.2} = 0.430$$

(9-8b)

$$P'_P(0.4) = \frac{0.05(0.4)}{0.2} = 0.100$$

(9-8c)

$$P'_P(0.6) = \frac{0.04(0.6)}{0.2} = 0.120$$

(9-8d)

$$P'_P(0.8) = \frac{0.02(0.8)}{0.2} = 0.080$$

(9-8e)

$$P_P'(0.9) = \frac{0.01(0.9)}{0.2} = 0.045 \tag{9-8f}$$

The resulting probabilities in Equations 9-8a to 9-8f add up to 1. Also, the average probability of 0.2 can be viewed as a normalizing factor for computing these probabilities. The mean defective component probability $\bar{p}(D)$ based on the posterior distribution is

$$\bar{p}(D) = 0.1(0.225) + 0.2(0.430) + 0.4(0.100) + 0.6(0.120) + 0.8(0.080) + 0.9(0.045) \tag{9-9a}$$

$$= 0.325$$

The posterior mean probability (0.325) is larger than the prior mean probability (0.200). The increase is due to the detected failure in the test. Now assume that a second component from line 3 was tested and found to be defective; the posterior distribution of Equations 9-8a to 9-8f needs to be revised to reflect the new (objective) information. The revised posterior distribution builds on the posterior distribution of Equations 9-8a to 9-8f, treating it as a prior distribution. Performing similar computations as in Equations 9-8 and 9-9 results in the posterior distribution shown in Table 9-1 in the column "Post. 2 D". The average defective component probability $\bar{p}(D)$ is also given in the table. The last row in the table is the average nondefective component probability [$\bar{p}(ND)$] for cases where a nondefective component results from a test. This value $\bar{p}(ND)$ can be computed similar to Equation 9-7 or 9-9. For example, the $\bar{p}(ND)$ in case of a nondefective test according to the prior distribution is

$$\bar{p}(ND) = (1-0.1)(0.225) + (1-0.2)(0.430) + (1-0.4)(0.100)$$

$$+ (1-0.6)(0.120) + (1-0.8)(0.080) + (1-0.9)(0.045) \tag{9-9b}$$

$$= 0.800$$

The computations for other cases are similarly performed as shown in Table 9-1. It should be noted that

$$\bar{p}(D) + \bar{p}(ND) = 1.0 \tag{9-10}$$

Now assume that a third component from line 3 was tested and found to be nondefective; the posterior distribution in column "Post. 2 D" of Table 9-1 needs to be revised to reflect the new (objective) information. The revised distribution is the posterior distribution [$P_P'(p)$] and can be computed using Equation 9-2 as follows:

$$P_P'(0.1) = \frac{0.0692(1-0.1)}{0.4883} = 0.1275 \tag{9-11a}$$

TABLE 9-1 Prior and Posterior Distributions for Line 3

Probability, p	P(p)	Post. 1 D	Post. 2 D	Post. 3 ND	Post. 4 D	Post. 5 D	Post. 6 D	Post. 7 D	Post. 8 D	Post. 9 D	Post. 10 D
0.1	0.45	0.225	0.0692308	0.127599244	0.0358090	0.0070718	0.0011355	0.0001638	2.22693E-05	2.912E-06	3.703E-07
0.2	0.43	0.430	0.2646154	0.433522369	0.2433245	0.0961062	0.0308633	0.0089068	0.002421135	0.0006332	0.0001611
0.4	0.05	0.100	0.1230769	0.151228733	0.1697613	0.1341016	0.0861300	0.0497125	0.027026626	0.0141359	0.0071914
0.6	0.04	0.120	0.2215385	0.181474480	0.3055703	0.3620744	0.3488266	0.3020033	0.246280127	0.1932203	0.1474458
0.8	0.02	0.080	0.1969231	0.080655325	0.1810787	0.2860835	0.3674882	0.4242131	0.461254413	0.4825060	0.4909318
0.9	0.01	0.045	0.1246154	0.025519849	0.0644562	0.1145626	0.1655564	0.2150004	0.262995429	0.3095016	0.3542696
$\bar{p}$ (D)	0.20	0.325	0.5116923	0.356332703	0.5063660	0.6227868	0.6930255	0.7357556	0.764764597	0.7862698	0.8029643
$\bar{p}$ (ND)	0.80	0.675	0.4883077	0.643667297	0.4936340	0.3772132	0.3069745	0.2642444	0.235235403	0.2137302	0.1970357

Similarly, the following posterior probabilities can be computed:

$$P_P'(0.2) = \frac{0.2646(1-0.2)}{0.4883} = 0.4335 \tag{9-11b}$$

$$P_P'(0.4) = \frac{0.1231(1-0.4)}{0.4883} = 0.1512 \tag{9-11c}$$

$$P_P'(0.6) = \frac{0.2215(1-0.6)}{0.4883} = 0.1815 \tag{9-11d}$$

$$P_P'(0.8) = \frac{0.1969(1-0.8)}{0.4883} = 0.0807 \tag{9-11e}$$

$$P_P'(0.9) = \frac{0.1246(1-0.9)}{0.4883} = 0.0255 \tag{9-11f}$$

The resulting probabilities in Equations 9-11a to 9-11f add up to 1. The probability $\bar{p}(ND)$ of 0.4883 was used in these calculations. The results of these calculations and the mean probability $\bar{p}(D)$ are shown in Table 9-1. It can be noted from the table that the mean defective component probability decreases as nondefective components are obtained through testing.

If the next seven tests result in defective components, the resulting posterior distributions are shown in Table 9-1. The results are also shown in Figure 9-3. It can be observed from the figure that the average probability is approaching one as more and more defective tests are obtained. Also, the effect of a nondefective component on the posterior probabilities can be seen in this figure.

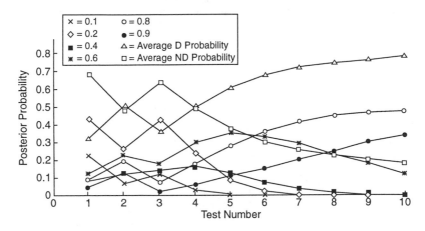

FIGURE 9-3 Posterior distributions for line 3.

9.3. BAYESIAN ESTIMATION OF PARAMETERS

9.3.1. Discrete Parameters

For an unknown parameter Θ, a prior distribution for the parameters can be subjectively determined and expressed using a probability mass function as

$$P_\Theta(\theta_i) = P(\Theta = \theta_i) \quad \text{for } i = 1, 2, \ldots, n \tag{9-12a}$$

or, in an abbreviated form, as

$$P_\Theta(\theta_i) = P(\theta_i) \quad \text{for } i = 1, 2, \ldots, n \tag{9-12b}$$

Therefore, the parameter Θ is assumed to take n discrete values with probabilities given by Equation 9-12. The distribution of Θ reflects the uncertainty in this parameter including its randomness. It is possible to have a parameter that is not random, but uncertain, therefore requiring its description through a distribution as given by Equation 9-12.

Now assume that new (objective) information ε was obtained. Using Equation 9-2, the posterior distribution of the parameter can be obtained as

$$P(\theta_i|\varepsilon) = \frac{P(\theta_i)\, P(\varepsilon|\theta_i)}{P(\theta_1)\, P(\varepsilon|\theta_1) + P(\theta_2)\, P(\varepsilon|\theta_2) + \ldots + P(\theta_n)\, P(\varepsilon|\theta_n)} \tag{9-13a}$$

where $P(\theta_i|\varepsilon)$ = the conditional probability of θ_i given ε, or the posterior probability for θ_i; $P(\theta_i)$ = prior probability as given by Equation 9-12; and $P(\varepsilon|\theta_i)$ = the probability of obtaining the new information (ε) given a certain value (θ_i) for the parameter. The following notation for the posterior distribution is also common:

$$P'(\theta_i) = \frac{P(\theta_i)\, P(\varepsilon|\theta_i)}{P(\theta_1)\, P(\varepsilon|\theta_1) + P(\theta_2)\, P(\varepsilon|\theta_2) + \ldots + P(\theta_n)\, P(\varepsilon|\theta_n)} \tag{9-13b}$$

where $P'(\theta_i)$ = the conditional probability of θ_i given ε, or the posterior probability for θ_i.

Using the prior distribution of the parameter Θ given by Equation 9-12, the expected value of the parameter can be computed as

$$E(\Theta) = \sum_{i=1}^{n} \theta_i\, P(\theta_i) \tag{9-14}$$

Based on the posterior distribution, the expected value of Θ can be computed as

$$E(\Theta|\varepsilon) = \sum_{i=1}^{n} \theta_i \, P'(\theta_i)$$
(9-15)

In many engineering problems, the parameter Θ can be used to define a probability distribution of a random variable X. The probability distribution of X can be either for a discrete random variable in the form of a probability mass function, $P_X(x)$, or for a continuous random variable in the form of a density function, $f_X(x)$. The Bayesian estimation of the parameter can be used to compute Bayesian probabilities that are obtained using the information gained about the parameters. For example, the probability that X is less than some value x_o can be computed using the prior distribution as

$$P(X < x_o) = \sum_{i=1}^{n} P(X < x_o|\theta_i) P(\theta_i)$$
(9-16)

or

$$F_X(x_o) = \sum_{i=1}^{n} F_X(x_o|\theta_i) P(\theta_i)$$
(9-17)

where $F_X(x)$ = the cumulative distribution function of X evaluated at x_o. Using the posterior distribution results in the following expressions:

$$P(X < x_o) = \sum_{i=1}^{n} P(X < x_o|\theta_i) P'(\theta_i)$$
(9-18)

or

$$F_X(x_o) = \sum_{i=1}^{n} F_X(x_o|\theta_i) P'(\theta_i)$$
(9-19)

Example 9-2: The Duration of a Construction Activity — Discrete Case

A construction engineer is interested in modeling the duration of a construction activity for the purpose of project scheduling and control. Based on previous experiences with similar activities, the average duration of the activity is uncertain and can be from 5 to 10 days. The engineer assumed a prior probability mass function for the average duration (D) as shown in Figure 9-4. The expected average duration based on this distribution according to Equation 9-14 is given by

$$E(D) = 5(0.1) + 6(0.1) + 7(0.2) + 8(0.3) + 9(0.2) + 10(0.1) = 7.7 \text{ days} \quad (9\text{-}20)$$

Assuming the duration of the activity (X) to have an exponential probability distribution, the probability of completing the activity within 8 days can be computed using Equation 9-17 as

$$P(X \le 8) = \sum_{i=1}^{n} F_X(8 \text{ days}|D_i) P(D_i) \quad (9\text{-}21a)$$

or

$$P(X \le 8) = \sum_{i=1}^{n} \left(1 - \exp\left(\frac{8}{D_i}\right)\right) P(D_i)$$

$$= 0.1\left(1 - \exp\left(-\frac{8}{5}\right)\right) + 0.1\left(1 - \exp\left(-\frac{8}{6}\right)\right) + 0.2\left(1 - \exp\left(-\frac{8}{7}\right)\right) \quad (9\text{-}21b)$$

$$+ 0.3\left(1 - \exp\left(-\frac{8}{8}\right)\right) + 0.2\left(1 - \exp\left(-\frac{8}{9}\right)\right) + 0.1\left(1 - \exp\left(-\frac{8}{10}\right)\right)$$

$$= 0.080 + 0.074 + 0.136 + 0.190 + 0.118 + 0.055 = 0.652$$

Now assume that an activity of this type was completed in another similar construction site in less than 8 days; the prior distribution in Figure 9-4 can be revised to reflect the new (objective) information. The revised (or posterior) distribution [$P'(d)$] can be computed using Equation 9-13 as follows:

$$P'(5) = \frac{0.080}{0.652} = 0.122 \qquad (9\text{-}22a)$$

Similarly, the following posterior probabilities can be computed:

$$P'(6) = \frac{0.074}{0.652} = 0.113 \qquad (9\text{-}22b)$$

$$P'(7) = \frac{0.136}{0.652} = 0.209 \qquad (9\text{-}22c)$$

$$P'(8) = \frac{0.190}{0.652} = 0.291 \qquad (9\text{-}22d)$$

$$P'(9) = \frac{0.118}{0.652} = 0.181 \qquad (9\text{-}22e)$$

$$P'(10) = \frac{0.055}{0.652} = 0.084 \qquad (9\text{-}22f)$$

The resulting probabilities in Equations 9-22a to 9-22f add up to 1 and are shown in Figure 9-5 with the prior probabilities. Also, the average probability of 0.652 can be viewed as a normalizing factor for computing these probabilities. The mean duration E(D) based on the posterior distribution is

$$E(D) = 5(0.122) + 6(0.113) + 7(0.209) + 8(0.291) + 9(0.181) + 10(0.084)$$
$$= 7.55 \text{ days} \qquad (9\text{-}23)$$

FIGURE 9-5 Prior and posterior probability distributions for average duration of a construction activity.

The resulting posterior average duration is smaller than the prior average duration. The probability of completing the activity within 8 days can be now be revised using the posterior distribution as

$$P(X \le 8) = \sum_{i=1}^{n} F_X(8 \text{ days}|D_i) P'(D_i) \qquad (9\text{-}24a)$$

or

$$P(X \le 8) = \sum_{i=1}^{n} \left(1 - \exp\left(\frac{8}{D_i}\right)\right) P'(D_i)$$

$$= 0.122\left(1 - \exp\left(-\frac{8}{5}\right)\right) + 0.113\left(1 - \exp\left(-\frac{8}{6}\right)\right) + 0.209\left(1 - \exp\left(-\frac{8}{7}\right)\right) \qquad (9\text{-}24b)$$

$$+ 0.291\left(1 - \exp\left(-\frac{8}{8}\right)\right) + 0.181\left(1 - \exp\left(-\frac{8}{9}\right)\right) + 0.084\left(1 - \exp\left(-\frac{8}{10}\right)\right)$$

$$= 0.098 + 0.083 + 0.142 + 0.184 + 0.106 + 0.046 = 0.660$$

9.3.2. Continuous Parameters

For an unknown parameter Θ, a prior distribution for the parameters can be subjectively determined and expressed using a probability density function $f_\Theta(\theta)$. The parameter Θ is assumed to be continuous with probabilities that can be computed based on its density function. Again, the distribution of Θ reflects the uncertainty in this parameter including its randomness.

Now assume that new (objective) information ε was obtained. Using Equation 9-2, the posterior distribution for the parameter can be obtained as

$$f_\Theta(\theta|\varepsilon) = \frac{f_\Theta(\theta) P(\varepsilon|\theta)}{\int_{-\infty}^{\infty} P(\varepsilon|\theta) f_\Theta(\theta) \, d\theta} \qquad (9\text{-}25)$$

where $f_\Theta(\theta)$ = the prior density function of Θ; $f_\Theta(\theta|\varepsilon)$ = the posterior density function of Θ; and $P(\varepsilon|\theta)$ = the probability of obtaining the new information (ε) given a certain value for the parameter (θ). The probability $P(\varepsilon|\theta)$ is called the likelihood function $L(\theta)$. The following notations for the posterior distribution are also common:

$$f'_\Theta(\theta) = \frac{f_\Theta(\theta) L(\theta)}{\int_{-\infty}^{\infty} L(\theta) f_\Theta(\theta) \, d\theta} \qquad (9\text{-}26)$$

where $f'_\Theta(\theta)$ = the conditional density function of θ given ε, or the posterior density function of Θ.

Using the prior density function of the parameter Θ, the expected value of the parameter can be computed as

$$E(\Theta) = \int_{-\infty}^{\infty} \theta f_\Theta(\theta)\, d\theta \qquad (9\text{-}27)$$

Based on the posterior distribution, the expected value of Θ can be computed as

$$E(\Theta|\varepsilon) = \int_{-\infty}^{\infty} \theta f'_\Theta(\theta)\, d\theta \qquad (9\text{-}28)$$

In many engineering problems, the parameter Θ can be used to define a probability distribution of a random variable X. The probability distribution of X can be either for a discrete random variable in the form of a probability mass function, $P_X(x)$, or for a continuous random variable in the form of a density function, $f_X(x)$. The Bayesian estimation of the parameter can be used to compute Bayesian probabilities that are obtained with the gained information about the parameters. For example, the probability that X is less than some value x_o can be computed using the prior distribution as

$$P(X < x_o) = \int_{-\infty}^{\infty} P(X < x_o|\theta) f_\Theta(\theta)\, d\theta \qquad (9\text{-}29)$$

or

$$F_X(x_o) = \int_{-\infty}^{\infty} F_X(x_o|\theta) f_\Theta(\theta)\, d\theta \qquad (9\text{-}30)$$

where $F_X(x_o)$ = the cumulative distribution function of X evaluated at x_o. Using the posterior distribution results in the following expression:

$$P(X < x_o) = \int_{-\infty}^{\infty} P(X < x_o|\theta) f'_\Theta(\theta)\, d\theta \qquad (9\text{-}31)$$

or

$$F_X(x_o) = \int_{-\infty}^{\infty} F_X(x_o|\theta) f'_\Theta(\theta)\, d0 \qquad (9\text{-}32)$$

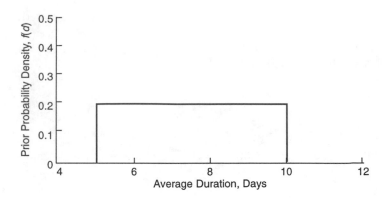

FIGURE 9-6 Prior probability distribution for average duration of a construction activity.

Example 9-3: The Duration of a Construction Activity — Continuous Case

The construction engineer in Example 9-2 is interested in modeling the duration of a construction activity for the purpose of project scheduling and control. Based on previous experiences with similar activities, the average duration of the activity is uncertain and is assumed again to be from 5 to 10 days. In this example, the engineer assumed a prior probability density function for the average duration (D) to be uniform, as shown in Figure 9-6. The expected average duration based on this distribution according to Equation 9-27 is given by

$$E(D) = \int_{-\infty}^{\infty} \theta f_\Theta(\theta)\, d\theta$$

$$= \int_5^{10} \theta(0.2)\, d\theta = 7.5 \text{ days}$$

(9-33)

Assuming the duration of the activity (X) to have an exponential probability distribution, the probability of completing the activity within 8 days can be computed using Equation 9-30 as

$$P(X \le 8) = \int_{-\infty}^{\infty} F_X(x_o|\theta) f_\Theta(\theta)\, d\theta \qquad (9\text{-}34a)$$

or

$$P(X \le 8) = \int_5^{10} \left(1 - \exp\left(-\frac{8}{\theta}\right)\right)(0.2)\, d\theta = 0.6626 \qquad (9\text{-}34b)$$

Now assume that an activity of this type was completed in another similar construction site in less than 8 days; the prior distribution in Figure 9-6 can be revised to reflect the new (objective) information. The revised (or posterior) distribution $[P'_D(d)]$ can be computed using Equation 9-26. The likelihood function $L(\theta)$ in Equation 9-26 is defined as the conditional probability of obtaining the outcome of the experiment given a value for the average activity duration D (i.e., the parameter θ). This conditional probability can be computed using the cumulative distribution function of the exponential distribution as

$$L(d) = 1 - \exp\left(-\frac{8}{\theta}\right) \tag{9-35}$$

Therefore, using Equation 9-26 results in the following:

$$f'_D(d) = \frac{f_D(d)\left[1 - \exp\left(-\frac{8}{d}\right)\right]}{\int_{-\infty}^{\infty}\left[1 - \exp\left(-\frac{8}{d}\right)\right]f_D(\theta)\,d\theta} \tag{9-36}$$

Substituting Equation 9-34b into Equation 9-36 produces

$$f'_D(d) = \frac{0.2}{0.6626}\left[1 - \exp\left(-\frac{8}{d}\right)\right] \quad \text{for } 5 \le d \le 10 \text{ days}$$

$$= 0.302\left[1 - \exp\left(-\frac{8}{d}\right)\right] \tag{9-37}$$

The area under the density function given by Equation 9-37 is 1. The prior and posterior density functions are shown in Figure 9-7. The mean duration $E(D)$ based on the posterior distribution is

$$E(D) = \int_{-\infty}^{\infty}\theta\, f'_\theta(\theta)\,d\theta$$

$$= \int_{5}^{10} 0.302\theta\left(1 - \exp\left(-\frac{8}{\theta}\right)\right)d\theta = 7.33 \text{ days} \tag{9-38}$$

The resulting posterior average duration is smaller than the prior average duration. The probability of completing the activity within 8 days can now be revised using the posterior distribution as

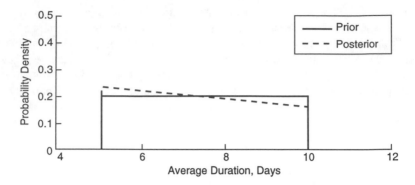

FIGURE 9-7 Prior and posterior probability density functions
for average duration of a construction activity.

$$P(X \le 8) = \int_{-\infty}^{\infty} F_X(x_o|\theta) f_\Theta'(\theta) \, d\theta \qquad (9\text{-}39a)$$

or

$$P(X \le 8) = \int_5^{10} 0.302 \left(1 - \exp\left(-\frac{8}{\theta}\right)\right)^2 d\theta = 0.671 \qquad (9\text{-}39b)$$

9.4. BAYESIAN STATISTICS

The Bayesian methods that were developed in the previous sections can be used in the statistical analysis of data. In this section, two cases are considered to illustrate their use in statistical analysis. The first case deals with a random variable X that is normally distributed with a known standard deviation. The mean value of the random variable is of interest and is estimated using Bayesian methods. In the second case, the random variable X is also normally distributed, but its standard deviation is unknown. In this case, both the mean value and the variance of the random variable are of interest and are estimated using Bayesian methods.

9.4.1. Mean Value with Known Variance

A random variable X is considered to be normally distributed with a known variance σ^2. The mean value of the random variable is of interest and is unknown. The prior distribution of the unknown mean (μ) is normal, with a mean value and variance, μ_o, and σ_o^2, respectively. New (objective) information was obtained by a sample of size n. The mean value based on the sample is $\overline{X}$. We are interested in determining the posterior distribution of the mean. Using Equation 9-26, the following expression can be established:

$$f'(\mu) = \frac{f(\mu)L(\mu)}{\int\limits_{-\infty}^{\infty} L(\mu)f(\mu)\,d\mu} \qquad (9\text{-}40)$$

where $f(\mu)$ = the prior density function of μ, which is normal with mean and variance of μ_o and σ_o^2, respectively [i.e., $N(\mu_o, \sigma_o^2)$]; $f'(\mu)$ = the posterior density function of the unknown mean μ; and $L(\mu)$ = the likelihood function for the sample of size n. The likelihood function can be computed as the product of n values of the density function of the normal distribution with a mean μ and standard deviation σ, each evaluated at a sampled value x_i. The product can be expressed as

$$L(\mu) = \frac{1}{(2\pi)^{n/2}\,\sigma^n}\exp\left[-\frac{1}{2}\sum_{i=1}^{n}\left(\frac{x_i - \mu}{\sigma}\right)^2\right] \qquad (9\text{-}41)$$

It can be shown that by substituting Equation 9-41 in Equation 9-40, the resulting $f'(\mu)$ is normally distributed with the following mean value and variance, respectively:

$$\mu' = \frac{n\overline{X}\sigma_o^2 + \mu_o\sigma^2}{n\sigma_o^2 + \sigma^2} \qquad (9\text{-}42a)$$

and

$$\sigma'^2 = \frac{\sigma_o^2\sigma^2}{n\sigma_o^2 + \sigma^2} \qquad (9\text{-}42b)$$

The resulting μ' and σ' are the posterior mean and standard deviation of the unknown mean value μ. Using the normal posterior distribution, any Bayesian probabilities of interest for the random variable X can be computed.

The prior and posterior mean values and variances can also be used in other aspects of statistical analysis such as confidence intervals and hypothesis testing. For example, they can be used to establish the following prior confidence interval on the mean:

$$\mu_o - z_{\alpha/2}\sigma \le \mu \le \mu_o + z_{\alpha/2}\sigma \qquad (9\text{-}43a)$$

Also, they can be used to establish the following posterior confidence interval:

$$\mu' - z_{\alpha/2}\sigma' \le \mu \le \mu' + z_{\alpha/2}\sigma' \qquad (9\text{-}43b)$$

where $(1 - \alpha)$ is the confidence level. In a similar approach, prior hypothesis testing and posterior hypothesis testing can be performed.

Example 9-4: Mean Concrete Strength

An existing reinforced concrete structure needs to be evaluated for its strength. The mean concrete strength is an important quantity for this evaluation. A review of the design and construction documents, which are about 15 years old, indicated that the mean concrete strength has a mean and variance as follows:

$$\mu_o = 3000 \text{ psi}$$

and

$$\sigma_o^2 = 90,000 \text{ (psi)}^2$$

These values can be considered to constitute prior information about the mean strength. Based on the construction practice during the period of construction of the building, it was common to produce concrete with a standard deviation (σ) of 400 psi.

Due to aging and environmental factors, μ_o and σ_o are uncertain. Therefore, it was decided to perform nondestructive testing of the concrete strength. A certified testing company produced a sample size (n) of 10, with a sample mean of 3500 psi. This information can be used to update the prior mean and variance of the average concrete strength using Equations 9-42a and 9-42b as follows:

$$\mu' = \frac{10(3500)(90,000) + (3000)(400)^2}{10(90,000) + (400)^2} = 3425 \text{ psi} \qquad (9\text{-}44\text{a})$$

and

$$\sigma'^2 = \frac{90,000(400)^2}{10(90,000) + (400)^2} = 13,585 \text{ (psi)}^2 \qquad (9\text{-}44\text{b})$$

Therefore, the posterior standard deviation is 117 psi.

The prior mean value and variance can be used to establish the following prior 95% confidence interval on the mean (using Equation 9-43a):

$$3000 - 1.96(300) \le \mu \le 3000 + 1.96(300) \qquad (9\text{-}45\text{a})$$

or

$$2412 \text{ psi} \le \mu \le 3588 \text{ psi} \qquad (9\text{-}45\text{b})$$

The posterior mean and variance can be used to establish the following posterior 95% confidence interval (using Equation 9-43b):

$$3425 - 1.96(117) \leq \mu \leq 3425 + 1.96(117) \tag{9-46a}$$

or

$$3197 \text{ psi} \leq \mu \leq 3653 \text{ psi} \tag{9-46b}$$

By comparing Equations 9-45 and 9-46, it can be observed that the nondestructive testing resulted in reducing the width of the confidence interval, i.e., reducing the uncertainty in the mean concrete strength.

9.4.2. Mean Value with Unknown Variance

The random variable X in this case is considered to be normally distributed with a unknown mean value (μ) and unknown variance (σ^2). Both the mean value and variance of the random variable are of interest and are unknown. The prior joint distribution of the unknown mean (μ) and unknown variance (σ^2) is assumed to be normal-gamma, which is defined as the product of a normal distribution for the mean (μ) and a gamma distribution for the variance (σ^2). The prior information about the mean and variance is based on a sample of size N with sample mean and variance of $\overline{X}_o$ and S_o^2, respectively. New (objective) information was obtained by a sample of size n. The mean value and variance based on the sample are $\overline{X}$ and S^2, respectively. We are interested in determining the posterior distribution of the mean and variance. Using Equation 9-26, it can be shown that the posterior distribution is also a normal-gamma. The posterior mean ($\overline{X}'$) and posterior variance (S'^2) can be shown to be

$$\overline{X}' = \frac{N\overline{X}_o + n\overline{X}}{n'} \tag{9-47a}$$

and

$$S'^2 = \frac{(N-1)S_o^2 + N\overline{X}_o^2 + (n-1)S^2 + n\overline{X}^2 - n'\overline{X}'^2}{n'-1} \tag{9-47b}$$

where

$$n' = N + n \tag{9-47c}$$

The resulting values from Equations 9-47a to 9-47c are the posterior mean and standard deviation of the unknown mean and variance.

Example 9-5: Mean and Variance of Concrete Strength

The existing reinforced concrete structure that was discussed in Example 9-4 is reexamined herein. In this example both the mean and variance of the concrete strength are of interest. Assume the following prior information about the concrete strength: mean strength, $\overline{X}_o$ = 3,000 psi; variance of strength, S_o^2, = 90,000 (psi)²; and sample size, N = 10. These values are considered to constitute prior information about the concrete strength. Due to aging and environmental factors, $\overline{X}_o$ and S_o^2 are uncertain. Therefore, it was decided to perform nondestructive testing of the concrete strength. A certified testing company produced a sample of size (n) 20, with a sample mean of 3500 psi and a sample standard deviation of 400 psi. This information can be used to update the prior mean and variance of the average concrete strength using Equations 9-47a to 9-47c as follows:

$$\overline{X}' = \frac{10(3000) + (20)(3500)}{10 + 20} = 3333 \text{ psi} \tag{9-48a}$$

and

$$S'^2 = \frac{(10-1)(90000) + 10(3000)^2 + (20-1)(400)^2 + (20)(3500)^2 - (10+20)(3333)^2}{10 + 20 - 1} \tag{9-48b}$$

$$= 190,230 \text{ (psi)}^2$$

where

$$n' = 10 + 20 = 30 \tag{9-48c}$$

Therefore, the posterior standard deviation is 436 psi.

9.5. APPLICATIONS

9.5.1. Sampling from an Assembly Line

An assembly line produces components of automobiles that have a defective probability distribution as shown in Figure 9-8. This prior distribution is called $P_P(p)$. The mean probability $\overline{P}$ based on this distribution is

$$\overline{P} = 0.1(0.6) + 0.2(0.4) \tag{9-49}$$

$$= 0.140$$

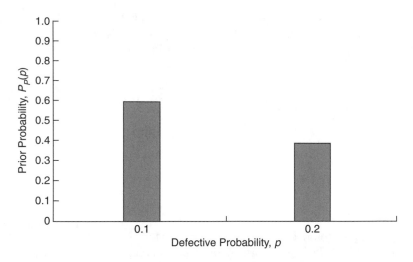

FIGURE 9-8 Prior probability distribution for defective probability of an assembly line.

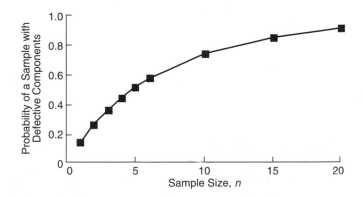

FIGURE 9-9 Probability of a sample with defective components.

For a sample of size n, the probability $P(D)$ of obtaining one or more defective components can be computed based on the binomial distribution as

$$P(D) = 0.6\left[1 - \binom{n}{0}(0.1)^0 (1-0.1)^n\right] + 0.4\left[1 - \binom{n}{0}(0.2)^0 (1-0.2)^n\right]$$

(9-50)

$$= 0.6\left[1 - (0.9)^n\right] + 0.4\left[1 - (0.8)^n\right]$$

The probability according to Equation 9-50 is shown in Figure 9-9 for different values of n.

Now assume that a sample of size $n = 4$ was tested and found to contain defective components; the subjective prior distribution of Figure 9-8 needs to be revised to reflect the new (objective) information. From Figure 9-9, the probability of the sample of size $n = 4$ containing defective components is 0.4425. The posterior distribution $[P'_P(p)]$ can then be computed using Equation 9-13 as follows:

$$P'_P(0.1) = \frac{0.6\left[1-(1-0.1)^4\right]}{0.4425} = 0.4663 \qquad (9\text{-}51a)$$

Similarly, the second posterior probability can be computed as follows:

$$P'_P(0.2) = \frac{0.4\left[1-(1-0.2)^4\right]}{0.4425} = 0.5337 \qquad (9\text{-}51b)$$

The resulting probabilities in Equation 9-51 add up to 1. Also, the average probability of 0.4425 can be viewed as a normalizing factor for computing these probabilities. The mean probability $\overline{P}$ based on the posterior distribution is

$$\overline{P} = 0.1(0.4663) + 0.2(0.5337)$$
$$= 0.1534 \qquad (9\text{-}52)$$

The result shows an increase from the average probability of 0.140 as given by Equation 9-49 due to the detected failure by testing.

9.5.2. Scour around Bridge Piers

The scour around a bridge pier develops due to the infrequent occurrence of extreme flow in a river and is of interest to engineers. The mean scour depth is an important quantity for the evaluation of bridge piers in streams. A review of several bridge piers in streams indicated that the mean annual scour has a mean and variance as follows:

$$\mu_o = 3 \text{ in.}$$

and

$$\sigma_o^2 = 1 \text{ in.}^2$$

These values can be considered to constitute prior information about the mean scour. Based on the physics of scour, it can be considered to be random variable with a high variability level with an assumed coefficient of variation of 0.5. Therefore, scour can be assumed to have, in this case, a known or assumed standard deviation (σ) of $0.5 \times 3 = 1.5$ in.

In order to improve estimates of the annual scour depth, a sample of size (n) five bridges was analyzed for estimating the annual scour. A sample mean of 2 in. was obtained. This information can be used to update the prior mean and variance of the average scour using Equations 9-42a and 9-42b as follows:

$$\mu' = \frac{5(2)(1)+(3)(1.5)^2}{5(1)+(1.5)^2} = 2.31 \text{ in.} \tag{9-53}$$

and

$$\sigma'^2 = \frac{(1)(1.5)^2}{5(1)+(1.5)^2} = 0.31 \text{ (in.)}^2 \tag{9-54}$$

Therefore, the posterior standard deviation is 0.56 in.

The prior mean value and variance can be used to establish the following prior 95% confidence interval on the mean (using Equation 9-43a):

$$3 - 1.96(1) \leq \mu \leq 3 + 1.96(1) \tag{9-55a}$$

or

$$1.04 \text{ in.} \leq \mu \leq 4.96 \text{ in.} \tag{9-55b}$$

The posterior mean and variance can be used to establish the following posterior 95% confidence interval (using Equation 9-43b):

$$2.31 - 1.96(0.56) \leq \mu \leq 2.31 + 1.96(0.56) \tag{9-56a}$$

or

$$1.21 \text{ in.} \leq \mu \leq 3.41 \text{ in.} \tag{9-56b}$$

By comparing Equations 9-55 and 9-56, it can be observed that the gained information by the sample resulted in reducing the width of the confidence interval, i.e., reducing the uncertainty in the mean concrete strength. However, the range might still be considered large, requiring the collection of more information.

9.6. PROBLEMS

9-1. The accident probability at a new intersection is of interest to a traffic engineer. The engineer subjectively estimated the weekly accident probability as follows:

Weekly accident probability	Subjective probability of accident probability
0.1	0.30
0.2	0.40
0.4	0.20
0.6	0.05
0.8	0.04
0.9	0.01

Solve the following:
a. What is the average accident probability based on the prior information?
b. Given an accident in the first week of traffic, update the distribution of the accident probability.
c. What is the new average accident probability based on the posterior information?
d. Given an accident in the first and second weeks and no accidents in the third week of traffic, update the distribution of the accident probability.
e. What is the average accident probability after the second week.
f. Given no additional accidents for the weeks 4, 5, 6, 7, 8, 9, and 10, update the distribution and the average accident probability. Plot your results.

9-2. The probability of back injuries of construction workers at XYZ Corporation is of interest to a safety engineer. The engineer subjectively estimated the annual probability as follows:

Annual probability	Subjective probability of annual probability
0.05	0.70
0.10	0.20
0.20	0.04
0.60	0.02
0.80	0.02
0.90	0.02

Solve the following:
a. What is the average annual probability based on the prior information?
b. Given no back injuries in a year, update the distribution of the annual probability.
c. What is the new average annual probability based on the posterior information?
d. Given injuries in the first and second years, and no injuries in the following year, update the distribution of the annual probability.
e. What is the average annual probability after the second year?
f. Given no additional injuries for the years 4, 5, 6, 7, 8, 9, and 10, update the distribution and the average annual probability. Plot your results.

9-3. In Problem 9-1 assume that a weekly rate is used with a uniform prior distribution over the range 0.1 to 0.5 per week. Use an exponential likelihood function. Determine the posterior distribution based on additional information that an accident occurred in less than 2 weeks.

9-4. In Problem 9-2 assume that an annual rate is used with a prior uniform distribution over the range 10 to 18 per year. Use an exponential likelihood function. Determine the posterior distribution based on additional information that an injury occurred in less than 2 weeks (i.e., 0.04 year).

9-5. Redo Problem 9-3 assuming the weekly rate to have a prior distribution that is exponential with a mean of 0.2 per week.

9-6. Redo Problem 9-4 assuming the annual rate to have a prior distribution that is exponential with a mean of 15 per year.

9-7. The mean thickness of a plate girder used in a suspension bridge needs to be estimated using ultrasonic testing. The mean thickness can be assumed to have a normal probability distribution with a mean (μ) and standard deviation (σ). The standard deviation (σ) is known to be 1.8 mm. The construction drawings indicate a mean thickness of 15 mm and a standard deviation of 2.0 mm. A sample of $n = 20$ was collected. The sample resulted in a mean of 16 mm. Determine (a) the posterior mean thickness; (b) the posterior standard deviation; and (c) the 95% confidence intervals for thickness based on prior and posterior information.

9-8. The mean thickness of the shell of a pressure vessel needs to be estimated using ultrasonic testing. The mean thickness can be assumed to have a normal probability distribution with a mean (μ) and standard deviation (σ). The standard deviation (σ) is known to be 0.5 mm. The construction drawings indicate a mean thickness of 635 mm and a standard deviation of 1.0 mm. A sample of $n = 20$ was collected. The sample resulted in a mean of 625 mm. Determine (a) the posterior mean thickness; (b) the posterior standard deviation; and (c) the 95% confidence intervals for the mean thickness based on prior and posterior information. Does the pressure vessel meet the specification of a mean thickness of at least 628 mm using two-sided hypothesis testing with a significance level of 5%?

9-9. The air quality in a building is of interest to an environmental engineer. The mean air quality is measured in the form of the amount of pollutant in the air in ppm. The mean amount of pollutant can be assumed to have a normal probability distribution with a mean (μ) and standard deviation (σ). The standard deviation of pollutant (σ) is known to be 0.2 ppm. The air quality based on the records of the building has a mean of 3 ppm and a standard deviation of 0.2 ppm. A sample of $n = 10$ was collected. The sample resulted in a mean of 2.7 ppm. Determine (a) the posterior mean pollutant concentration; (b) the posterior standard deviation; and (c) the 95% confidence intervals for the mean concentration based on prior and posterior information.

9-10. Redo Problem 9-9 using a sample size $n = 100$ with the same mean and standard deviation. Compare and discuss your results.

Appendix A
Probability and Statistics Tables

TABLE OF CONTENTS

A-1. Cumulative Distribution Function of Standard Normal [$\Phi(z)$]

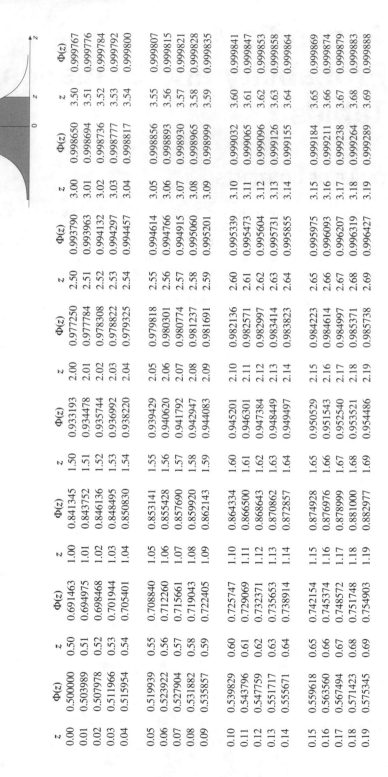

z	$\Phi(z)$	z	$\Phi(z)$	z	$\Phi(z)$	z	$\Phi(z)$	z	$\Phi(z)$	z	$\Phi(z)$	z	$\Phi(z)$	z	$\Phi(z)$
0.00	0.500000	0.50	0.691463	1.00	0.841345	1.50	0.933193	2.00	0.977250	2.50	0.993790	3.00	0.998650	3.50	0.999767
0.01	0.503989	0.51	0.694975	1.01	0.843752	1.51	0.934478	2.01	0.977784	2.51	0.993963	3.01	0.998694	3.51	0.999776
0.02	0.507978	0.52	0.698468	1.02	0.846136	1.52	0.935744	2.02	0.978308	2.52	0.994132	3.02	0.998736	3.52	0.999784
0.03	0.511966	0.53	0.701944	1.03	0.848495	1.53	0.936992	2.03	0.978822	2.53	0.994297	3.03	0.998777	3.53	0.999792
0.04	0.515954	0.54	0.705401	1.04	0.850830	1.54	0.938220	2.04	0.979325	2.54	0.994457	3.04	0.998817	3.54	0.999800
0.05	0.519939	0.55	0.708840	1.05	0.853141	1.55	0.939429	2.05	0.979818	2.55	0.994614	3.05	0.998856	3.55	0.999807
0.06	0.523922	0.56	0.712260	1.06	0.855428	1.56	0.940620	2.06	0.980301	2.56	0.994766	3.06	0.998893	3.56	0.999815
0.07	0.527904	0.57	0.715661	1.07	0.857690	1.57	0.941792	2.07	0.980774	2.57	0.994915	3.07	0.998930	3.57	0.999821
0.08	0.531882	0.58	0.719043	1.08	0.859920	1.58	0.942947	2.08	0.981237	2.58	0.995060	3.08	0.998965	3.58	0.999828
0.09	0.535857	0.59	0.722405	1.09	0.862143	1.59	0.944083	2.09	0.981691	2.59	0.995201	3.09	0.998999	3.59	0.999835
0.10	0.539829	0.60	0.725747	1.10	0.864334	1.60	0.945201	2.10	0.982136	2.60	0.995339	3.10	0.999032	3.60	0.999841
0.11	0.543796	0.61	0.729069	1.11	0.866500	1.61	0.946301	2.11	0.982571	2.61	0.995473	3.11	0.999065	3.61	0.999847
0.12	0.547759	0.62	0.732371	1.12	0.868643	1.62	0.947384	2.12	0.982997	2.62	0.995604	3.12	0.999096	3.62	0.999853
0.13	0.551717	0.63	0.735653	1.13	0.870862	1.63	0.948449	2.13	0.983414	2.63	0.995731	3.13	0.999126	3.63	0.999858
0.14	0.555671	0.64	0.738914	1.14	0.872857	1.64	0.949497	2.14	0.983823	2.64	0.995855	3.14	0.999155	3.64	0.999864
0.15	0.559618	0.65	0.742154	1.15	0.874928	1.65	0.950529	2.15	0.984223	2.65	0.995975	3.15	0.999184	3.65	0.999869
0.16	0.563560	0.66	0.745374	1.16	0.876976	1.66	0.951543	2.16	0.984614	2.66	0.996093	3.16	0.999211	3.66	0.999874
0.17	0.567494	0.67	0.748572	1.17	0.878999	1.67	0.952540	2.17	0.984997	2.67	0.996207	3.17	0.999238	3.67	0.999879
0.18	0.571423	0.68	0.751748	1.18	0.881000	1.68	0.953521	2.18	0.985371	2.68	0.996319	3.18	0.999264	3.68	0.999883
0.19	0.575345	0.69	0.754903	1.19	0.882977	1.69	0.954486	2.19	0.985738	2.69	0.996427	3.19	0.999289	3.69	0.999888

$P(Z < z) = \Phi(z) = area$

z	P	z	P	z	P	z	P	z	P	z	P	z	P	z	P
0.20	0.579260	0.70	0.758036	1.20	0.884930	1.70	0.955435	2.20	0.986097	2.70	0.996533	3.20	0.999313	3.70	0.999892
0.21	0.583166	0.71	0.761148	1.21	0.886860	1.71	0.956367	2.21	0.986447	2.71	0.996636	3.21	0.999336	3.71	0.999896
0.22	0.587064	0.72	0.764238	1.22	0.888767	1.72	0.957284	2.22	0.986791	2.72	0.996736	3.22	0.999359	3.72	0.999900
0.23	0.590954	0.73	0.767305	1.23	0.890651	1.73	0.958185	2.23	0.987126	2.73	0.996833	3.23	0.999381	3.73	0.999904
0.24	0.594835	0.74	0.770350	1.24	0.892512	1.74	0.959071	2.24	0.987455	2.74	0.996928	3.24	0.999402	3.74	0.999908
0.25	0.598706	0.75	0.773373	1.25	0.894350	1.75	0.959941	2.25	0.987776	2.75	0.997020	3.25	0.999423	3.75	0.999912
0.26	0.602568	0.76	0.776373	1.26	0.896165	1.76	0.960796	2.26	0.988089	2.76	0.997110	3.26	0.999443	3.76	0.999915
0.27	0.606420	0.77	0.779350	1.27	0.897958	1.77	0.961636	2.27	0.988396	2.77	0.997197	3.27	0.999462	3.77	0.999918
0.28	0.610262	0.78	0.782305	1.28	0.899727	1.78	0.962462	2.28	0.988696	2.78	0.997282	3.28	0.999481	3.78	0.999922
0.29	0.614092	0.79	0.785236	1.29	0.901475	1.79	0.963273	2.29	0.988989	2.79	0.997365	3.29	0.999499	3.79	0.999925
0.30	0.617912	0.8	0.788145	1.30	0.903199	1.80	0.964070	2.30	0.989276	2.80	0.997445	3.30	0.999516	3.80	0.999928
0.31	0.621720	0.81	0.791030	1.31	0.904902	1.81	0.964852	2.31	0.989556	2.81	0.997523	3.31	0.999533	3.81	0.999931
0.32	0.625517	0.82	0.793892	1.32	0.906583	1.82	0.965621	2.32	0.989830	2.82	0.997599	3.32	0.99955	3.82	0.999933
0.33	0.629301	0.83	0.796731	1.33	0.908241	1.83	0.966375	2.33	0.990097	2.83	0.997673	3.33	0.999566	3.83	0.999936
0.34	0.633072	0.84	0.799546	1.34	0.909877	1.84	0.967116	2.34	0.990358	2.84	0.997744	3.34	0.999581	3.84	0.999938
0.35	0.636831	0.85	0.802337	1.35	0.911492	1.85	0.967843	2.35	0.990613	2.85	0.997814	3.35	0.999596	3.85	0.999941
0.36	0.640576	0.86	0.805105	1.36	0.913085	1.86	0.968557	2.36	0.990863	2.86	0.997882	3.36	0.999610	3.86	0.999943
0.37	0.644309	0.87	0.807850	1.37	0.914656	1.87	0.969258	2.37	0.991106	2.87	0.997948	3.37	0.999624	3.87	0.999946
0.38	0.648027	0.88	0.810570	1.38	0.916207	1.88	0.969946	2.38	0.991344	2.88	0.998012	3.38	0.999637	3.88	0.999948
0.39	0.651732	0.89	0.813267	1.39	0.917735	1.89	0.970621	2.39	0.991576	2.89	0.998074	3.39	0.99965	3.89	0.999950
0.40	0.655422	0.9	0.815940	1.40	0.919243	1.90	0.971284	2.40	0.991802	2.90	0.998134	3.40	0.999663	3.90	0.999952
0.41	0.659097	0.91	0.818589	1.41	0.920730	1.91	0.971933	2.41	0.992024	2.91	0.998193	3.41	0.999675	3.91	0.999954
0.42	0.662757	0.92	0.821214	1.42	0.922196	1.92	0.972571	2.42	0.992240	2.92	0.998250	3.42	0.999687	3.92	0.999956
0.43	0.666402	0.93	0.823815	1.43	0.923641	1.93	0.973197	2.43	0.992451	2.93	0.998305	3.43	0.999698	3.93	0.999958
0.44	0.670032	0.94	0.826391	1.44	0.925066	1.94	0.973810	2.44	0.992656	2.94	0.998359	3.44	0.999709	3.94	0.999959
0.45	0.673645	0.95	0.828944	1.45	0.926471	1.95	0.974412	2.45	0.992857	2.95	0.998411	3.45	0.999720	3.95	0.999961
0.46	0.677242	0.96	0.831473	1.46	0.927855	1.96	0.975002	2.46	0.993053	2.96	0.998462	3.46	0.999730	3.96	0.999963
0.47	0.680823	0.97	0.833977	1.47	0.929219	1.97	0.975581	2.47	0.993244	2.97	0.998511	3.47	0.999740	3.97	0.999964
0.48	0.684387	0.98	0.836457	1.48	0.930563	1.98	0.976148	2.48	0.993431	2.98	0.998559	3.48	0.999749	3.98	0.999966
0.49	0.687933	0.99	0.838913	1.49	0.931888	1.99	0.976705	2.49	0.993613	2.99	0.998605	3.49	0.999758	3.99	0.999967

A-1. Cumulative Distribution Function of Standard Normal [$\Phi(z)$] (continued)

z	$1 - \Phi(z)$	z	$1 - \Phi(z)$	z	$1 - \Phi(z)$	z	$1 - \Phi(z)$	z	$1 - \Phi(z)$	z	$1 - \Phi(z)$
4.00	3.17E-05	4.50	3.40E-06	5.00	2.87E-07	6.00	9.87E-10	7.00	1.28E-12	8.00	6.66E-16
4.05	2.56E-05	4.55	2.68E-06	5.10	1.70E-07	6.10	5.30E-10	7.10	6.24E-13	8.10	2.22E-16
4.10	2.07E-05	4.60	2.11E-06	5.20	9.96E-08	6.20	2.82E-10	7.20	3.01E-13	8.20	1.11E-16
4.15	1.66E-05	4.65	1.66E-06	5.30	5.79E-08	6.30	1.49E-10	7.30	1.44E-13		
4.20	1.33E-05	4.70	1.30E-06	5.40	3.33E-08	6.40	7.77E-11	7.40	6.80E-14		
4.25	1.07E-05	4.75	1.02E-06	5.50	1.90E-08	6.50	4.02E-11	7.50	3.20E-14		
4.30	8.54E-06	4.80	7.93E-07	5.60	1.07E-08	6.60	2.06E-11	7.60	1.50E-14		
4.35	6.81E-06	4.85	6.17E-07	5.70	5.99E-09	6.70	1.04E-11	7.70	7.00E-15		
4.40	5.41E-06	4.90	4.79E-07	5.80	3.32E-09	6.80	5.23E-12	7.80	3.00E-15		
4.45	4.29E-06	4.95	3.71E-07	5.90	1.82E-09	6.90	2.60E-12	7.90	1.50E-15		

A-2. Critical Values for the Student-*t* Distribution ($t_{\alpha,k}$)

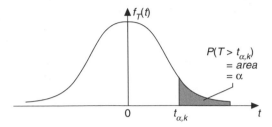

For Upper Tail Values

Degrees of freedom, *k*	Level of significance, α							
	0.2500	0.1000	0.0500	0.0250	0.0100	0.0050	0.0025	0.0005
1	1.000	3.078	6.314	12.706	31.821	63.657	127.321	536.627
2	0.816	1.886	2.920	4.303	6.965	9.925	14.089	31.599
3	0.765	1.638	2.353	3.182	4.541	5.841	7.453	12.924
4	0.741	1.533	2.132	2.776	3.747	4.604	5.598	8.610
5	0.727	1.476	2.015	2.571	3.365	4.032	4.773	6.869
6	0.718	1.440	1.943	2.447	3.143	3.707	4.317	5.959
7	0.711	1.415	1.895	2.365	2.998	3.499	4.029	5.408
8	0.706	1.397	1.860	2.306	2.896	3.355	3.833	5.041
9	0.703	1.383	1.833	2.262	2.821	3.250	3.690	4.781
10	0.700	1.372	1.812	2.228	2.764	3.169	3.581	4.587
11	0.697	1.363	1.796	2.201	2.718	3.106	3.497	4.437
12	0.695	1.356	1.782	2.179	2.681	3.055	3.428	4.318
13	0.694	1.350	1.771	2.160	2.650	3.012	3.372	4.221
14	0.692	1.345	1.761	2.145	2.624	2.977	3.326	4.140
15	0.691	1.341	1.753	2.131	2.602	2.947	3.286	4.073
16	0.690	1.337	1.746	2.120	2.583	2.921	3.252	4.015
17	0.689	1.333	1.740	2.110	2.567	2.898	3.222	3.965
18	0.688	1.330	1.734	2.101	2.552	2.878	3.197	3.922
19	0.688	1.328	1.729	2.093	2.539	2.861	3.174	3.883
20	0.687	1.325	1.725	2.086	2.528	2.845	3.153	3.850
21	0.686	1.323	1.721	2.080	2.518	2.831	3.135	3.819
22	0.686	1.321	1.717	2.074	2.508	2.819	3.119	3.792
23	0.685	1.319	1.714	2.069	2.500	2.807	3.104	3.768
24	0.685	1.318	1.711	2.064	2.492	2.797	3.091	3.745
25	0.684	1.316	1.708	2.060	2.485	2.787	3.078	3.725
26	0.684	1.315	1.706	2.056	2.479	2.779	3.067	3.707
27	0.684	1.314	1.703	2.052	2.473	2.771	3.057	3.690
28	0.683	1.313	1.701	2.048	2.467	2.763	3.047	3.674
29	0.683	1.311	1.699	2.045	2.462	2.756	3.038	3.659
30	0.683	1.310	1.697	2.042	2.457	2.750	3.030	3.646
35	0.682	1.306	1.690	2.030	2.438	2.724	2.996	3.591
40	0.681	1.303	1.684	2.021	2.423	2.704	2.971	3.551
45	0.680	1.301	1.679	2.014	2.412	2.690	2.952	3.520
50	0.679	1.299	1.676	2.009	2.403	2.678	2.937	3.496

A-2. Critical Values for the Student-*t* Distribution ($t_{\alpha,k}$) (continued)

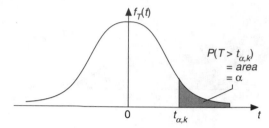

For Upper Tail Values

Degrees of freedom, *k*	Level of significance, α							
	0.2500	0.1000	0.0500	0.0250	0.0100	0.0050	0.0025	0.0005
55	0.679	1.297	1.673	2.004	2.396	2.668	2.925	3.476
60	0.679	1.296	1.671	2.000	2.390	2.660	2.915	3.460
65	0.678	1.295	1.669	1.997	2.385	2.654	2.906	3.447
70	0.678	1.294	1.667	1.994	2.381	2.648	2.899	3.435
80	0.678	1.292	1.664	1.990	2.374	2.639	2.887	3.416
90	0.677	1.291	1.662	1.987	2.368	2.632	2.878	3.402
100	0.677	1.290	1.660	1.984	2.364	2.626	2.871	3.390
125	0.676	1.288	1.657	1.979	2.357	2.616	2.858	3.370
150	0.676	1.287	1.655	1.976	2.351	2.609	2.849	3.357
200	0.676	1.286	1.653	1.972	2.345	2.601	2.839	3.340
∞	0.6745	1.2816	1.6448	1.9600	2.3267	2.5758	2.8070	3.2905

A-3. Critical Values for the Chi-Square Distribution ($c_{\alpha,k} = \chi^2_{\alpha,k}$)

$P(C > c_{\alpha,k}) = area = \alpha$

For Upper Tail Values

Degrees of freedom, k	Level of significance, α																	
	0.9950	0.9900	0.9750	0.9500	0.9000	0.7000	0.5000	0.2000	0.1000	0.0500	0.0200	0.0100	0.0050	0.0010	0.9990
1	0.000	0.000	0.000	0.001	0.004	0.016	0.148	0.455	1.642	2.706	3.842	5.405	6.637	7.905	10.809
2	0.002	0.010	0.020	0.051	0.102	0.211	0.713	1.386	3.219	4.504	5.995	7.822	9.221	10.589	13.691
3	0.024	0.071	0.115	0.216	0.352	0.584	1.424	2.366	4.642	6.252	7.817	9.841	11.325	12.819	16.292
4	0.090	0.205	0.297	0.484	0.711	1.064	2.195	3.357	5.989	7.782	9.492	11.660	13.280	14.824	18.432
5	0.209	0.411	0.553	0.831	1.145	1.610	3.000	4.352	7.291	9.237	11.073	13.385	15.088	16.762	20.751
6	0.377	0.673	0.871	1.236	1.635	2.204	3.828	5.349	8.559	10.646	12.596	15.033	16.810	18.550	22.677
7	0.597	0.988	1.237	1.688	2.167	2.833	4.671	6.346	9.804	12.070	14.070	16.624	18.471	20.270	24.527
8	0.850	1.341	1.642	2.179	2.732	3.489	5.527	7.344	11.031	13.363	15.512	18.171	20.082	21.938	26.318
9	1.135	1.728	2.086	2.699	3.324	4.168	6.393	8.343	12.243	14.686	16.925	19.683	21.654	23.563	28.061
10	1.446	2.152	2.555	3.244	3.938	4.864	7.266	9.342	13.443	15.990	18.311	21.165	23.194	25.154	29.763
11	1.819	2.597	3.047	3.815	4.574	5.576	8.148	10.341	14.633	17.278	19.681	22.623	24.755	26.714	31.431
12	2.188	3.064	3.568	4.402	5.225	6.303	9.034	11.340	15.813	18.551	21.030	24.059	26.246	28.249	33.070
13	2.577	3.560	4.102	5.006	5.890	7.041	9.926	12.340	16.986	19.814	22.367	25.477	27.717	29.878	34.683
14	3.018	4.066	4.653	5.624	6.568	7.789	10.821	13.339	18.152	21.067	23.691	26.879	29.169	31.376	36.272
15	3.449	4.588	5.226	6.260	7.260	8.546	11.721	14.339	19.312	22.310	25.000	28.266	30.605	32.857	37.842
16	3.894	5.135	5.807	6.905	7.960	9.311	12.624	15.339	20.466	23.546	26.301	29.640	32.027	34.321	39.392
17	4.350	5.687	6.400	7.560	8.670	10.083	13.530	16.338	21.616	24.771	27.593	31.002	33.435	35.771	40.926
18	4.864	6.251	7.004	8.225	9.388	10.864	14.440	17.337	22.761	25.992	28.877	32.353	34.831	37.208	42.444
19	5.351	6.825	7.627	8.904	10.114	11.650	15.351	18.338	23.902	27.206	30.148	33.694	36.216	38.633	43.949
20	5.848	7.422	8.252	9.587	10.849	12.442	16.625	19.337	25.039	28.415	31.416	35.026	37.591	40.046	45.440
21	6.398	8.018	8.886	10.278	11.590	13.238	17.182	20.337	26.173	29.619	32.678	36.350	38.957	41.449	46.919
22	6.919	8.622	9.528	10.976	12.336	14.040	18.100	21.337	27.304	30.817	33.933	37.666	40.314	42.843	48.387

A-3. Critical Values for the Chi-Square Distribution ($c_{\alpha,k} = \chi^2_{\alpha,k}$) (continued)

$P(C > c_{\alpha,k}) = area = \alpha$

For Upper Tail Values

Degrees of freedom, k	0.9990	0.9950	0.9900	0.9750	0.9500	0.9000	0.7000	0.5000	0.2000	0.1000	0.0500	0.0200	0.0100	0.0050	0.0010
								Level of significance, α							
23	7.447	9.247	10.187	11.685	13.088	14.846	19.020	22.337	28.431	32.012	35.178	38.975	41.662	44.228	49.845
24	8.027	9.869	10.846	12.397	13.845	15.657	19.943	23.337	29.556	33.199	36.421	40.277	43.004	45.604	51.293
25	8.576	10.498	11.510	13.115	14.607	16.471	20.866	24.337	30.678	34.384	37.660	41.573	44.338	46.973	52.732
26	9.130	11.132	12.190	13.837	15.377	17.291	21.792	25.337	31.796	35.566	38.894	42.863	45.665	48.334	54.162
27	9.735	11.789	12.868	14.565	16.149	18.113	22.718	26.336	32.913	36.745	40.119	44.147	46.986	49.688	55.584
28	10.306	12.348	13.551	15.304	16.925	18.938	23.646	27.336	34.028	37.920	41.344	45.426	48.301	51.036	56.998
29	10.882	13.092	14.240	16.042	17.705	19.766	24.576	28.336	35.140	39.092	42.565	46.699	49.610	52.378	58.405
30	11.509	13.767	14.943	16.784	18.488	20.598	25.507	29.336	36.251	40.261	43.782	47.968	50.914	53.713	59.805
35	14.554	17.160	18.494	20.562	22.461	24.793	30.176	34.336	41.780	46.063	49.811	54.250	57.363	60.313	66.714
40	17.846	20.669	22.139	24.423	26.508	29.055	34.879	39.337	47.261	51.796	55.753	60.443	63.710	66.802	73.490
45	21.183	24.275	25.878	28.357	30.611	33.355	39.591	44.337	52.721	57.497	61.652	66.562	69.976	73.201	80.160
50	24.609	27.957	29.685	32.349	34.763	37.693	44.319	49.336	58.157	63.159	67.501	72.619	76.172	79.523	86.740
55	28.111	31.702	33.549	36.390	38.957	42.064	49.061	54.336	63.571	68.789	73.308	78.625	82.309	85.780	93.243
60	31.678	35.503	37.465	40.474	43.187	46.463	53.814	59.336	68.966	74.390	79.078	84.586	88.396	91.982	99.679
65	35.303	39.353	41.424	44.596	47.449	50.887	58.578	64.336	74.344	79.966	84.817	90.507	94.438	98.134	106.057
70	38.980	43.246	45.423	48.750	51.739	55.333	63.351	69.335	79.709	85.521	90.528	96.393	100.441	104.243	112.383
80	46.466	51.145	53.523	57.147	60.391	64.282	72.920	79.335	90.400	96.572	101.876	108.075	112.344	116.348	124.901
90	54.104	59.171	61.738	65.641	69.126	73.295	82.515	89.335	101.048	107.559	113.143	119.654	124.130	128.324	137.267
100	61.869	67.303	70.049	74.216	77.929	82.362	92.133	99.335	111.662	118.493	124.340	131.147	135.820	140.193	149.505
125	81.726	88.007	91.166	95.941	100.178	105.217	116.254	124.335	138.071	145.638	152.092	159.580	164.706	169.493	179.653
150	102.073	109.122	112.655	117.980	122.692	128.278	140.460	149.334	164.345	172.577	179.579	187.683	193.219	198.380	209.310
200	143.807	152.224	156.421	162.724	168.279	174.838	189.051	199.334	216.605	226.017	233.993	243.191	249.455	255.281	267.579

A-4. Critical Values for the F Distribution ($f_{\alpha,k,u}$)

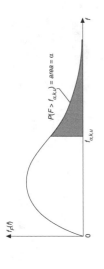

$f_F(f)$

$P(F > f_{\alpha,k,u}) = area = \alpha$

$f_{\alpha,k,u}$

Upper Values for 5% (in First Row) and 1% (in Second Row) Significance Levels, α

First degrees of freedom[1], k

Second degrees of freedom, u	1	2	3	4	5	6	7	8	9	10	11	12	13	14	15
	161.	200.	216.	225.	230.	234.	237.	239.	241.	242.	243.	244.	245.	245.	246.
	4052.	5000.	5403.	5625.	5764.	5859.	5928.	5981.	6022.	6056.	6083.	6106.	6126.	6143.	6157.
2	18.51	19.00	19.16	19.25	19.30	19.33	19.35	19.37	19.38	19.40	19.41	19.41	19.42	19.42	19.43
	98.50	99.00	99.17	99.25	99.30	99.33	99.36	99.37	99.39	99.40	99.41	99.42	99.42	99.43	99.43
3	10.13	9.55	9.28	9.12	9.01	8.94	8.89	8.85	8.81	8.79	8.76	8.74	8.73	8.71	8.70
	34.12	30.82	29.46	28.71	28.24	27.91	27.67	27.49	27.34	27.23	27.13	27.05	26.99	26.93	26.87
4	7.71	6.94	6.59	6.39	6.26	6.16	6.09	6.04	6.00	5.96	5.94	5.91	5.89	5.87	5.86
	21.20	18.00	16.69	15.98	15.52	15.21	14.98	14.80	14.66	14.55	14.45	14.37	14.31	14.25	14.20
5	6.61	5.79	5.41	5.19	5.05	4.95	4.88	4.82	4.77	4.74	4.70	4.68	4.66	4.64	4.62
	16.26	13.27	12.06	11.39	10.97	10.67	10.46	10.29	10.16	10.05	9.96	9.89	9.82	9.77	9.72
6	5.99	5.14	4.76	4.53	4.39	4.28	4.21	4.15	4.10	4.06	4.03	4.00	3.98	3.96	3.94
	13.75	10.92	9.78	9.15	8.75	8.47	8.26	8.10	7.98	7.87	7.79	7.72	7.66	7.60	7.56
7	5.59	4.74	4.35	4.12	3.97	3.87	3.79	3.73	3.68	3.64	3.60	3.57	3.55	3.53	3.51
	12.25	9.55	8.45	7.85	7.46	7.19	6.99	6.84	6.72	6.62	6.54	6.47	6.41	6.36	6.31
8	5.32	4.46	4.07	3.84	3.69	3.58	3.50	3.44	3.39	3.35	3.31	3.28	3.26	3.24	3.22
	11.26	8.65	7.59	7.01	6.63	6.37	6.18	6.03	5.91	5.81	5.73	5.67	5.61	5.56	5.52
9	5.12	4.26	3.86	3.63	3.48	3.37	3.29	3.23	3.18	3.14	3.10	3.07	3.05	3.03	3.01
	10.56	8.02	6.99	6.42	6.06	5.80	5.61	5.47	5.35	5.26	5.18	5.11	5.05	5.01	4.96
10	4.96	4.10	3.71	3.48	3.33	3.22	3.14	3.07	3.02	2.98	2.94	2.91	2.89	2.86	2.85
	10.04	7.56	6.55	5.99	5.64	5.39	5.20	5.06	4.94	4.85	4.77	4.71	4.65	4.60	4.56

[1] For k values larger than 15, refer to pages 495 to 497.

A-4. Critical Values for the *F* Distribution ($f_{\alpha,k,u}$) (continued)

Upper Values for 5% (in First Row) and 1% (in Second Row) Significance Levels, α

Second degrees of freedom, u		First degrees of freedom, k													
	1	2	3	4	5	6	7	8	9	10	11	12	13	14	15
11	4.84	3.98	3.59	3.36	3.20	3.09	3.01	2.95	2.90	2.85	2.82	2.79	2.76	2.74	2.72
	9.65	7.21	6.22	5.67	5.32	5.07	4.89	4.74	4.63	4.54	4.46	4.40	4.34	4.29	4.25
12	4.75	3.89	3.49	3.26	3.11	3.00	2.91	2.85	2.80	2.75	2.72	2.69	2.66	2.64	2.62
	9.33	6.93	5.95	5.41	5.06	4.82	4.64	4.50	4.39	4.30	4.22	4.16	4.10	4.05	4.01
13	4.67	3.81	3.41	3.18	3.03	2.92	2.83	2.77	2.71	2.67	2.63	2.60	2.58	2.55	2.53
	9.07	6.70	5.74	5.21	4.86	4.62	4.44	4.30	4.19	4.10	4.02	3.96	3.91	3.86	3.82
14	4.60	3.74	3.34	3.11	2.96	2.85	2.76	2.70	2.65	2.60	2.57	2.53	2.51	2.48	2.46
	8.86	6.51	5.56	5.04	4.69	4.46	4.28	4.14	4.03	3.94	3.86	3.80	3.75	3.70	3.66
15	4.54	3.68	3.29	3.06	2.90	2.79	2.71	2.64	2.59	2.54	2.51	2.48	2.45	2.42	2.40
	8.68	6.36	5.42	4.89	4.56	4.32	4.14	4.00	3.89	3.80	3.73	3.67	3.61	3.56	3.52
17	4.45	3.59	3.20	2.96	2.81	2.70	2.61	2.55	2.49	2.45	2.41	2.38	2.35	2.33	2.31
	8.40	6.11	5.19	4.67	4.34	4.10	3.93	3.79	3.68	3.59	3.52	3.46	3.40	3.35	3.31
20	4.35	3.49	3.10	2.87	2.71	2.60	2.51	2.45	2.39	2.35	2.31	2.28	2.25	2.22	2.20
	8.10	5.85	4.94	4.43	4.10	3.87	3.70	3.56	3.46	3.37	3.29	3.23	3.18	3.13	3.09
25	4.24	3.39	2.99	2.76	2.60	2.49	2.40	2.34	2.28	2.24	2.20	2.16	2.14	2.11	2.09
	7.77	5.57	4.68	4.18	3.85	3.63	3.46	3.32	3.22	3.13	3.06	2.99	2.94	2.89	2.85
30	4.17	3.32	2.92	2.69	2.53	2.42	2.33	2.27	2.21	2.16	2.13	2.09	2.06	2.04	2.01
	7.56	5.39	4.51	4.02	3.70	3.47	3.30	3.17	3.07	2.98	2.91	2.84	2.79	2.74	2.70
40	4.08	3.23	2.84	2.61	2.45	2.34	2.25	2.18	2.12	2.08	2.04	2.00	1.97	1.95	1.92
	7.31	5.18	4.31	3.83	3.51	3.29	3.12	2.99	2.89	2.80	2.73	2.66	2.61	2.56	2.52

Table values are given as pairs (upper value / lower value).

(continuation — First degrees of freedom, k)

Second degrees of freedom, u	1	2	3	4	5	6	7	8	9	10	11	12	13	14	15
50	4.03 / 7.17	3.18 / 5.03	2.79 / 4.20	2.56 / 3.72	2.40 / 3.41	2.29 / 3.19	2.20 / 3.02	2.13 / 2.89	2.07 / 2.78	2.03 / 2.76	1.99 / 2.63	1.95 / 2.56	1.92 / 2.51	1.89 / 2.46	1.87 / 2.42
75	3.97 / 6.99	3.12 / 4.90	2.73 / 4.05	2.49 / 3.58	2.34 / 3.27	2.22 / 3.05	2.13 / 2.89	2.06 / 2.76	2.01 / 2.65	1.96 / 2.57	1.92 / 2.49	1.88 / 2.43	1.85 / 2.38	1.83 / 2.33	1.80 / 2.29
100	3.94 / 6.90	3.09 / 4.82	2.70 / 3.98	2.48 / 3.51	2.31 / 3.21	2.19 / 2.99	2.10 / 2.82	2.03 / 2.69	1.97 / 2.59	1.93 / 2.50	1.89 / 2.43	1.85 / 2.37	1.82 / 2.31	1.79 / 2.27	1.77 / 2.22
200	3.89 / 6.76	3.04 / 4.71	2.65 / 3.88	2.42 / 3.41	2.26 / 3.11	2.14 / 2.89	2.06 / 2.73	1.98 / 2.60	1.93 / 2.50	1.88 / 2.41	1.84 / 2.34	1.80 / 2.27	1.77 / 2.22	1.74 / 2.17	1.72 / 2.13
500	3.86 / 6.69	3.01 / 4.65	2.62 / 3.82	2.39 / 3.36	2.23 / 3.05	2.12 / 2.84	2.03 / 2.68	1.96 / 2.55	1.90 / 2.44	1.85 / 2.36	1.81 / 2.28	1.77 / 2.22	1.74 / 2.17	1.71 / 2.12	1.69 / 2.07
1,000	3.85 / 6.66	3.00 / 4.63	2.61 / 3.80	2.38 / 3.34	2.22 / 3.04	2.11 / 2.82	2.02 / 2.66	1.98 / 2.53	1.89 / 2.43	1.84 / 2.34	1.80 / 2.27	1.76 / 2.20	1.73 / 2.15	1.70 / 2.10	1.68 / 2.06
10,000	3.84 / 6.64	3.00 / 4.61	2.61 / 3.78	2.37 / 3.32	2.21 / 3.02	2.10 / 2.80	2.01 / 2.64	1.94 / 2.51	1.88 / 2.41	1.83 / 2.32	1.79 / 2.25	1.75 / 2.19	1.72 / 2.13	1.69 / 2.08	1.67 / 2.04

First degrees of freedom, k

Second degrees of freedom, u	17	20	25	30	40	50	75	100	200	500	1,000	10,000
1	247. / 6181.	248. / 6209.	249. / 6240.	250. / 6261.	251. / 6287.	252. / 6303.	253. / 6324.	253. / 6334.	254. / 6350.	254. / 6360.	254. / 6363.	254. / 6364.
2	19.44 / 99.44	19.45 / 99.45	19.46 / 99.46	19.46 / 99.47	18.47 / 99.47	19.48 / 99.48	19.48 / 99.49	19.49 / 99.49	19.49 / 99.49	19.49 / 99.49	19.49 / 99.50	19.49 / 99.49
3	8.68 / 26.79	8.66 / 26.69	8.63 / 26.58	8.62 / 26.50	8.59 / 26.41	8.58 / 26.35	8.56 / 26.28	8.55 / 26.24	8.54 / 26.18	8.53 / 26.15	8.53 / 26.14	8.52 / 26.12
4	5.83 / 14.11	5.80 / 14.02	5.77 / 13.91	5.75 / 13.84	5.72 / 13.75	5.70 / 13.69	5.68 / 13.61	5.66 / 13.58	5.65 / 13.52	5.64 / 13.49	5.63 / 13.47	5.63 / 13.46
5	4.59 / 9.64	4.56 / 9.55	4.52 / 9.45	4.50 / 9.38	4.46 / 9.29	4.44 / 9.24	4.42 / 9.17	4.41 / 9.13	4.39 / 9.08	4.37 / 9.04	4.37 / 9.03	4.37 / 9.02
6	3.91 / 7.48	3.87 / 7.40	3.83 / 7.30	3.81 / 7.23	3.77 / 7.14	3.75 / 7.09	3.73 / 7.02	3.71 / 6.99	3.69 / 6.93	3.68 / 6.90	3.67 / 6.89	3.67 / 6.88
7	3.48 / 6.24	3.44 / 6.16	3.40 / 6.06	3.38 / 5.99	3.34 / 5.91	3.32 / 5.86	3.29 / 5.79	3.27 / 5.75	3.25 / 5.70	3.24 / 5.67	3.23 / 5.66	3.23 / 5.65

A-4. Critical Values for the F Distribution ($f_{\alpha,k,u}$) (continued)

Upper Values for 5% (in First Row) and 1% (in Second Row) Significance Levels, α

Second degrees of freedom, u	First degrees of freedom, k											
	17	20	25	30	40	50	75	100	200	500	1,000	10,000
8	3.19	3.15	3.11	3.08	3.04	3.02	2.99	2.97	2.95	2.94	2.93	2.93
	5.44	5.36	5.26	5.20	5.12	5.07	5.00	4.96	4.91	4.88	4.87	4.86
9	2.97	2.94	2.89	2.86	2.83	2.80	2.77	2.76	2.73	2.72	2.71	2.71
	4.89	4.81	4.71	4.65	4.57	4.52	4.45	4.41	4.36	4.33	4.32	4.31
10	2.81	2.77	2.73	2.70	2.66	2.64	2.60	2.59	2.56	2.55	2.54	2.54
	4.49	4.41	4.31	4.25	4.17	4.12	4.05	4.01	3.96	3.93	3.92	3.91
11	2.69	2.65	2.60	2.57	2.53	2.51	2.47	2.46	2.43	2.42	2.41	2.41
	4.18	4.10	4.01	3.94	3.86	3.81	3.74	3.71	3.66	3.62	3.61	3.60
12	2.58	2.54	2.50	2.47	2.43	2.40	2.37	2.35	2.32	2.31	2.30	2.30
	3.94	3.86	3.76	3.70	3.62	3.57	3.50	3.47	3.41	3.38	3.37	3.36
13	2.50	2.46	2.41	2.38	2.34	2.31	2.28	2.26	2.23	2.22	2.21	2.21
	3.75	3.66	3.57	3.51	3.43	3.38	3.31	3.27	3.22	3.19	3.18	3.17
14	2.43	2.39	2.34	2.31	2.27	2.24	2.21	2.19	2.16	2.14	2.14	2.13
	3.59	3.51	3.41	3.35	3.27	3.22	3.15	3.11	3.06	3.03	3.02	3.00
15	2.37	2.33	2.28	2.25	2.20	2.18	2.14	2.12	2.10	2.08	2.07	2.07
	3.45	3.37	3.28	3.21	3.13	3.08	3.01	2.98	2.92	2.89	2.88	2.87
17	2.27	2.23	2.18	2.15	2.10	2.08	2.04	2.02	1.99	1.97	1.97	1.96
	3.24	3.16	3.07	3.00	2.92	2.87	2.80	2.76	2.71	2.68	2.66	2.65
20	2.17	2.12	2.07	2.04	1.99	1.97	1.93	1.91	1.88	1.86	1.85	1.85
	3.02	2.94	2.84	2.78	2.69	2.64	2.57	2.54	2.48	2.44	2.43	2.42

n												
25	1.72	1.72	1.73	1.75	1.78	1.80	1.84	1.87	1.92	1.96	2.01	2.05
	2.17	2.18	2.19	2.23	2.29	2.33	2.40	2.45	2.54	2.60	2.70	2.78
30	1.62	1.63	1.64	1.66	1.70	1.74	1.76	1.79	1.84	1.88	1.93	1.98
	2.01	2.02	2.03	2.07	2.13	2.17	2.25	2.30	2.39	2.45	2.55	2.63
40	1.51	1.52	1.53	1.55	1.59	1.61	1.66	1.69	1.74	1.78	1.84	1.89
	0.18	1.82	1.83	1.87	1.94	1.98	2.06	2.11	2.20	2.27	2.37	2.45
50	1.44	1.45	1.46	1.48	1.52	1.55	1.60	1.63	1.69	1.73	1.78	1.83
	1.68	1.70	1.71	1.76	1.82	1.87	1.95	2.01	2.10	2.17	2.27	2.35
75	1.34	1.35	1.36	1.39	1.44	1.47	1.52	1.55	1.61	1.65	1.71	1.76
	1.52	1.53	1.55	1.60	1.67	1.72	1.81	1.87	1.96	2.03	2.13	2.22
100	1.28	0.30	1.31	1.34	1.39	1.42	1.48	1.52	1.57	1.62	1.68	1.73
	1.43	1.45	1.47	1.52	1.60	1.65	1.74	1.80	1.89	1.97	2.07	2.15
200	1.19	1.21	1.22	1.26	1.32	1.35	1.41	1.46	1.52	1.56	1.62	1.67
	1.28	1.30	1.33	1.39	1.48	1.53	1.63	1.69	1.79	1.87	1.97	2.06
500	1.12	1.14	1.16	1.21	1.28	1.31	1.38	1.42	1.48	1.53	1.59	1.64
	1.17	1.20	1.23	1.31	1.41	1.47	1.57	1.63	1.74	1.81	1.92	2.00
1,000	1.08	1.11	1.13	1.19	1.26	1.30	1.36	1.41	1.47	1.52	1.58	1.63
	1.12	1.16	1.19	1.28	1.38	1.44	1.54	1.61	1.72	1.79	1.90	1.98
10,000	1.03	1.08	1.11	1.17	1.25	1.28	1.35	1.40	1.46	1.51	1.57	1.62
	1.05	1.11	1.16	1.25	1.36	1.42	1.53	1.59	1.70	1.77	1.88	1.97

A-5. Critical Values for the Pearson Correlation Coefficient for the Null Hypothesis H_0: $\rho = 0$ and Both One-Tailed Alternative H_A: $|\rho| > 0$ and the Two-Tailed Alternative H_A: $\rho \neq 0$

	Level of significance for a one-tailed test					
	0.1000	0.0500	0.0250	0.0100	0.0050	0.0005
Degrees of	Level of significance for a two-tailed test					
freedom	0.2000	0.1000	0.0500	0.0200	0.0100	0.0010
1	0.9511	0.9877	0.9969	0.9995	0.9999	1.0000
2	0.8001	0.9000	0.9500	0.9800	0.9900	0.9990
3	0.6871	0.8053	0.8783	0.9343	0.9587	0.9912
4	0.6083	0.7293	0.8114	0.8822	0.9172	0.9741
5	0.5509	0.6694	0.7545	0.8329	0.8745	0.9508
6	0.5068	0.6215	0.7067	0.7888	0.8343	0.9249
7	0.4716	0.5823	0.6664	0.7498	0.7976	0.8982
8	0.4428	0.5495	0.6319	0.7154	0.7646	0.8721
9	0.4187	0.5214	0.6020	0.6850	0.7348	0.8471
10	0.3980	0.4972	0.5760	0.6581	0.7079	0.8233
11	0.3801	0.4762	0.5529	0.6339	0.6836	0.8010
12	0.3645	0.4574	0.5324	0.6120	0.6614	0.7800
13	0.3506	0.4409	0.5139	0.5922	0.6411	0.7604
14	0.3383	0.4258	0.4973	0.5742	0.6226	0.7419
15	0.3272	0.4124	0.4821	0.5577	0.6055	0.7247
16	0.3170	0.4000	0.4683	0.5425	0.5897	0.7084
17	0.3076	0.3888	0.4556	0.5285	0.5750	0.6932
18	0.2991	0.3783	0.4438	0.5154	0.5614	0.6853
19	0.2914	0.3687	0.4329	0.5033	0.5487	0.6652
20	0.2841	0.3599	0.4227	0.4921	0.5368	0.6524
21	0.2774	0.3516	0.4133	0.4816	0.5256	0.6402
22	0.2711	0.3438	0.4044	0.4715	0.5151	0.6287
23	0.2652	0.3365	0.3961	0.4623	0.5051	0.6177
24	0.2598	0.3297	0.3883	0.4534	0.4958	0.6073
25	0.2545	0.3233	0.3809	0.4451	0.4869	0.5974
26	0.2497	0.3173	0.3740	0.4372	0.4785	0.5880
27	0.2452	0.3114	0.3673	0.4297	0.4706	0.5790
28	0.2408	0.3060	0.3609	0.4226	0.4629	0.5703
29	0.2365	0.3009	0.3550	0.4158	0.4556	0.5620
30	0.2326	0.2959	0.3493	0.4093	0.4487	0.5541
40	0.2018	0.2573	0.3044	0.3578	0.3931	0.4896
60	0.1650	0.2109	0.2500	0.2948	0.3248	0.4078
120	0.1169	0.1496	0.1779	0.2104	0.2324	0.2943
250	0.0808	0.1035	0.1230	0.1455	0.1608	0.2038

A-6. Random Numbers

0.538246	0.181648	0.172614	0.450166	0.293027	0.030195	0.757836	0.915061
0.663357	0.368934	0.516388	0.656254	0.284258	0.906335	0.329788	0.054487
0.035771	0.053784	0.424573	0.942479	0.293872	0.326815	0.862351	0.358055
0.513560	0.165508	0.667312	0.878444	0.414203	0.100839	0.555287	0.685601
0.880006	0.069305	0.854410	0.371911	0.751341	0.128446	0.678679	0.514995
0.748794	0.902497	0.629615	0.662531	0.932879	0.018376	0.683876	0.554810
0.115441	0.207278	0.887853	0.812124	0.082143	0.939258	0.666874	0.582525
0.953369	0.543997	0.806486	0.707493	0.503949	0.489926	0.774467	0.248617
0.243600	0.537111	0.181388	0.619277	0.131852	0.131876	0.361814	0.582682
0.610186	0.411580	0.339972	0.080869	0.429448	0.822770	0.632690	0.863227
0.848375	0.043973	0.071429	0.713405	0.562010	0.716050	0.536620	0.357681
0.102922	0.201752	0.617270	0.416471	0.371492	0.633301	0.857578	0.483474
0.009326	0.912932	0.113850	0.331600	0.852807	0.626191	0.035676	0.581386
0.801494	0.365068	0.548750	0.480788	0.032959	0.906331	0.291263	0.706212
0.682049	0.946008	0.960047	0.830463	0.186225	0.123762	0.674147	0.012839
0.610019	0.495159	0.165360	0.207562	0.676065	0.843231	0.581861	0.669062
0.203327	0.400473	0.257740	0.474822	0.439519	0.722845	0.274275	0.862333
0.081627	0.889037	0.446990	0.102859	0.384948	0.785505	0.655849	0.305504
0.866440	0.209892	0.558315	0.422571	0.277399	0.548600	0.009868	0.121337
0.514767	0.666866	0.498019	0.253118	0.081162	0.306135	0.762659	0.284422
0.739733	0.030429	0.054100	0.524432	0.654229	0.393821	0.928477	0.784502
0.131999	0.540034	0.884054	0.508152	0.767278	0.646574	0.211862	0.912805
0.569512	0.983024	0.488274	0.554083	0.860977	0.868964	0.304604	0.729149
0.388688	0.592203	0.241637	0.331246	0.380662	0.696245	0.029850	0.363413
0.518582	0.634681	0.409353	0.468436	0.724978	0.222090	0.357835	0.507892
0.204704	0.364712	0.046972	0.526575	0.705034	0.352802	0.077436	0.715174
0.613146	0.281786	0.635090	0.433333	0.976778	0.148786	0.442014	0.232516
0.914132	0.089797	0.111000	0.346466	0.855181	0.945228	0.293689	0.531055
0.898415	0.959250	0.005809	0.260049	0.150843	0.637507	0.171769	0.151870
0.543517	0.275269	0.629999	0.428829	0.744025	0.279269	0.913490	0.597542
0.091718	0.646875	0.173754	0.647365	0.987107	0.588587	0.721938	0.361474

A-7. Critical Values for the Kolmogorov-Smirnov One-Sample Test

Sample size (n)	Level of Significance α				
	.20	**.15**	**.10**	**.05**	**.01**
1	.900	.925	.950	.975	.995
2	.684	.726	.776	.842	.929
3	.565	.597	.642	.708	.828
4	.494	.525	.564	.624	.733
5	.446	.474	.510	.565	.669
6	.410	.436	.470	.521	.618
7	.381	.405	.438	.486	.577
8	.358	.381	.411	.457	.543
9	.339	.360	.388	.432	.514
10	.322	.342	.368	.410	.490
11	.307	.326	.352	.391	.468
12	.295	.313	.338	.375	.450
13	.284	.302	.325	.361	.433
14	.274	.292	.314	.349	.418
15	.266	.283	.304	.338	.404
16	.258	.274	.295	.328	.392
17	.250	.266	.286	.318	.381
18	.244	.259	.278	.309	.371
19	.237	.252	.272	.301	.363
20	.231	.246	.264	.294	.356
25	.21	.22	.24	.27	.32
30	.19	.20	.22	.24	.29
35	.18	.19	.21	.23	.27
Over 35	$\dfrac{1.07}{\sqrt{n}}$	$\dfrac{1.14}{\sqrt{n}}$	$\dfrac{1.22}{\sqrt{n}}$	$\dfrac{1.36}{\sqrt{n}}$	$\dfrac{1.63}{\sqrt{n}}$

Appendix B
Values of the Gamma Function

$$\Gamma(t) = \int_0^\infty r^{t-1} \exp(-r)\, dr$$

t	$\Gamma(t)$	t	$\Gamma(t)$	t	$\Gamma(t)$	t	$\Gamma(t)$
1.000	1.00000	1.250	0.90640	1.500	0.88623	1.750	0.91906
1.005	0.99714	1.255	0.90539	1.505	0.88640	1.755	0.92021
1.010	0.99433	1.260	0.90440	1.510	0.88659	1.760	0.92137
1.015	0.99156	1.265	0.90344	1.515	0.88680	1.765	0.92256
1.020	0.98884	1.270	0.90250	1.520	0.88704	1.770	0.92376
1.025	0.98617	1.275	0.90160	1.525	0.88729	1.775	0.92499
1.030	0.98355	1.280	0.90072	1.530	0.88757	1.780	0.92623
1.035	0.98097	1.285	0.89987	1.535	0.88786	1.785	0.92749
1.040	0.97844	1.290	0.89904	1.540	0.88818	1.790	0.92877
1.045	0.97595	1.295	0.89824	1.545	0.88851	1.795	0.93007
1.050	0.97350	1.300	0.89747	1.550	0.88887	1.800	0.93138
1.055	0.97110	1.305	0.89672	1.555	0.88924	1.805	0.93272
1.060	0.96874	1.310	0.89600	1.560	0.88964	1.810	0.93408
1.065	0.96643	1.315	0.89531	1.565	0.89005	1.815	0.93545
1.070	0.96415	1.320	0.89464	1.570	0.89049	1.820	0.93685
1.075	0.96192	1.325	0.89400	1.575	0.89094	1.825	0.93826
1.080	0.95973	1.330	0.89338	1.580	0.89142	1.830	0.93969
1.085	0.95757	1.335	0.89278	1.585	0.89191	1.835	0.94114
1.090	0.95546	1.340	0.89222	1.590	0.89243	1.840	0.94261
1.095	0.95339	1.345	0.89167	1.595	0.89296	1.845	0.94410
1.100	0.95135	1.350	0.89115	1.600	0.89352	1.850	0.94561
1.105	0.94935	1.355	0.89066	1.605	0.89409	1.855	0.94714
1.110	0.94740	1.360	0.89018	1.610	0.89468	1.860	0.94869
1.115	0.94547	1.365	0.88974	1.615	0.89529	1.865	0.95025
1.120	0.94359	1.370	0.88931	1.620	0.89592	1.870	0.95184
1.125	0.94174	1.375	0.88891	1.625	0.89657	1.875	0.95345
1.130	0.93993	1.380	0.88854	1.630	0.89724	1.880	0.95507
1.135	0.93816	1.385	0.88818	1.635	0.89793	1.885	0.95672
1.140	0.93642	1.390	0.88785	1.640	0.89864	1.890	0.95838
1.145	0.93471	1.395	0.88755	1.645	0.89937	1.895	0.96006

t	$\Gamma(t)$	t	$\Gamma(t)$	t	$\Gamma(t)$	t	$\Gamma(t)$
1.150	0.93304	1.400	0.88726	1.650	0.90012	1.900	0.96177
1.155	0.93141	1.405	0.88700	1.655	0.90088	1.905	0.96349
1.160	0.92980	1.410	0.88676	1.660	0.90167	1.910	0.96523
1.165	0.92823	1.415	0.88655	1.665	0.90247	1.915	0.96699
1.170	0.92670	1.420	0.88636	1.670	0.90330	1.920	0.96877
1.175	0.92520	1.425	0.88618	1.675	0.90414	1.925	0.97058
1.180	0.92373	1.430	0.88604	1.680	0.90500	1.930	0.97240
1.185	0.92229	1.435	0.88591	1.685	0.90588	1.935	0.97424
1.190	0.92089	1.440	0.88581	1.690	0.90678	1.940	0.97610
1.195	0.91951	1.445	0.88572	1.695	0.90770	1.945	0.97798
1.200	0.91817	1.450	0.88566	1.700	0.90864	1.950	0.97988
1.205	0.91686	1.455	0.88562	1.705	0.90960	1.955	0.98180
1.210	0.91558	1.460	0.88560	1.710	0.91057	1.960	0.98374
1.215	0.91433	1.465	0.88561	1.715	0.91157	1.965	0.98570
1.220	0.91311	1.470	0.88563	1.720	0.91258	1.970	0.98768
1.225	0.91192	1.475	0.88568	1.725	0.91361	1.975	0.98969
1.230	0.91075	1.480	0.88575	1.730	0.91467	1.980	0.99171
1.235	0.90962	1.485	0.88584	1.735	0.91574	1.985	0.99375
1.240	0.90852	1.490	0.88595	1.740	0.91683	1.990	0.99581
1.245	0.90745	1.495	0.88608	1.745	0.91793	1.995	0.99790
1.250	0.90640	1.500	0.88623	1.750	0.91606	2.000	1.00000

INDEX

S